HUMAN RIGHTS, VIRTUE, AND THE COMMON GOOD

ERNEST FORTIN: COLLECTED ESSAYS
Edited by J. Brian Benestad

Volume 1
The Birth of Philosophic Christianity: Studies
in Early Christian and Medieval Thought

Foreword by Ernest L. Fortin

Volume 2
Classical Christianity and the Political Order:
Reflections on the Theologico-Political Problem

Foreword by Dan Mahoney

Volume 3
Human Rights, Virtue, and the Common Good:
Untimely Meditations on Religion and Politics

Foreword by J. Brian Benestad

HUMAN RIGHTS, VIRTUE, AND THE COMMON GOOD

Untimely Meditations on Religion and Politics

ERNEST L. FORTIN

Edited by
J. Brian Benestad

ROWMAN & LITTLEFIELD PUBLISHERS, INC.
Lanham • Boulder • New York • London

ROWMAN & LITTLEFIELD PUBLISHERS, INC.

Published in the United States of America
by Rowman & Littlefield Publishers, Inc.
4720 Boston Way, Lanham, Maryland 20706

3 Henrietta Street
London WC2E 8LU, England

Copyright © 1996 by Rowman & Littlefield Publishers, Inc.

All rights reserved. No part of this publication may be reproduced,
stored in a retrieval system, or transmitted in any form or by any
means, electronic, mechanical, photocopying, recording, or otherwise,
without the prior permission of the publisher.

British Cataloging in Publication Information Available

Library of Congress Cataloging-in-Publication Data

Fortin, Ernest L.
Human rights, virtue, and the common good : untimely meditations on
religion and politics / Ernest L. Fortin ; edited by J. Brian Benestad.
p.   cm. — (Ernest Fortin, collected essays ; v. 3)
Includes bibliographical references and index.
1.Christianity and politics—History. 2. Christianity and politics.
I. Benestad, J. Brian. II. Title. III. Series: Fortin, Ernest L. Essays ; v. 3.
BR115.P7F676   1996   261.7—dc20   96–23666   CIP

ISBN 0–8476–8278–1 (cloth : alk. paper)
ISBN 0–8476–8279–X (pbk. : alk. paper)

Printed in the United States of America

∞™ The paper used in this publication meets the minimum requirements of
American National Standard for Information Sciences—Permanence of
Paper for Printed Library Materials, ANSI Z39.48–1984.

CONTENTS

III. ANCIENTS AND MODERNS

IV. PAPAL SOCIAL THOUGHT, VIRTUE, AND LIBERALISM

V. PAGAN AND CHRISTIAN VIRTUE

VI. THE AMERICAN CATHOLIC CHURCH AND POLITICS

Contents

FOREWORD

The very first paragraph of Fortin's own foreword to volume 1 provides the best hermeneutical guide to all his essays. Like St. Augustine and "the intellectual tradition to which he was heir" Fortin is persuaded that "anybody who wishes to learn has to start by unlearning—*dediscere.*" This work of discarding opinions is necessary because we all absorb prejudices from the historical cave in which we happen to live. Fortin's essays reveal a scholar who realizes that acquiring self-knowledge—an important goal of any genuine liberal education—is a laborious task. Fortin writes so that he and his readers might see more clearly. In his words:

> [All my essays] were born of a desire to know more about the world in which we live, the philosophic, religious and political forces that shaped it, and the type of human being it tends to produce. As such, they are more preoccupied with achieving a measure of clarity about the ends of human existence than with offering made-to-order solutions to problems to which the pursuit of those ends give rise.[1]

In teaching St. Augustine's *Confessions* I have come to appreciate more and more how helpful Fortin's style of writing really is. Augustine is very persuasive in arguing that self-knowledge is elusive because human beings run from the truth: "They hate the truth for the sake of that other thing which they love because they take it for truth. They love truth when it enlightens them, they hate truth when it accuses them."[2] In other words, people want the object of their love to be the truth. Consequently, they see things the way they *want* to see them, rather than as they present themselves to the unbiased eye.

In my view, the third volume of Ernest Fortin's *Collected Essays* is an important contribution both to the growing literature on the renewal of liberalism and to the few voices calling for new approaches to Christian social ethics and social activism. Fortin's meditations on religion and politics in the contemporary era are only "untimely" because they reflect perspectives that don't fit into conventional categories. His writings reveal a scholar who has meditated on classic texts and digested the most current scholarship. Fortin is one of those rare theologians who has a thorough knowledge of theology, political theory, literature and languages. His perspicacity should appeal to any thoughtful reader.

One of the helpful categories reintroduced into the contemporary discussion of political philosophy is the distinction between the Ancients and the Moderns. Fortin reminds us of this distinction not to create nostalgia for the ancient city or medieval Christendom, but to shed light on contemporary problems. Modernity has been remarkably successful in providing health, prosperity, democracy and freedom, but has generated a host of social problems that need immediate attention. However, it has been difficult to find an approach to such problems as the erosion of religious belief and practice, the breakup of the family, greed, drug and alcohol abuse, and the culture of death (especially the fascination with physician-assisted suicide), as well as ignorance and character defects in people occupying leadership positions in all areas of society. The language of rights—the moral language of America—is not able to describe all the problems, much less come up with solutions. The language of values has been used to pick up the slack, but with mixed success. How can a word that implies the subjectivity of belief about good and evil address real human problems? Is there any precise meaning to such terms as family values or gospel values?

By drawing a distinction between Ancients and Moderns, discussing the tensions between Faith and reason, emphasizing the importance of political philosophy, and even reminding readers that "esoteric writing" does exist, Fortin has contributed to the recovery of perspectives on both the premodern and modern traditions that could promote a deeper understanding of religion and politics. His writings give some indication of how premodern insights can be incorporated into a liberal framework without bringing back undesirable aspects of the ancient and medieval world. For example, Jewish, Christian and classical understandings of virtue help Americans think about the meaning of character and the effect of good and bad character on political and religious leadership. Ironically, the study of virtue could also help Christian churches better understand their own moral teachings. It is also ironic that the study of the Ancients might help Americans understand how to preserve freedom in the modern world. As Fortin notes, Tocqueville's comment that there is "no freedom of the spirit in America" and very little "real freedom of discussion"[3] may still be more true than we are willing to admit.

Fortin's essays describe the interaction between liberalism and Christianity. As everyone knows, liberalism in the United States has taken a decidedly secular turn in recent years. In a recent article James Q. Wilson even speaks of "the social marginalization of religious believers."[4] Their input is just not welcome in the political and cultural realm. Less well known is Fortin's point that Christianity has contributed to its own difficulties by allowing the world to set the agenda for the Church. Consequently, the Church is not only hard pressed to educate its members, but also unable to effectively counter destructive forces in contemporary culture.

A BRIEF NOTE ON MODERNITY

Against the widespread scholarly opinion that Western thought is continuous from Socrates to the nineteenth century, Fortin defends the view that modernity represents a break with classical philosophy and Christianity. "Classical philosophy," writes Fortin, "studies human behavior . . . , in the light of virtue, and it claims to be able to show the way to these goals. It culminates in a discussion of the best life and, on the political level, of the best 'regime' or the type of rule that is most conducive to the best life."[5] The best life is the life of virtue and the best regime promotes the practice of virtue in the lives of citizens. Good government, then, makes for good human beings. Of course, the converse is also true: the predominance of good human beings, especially among those who set the tone for society, makes for good government. Plato and Aristotle were well aware that the best regime would exist only in speech, as a model for actual regimes to imitate as closely as possible. They knew that a great number of circumstances would have to be favorable for the emergence of good regimes, much less the "best regime."

Machiavelli was the first philosopher to break with the classical view. A passage from chapter 15 of *The Prince* reveals his quarrel with the classical orientation:

> And many have imagined republics and principalities that have never been
> seen or known to exist in truth: for it is so far from how one lives to how
> one should live that he who lets go of what is done for what should be
> done learns his ruin rather than his preservation. For a man who wants to
> make a profession of good in all regards must come to ruin among so many
> who are not good.[6]

What has never existed in reality is, for example, the best regime as
described by Plato in the *Republic*. The ideal cannot be realized because
most people do not seek excellence as they ought. Seeing things the way
they are, Machiavelli proposes a political philosophy that abandons the
ideal and takes its bearings by the way most people tend to behave most
of the time. In line with his new orientation Machiavelli proposes a new
"ought." Henceforth, rulers should not rule in such a way as to help
people live as they ought in some ideal sense. Rather, they ought to be
good or bad according to the needs of the situation. In other words, they
should not hesitate to use evil means in order to achieve decent political
goals.

The form in which the trend inaugurated by Machiavelli became
respectable and attractive was through the invention of universal natural
rights by Thomas Hobbes and their elaboration by John Locke. The latter
proposed a beguiling alternative to the classical position that only the
practice of virtue could bring about harmony between the individual and
society or between self-interest and the common good. Locke held that
things could be so arranged in society that "unlimited appropriation with
no concern for the need of others [would be] true charity."[7] The key is to
unleash people's acquisitive passions. Even if entrepreneurs are focused
exclusively on their well-being, their genius creates many well-paying jobs
for others. In Locke's perspective unlimited acquisitiveness is more effective in promoting the common good of society than moderation or charity
because it is more reliable. Simply by pursuing their own selfish goals
people automatically contribute to the realization of the common good.

Locke's political philosophy is largely responsible for the acceptance
of natural rights doctrines. Locke convinced people that they are by nature
free and equal and have rights to life, liberty, and property. I believe that
looking at life through the prism of rights initiated a quiet revolution in
the way people understood the purpose of their own lives as well as the
end of society. It was a quiet revolution because many citizens, including
Church leaders, do not realize that rights are not simply another way of

talking about classical virtue or the teaching of Jesus Christ. In fact, the doctrine of rights presupposes an understanding of human nature "which is no longer defined in terms of its highest aspirations," but rather assumes that people cannot really rise above preoccupation with their own interests. The language and perspective of rights, then, constitutes a sharp break with the idealism of the Ancients. Fortin summarizes this shift in perspective in a way that illustrates the difference between Ancients and Moderns.

> The passage from natural law to natural rights and later (once nature had fallen into disrepute) human rights represents a major shift in our understanding of justice and moral phenomena generally. Prior to that time the emphasis was on virtue and duty, that is, to what human beings owe to other human beings or to society at large rather than on what they can claim from them.[8]

A lawyer's perspective on the revolution effected by the passage from natural law to human rights is provided by Mary Ann Glendon's book *Rights Talk: The Impoverishment of Political Discourse.* Glendon, a law school professor, explains why the preoccupation with creating and asserting rights in the United States undermines public morality. She argues that the pervasive presence of rights talk in political, social, and cultural life causes difficulty in defining critical questions, finding common ground for discussion, and arriving at compromises in the face of intractable differences. Rights talk is silent "with respect to personal, civic, and collective responsibilities."[9] Furthermore, "simplistic rights talk," says Glendon, "simultaneously reflects and distorts American culture. It captures our devotion to individualism and liberty, but omits our traditions of hospitality and care for the community."[10] Americans very naturally express what is important to them in terms of rights, and "frame nearly every social controversy as a clash of rights."[11] These two tendencies lead Americans to misperceive the social dimension of the human person. Under the influence of the rights paradigm they fail to recognize personal and collective duties.

THE EFFECT OF MODERNITY ON CHRISTIANITY

Liberalism and liberal democracy, the contemporary offspring of modernity, incline Americans to think about morality almost exclusively in terms of rights. This in turn leads to a preoccupation with choice and freedom as ends in themselves, and about the sovereignty of the individual and the goods of the body: safety, health, pleasure, and prosperity. The liberal temper is anything but neutral in the moral tone it sets for citizens. It does not encourage the practice of virtue, but rather an unprecedented openness to all human possibilities. Fortin explains:

> What this leads to most of the time is neither Nietzschean creativity, nor a noble dedication to some pregiven ideal, nor a deeper religious life, nor a rich and diversified society, but easygoing indifference and mindless conformism.[12]

In other words, today's version of openness encourages not the pursuit of truth, but rather subservience to public opinion, preoccupation with material things, and a reshaping of religion to suit the temper of the times.

Liberal democracy in the United States has provided significant protection for religion through constitutional guarantees of religious liberty. The churches of America, then, would seem to be in a good position to go about their business of working for the salvation of souls and the promotion of justice in civil society. In fact, public opinion polls still reveal that most Americans identify themselves as religious.

Thoughtful observers of religion in America like Professor Fortin, however, note that a burgeoning religiosity is compatible with the death of God. Religion may continue to exist but remain limited in its ability to inspire people and to educate their minds. In fact, many churches in the United States have been unduly influenced by public opinion, contemporary culture, and modern philosophy, and are now hampered in their ability to make religion a vital force in people's lives. Even non-Catholic scholars, I see, have understood that the Catholic Church is greatly influenced by liberal political principles. Consider the following observation made by William Galston, a professor of political theory, on a meeting that took place in the Vatican between U.S. Catholic bishops and representatives of Pope John Paul II:

> The Roman prelates inveighed against what they saw as the laxity of the American Church. American bishops responded with a fascinating disquisition in which they pointed out, inter alia, that liberal political culture encourages rational criticism of all forms of authority, a tendency the American Church is not free to disregard. The notion of unquestioned authority that seems unimpeachable to a Polish Pope is almost unintelligible to U.S. Catholics, and the American Church has been significantly reconstructed in response to the influence of liberal public culture.[13]

In his chapter, "Rome and the Theologians," Fortin gives a revealing example of how Catholics actually embrace the spirit of the age on principle. He quotes the reaction of *Commonweal* magazine to an important document issued in June of 1990 by the Vatican Congregation for the Doctrine of the Faith under the title "Instruction on the Ecclesial Vocation of the Theologian."[14] "*Commonweal*," he says, ". . . read it as a 'sweeping rejection of what constitutes much of the framework of human consciousness in the twentieth century,' thereby implying that this framework should have the same normative value for the Catholic Church as it supposedly does for everyone else in our time."[15] Embracing contemporary consciousness as normative is not atypical of Catholics in the United States. The fact that a number of people feel or think a certain way about a particular matter becomes a *locus theologicus*, that is, an authoritative criterion for making judgments about the validity of Christian teachings. Disagreements among theologians, among bishops, and between bishops and theologians, are sometimes taken as a sign from the Holy Spirit that a particular question should remain open. It really does not matter that one side may have weaker arguments; as long as it has numbers or influence, the quality of reasoning is not decisive. In other words, it is human consciousness which is decisive.

Trends in Catholic religious education confirm Fortin's insight into the effect of modernity on Christianity. A number of religious educators have followed the signs of the times and devalued the importance of acquiring religious knowledge. Not a few religious education experts prefer to discuss the experiences of students and to create an atmosphere in which they might have "faith experiences" rather than explain the Christian faith in any depth. As a result, many young Catholics have only a meager knowledge of their faith. The typical undergraduate at a Catholic college can hardly recite or explain the Apostles' Creed, or even name a cardinal or theological virtue.

The international commission responsible for putting together the new *Catechism of the Catholic Church*, however, rejected human experience as the primary focus of religious education as, in the words of Avery Dulles, "too arbitrary, . . . too various and ephemeral to offer a solid platform that would apply today and tomorrow, in New York and Madagascar, Bangladesh and Moscow." Instead, adds Dulles, the commission focused the *Catechism* on "the Church's patrimony of faith," rejecting any catechetical method "that would reduce faith to personal experience." The attachment of religious educators to an improper reliance on experience is so strong that Dulles felt compelled to write what should be obvious to the ordinary Christian:

No analysis of contemporary experience can by itself disclose the contents of Christian faith, such as the Trinity, the Incarnation, and the Resurrection, which are known only from revelation. Cardinal Ratzinger rightly finds fault with a kind of "theological empiricism" in which present-day experience is allowed to block the dynamism of the original sources. Speaking of certain European catechetical programs, he remarks that they emphasize experience and method to the detriment of faith and content. Such instruction, he observes, has proved itself incapable of arousing interest. The word of God must be allowed to shine forth again as the power of salvation. The truths of revelation must be presented in their organic unity, apart from which they can seem meaningless.[16]

Catholic universities have also been adversely affected by the reigning secular culture. For example, a number of Catholic universities are more interested in making a core requirement of cultural-diversity courses than in requiring the study of Plato, Augustine and Shakespeare. Even Catholic universities are reluctant to require the reading of the masterworks that have shaped Christianity and Western civilization. Fortin explains,

In true Nietzschean fashion, we are all imbued with the idea that they should be approached as documents of the age that produced them and whose prejudices they reflect. Any thought that they might have something of ultimate importance to teach us about ourselves and the way we should live is out of the question.[17]

Acceptance of this view largely explains why many Catholic universities no longer have the self-confidence to proclaim from the rooftops that all their students will carefully read great works of philosophy, literature, and theology. Administrators have little or no incentive to hold fast before the pressure of the marketplace or the wishes of faculty who argue that their students don't have room in their schedule for getting a thorough liberal arts education.

A third area where modern modes of thought have caused the Catholic Church to modify its teaching is in its work for justice. For the past thirty years the Catholic bishops of the United States and American Catholic theologians have been routinely addressing public issues through the categories of social justice and rights rather than virtue and the common good. As a result they often end up focusing on the needs of the body rather than the transcendent goals of Christianity. Fortin explains:

It surprises [my students] that the bishops, to whom they still look up, spend so much time talking about self-preservation (as they do in the letter on nuclear warfare) and about comfortable self-preservation (as they do in the letter on the economy) and so little time talking about the love of the good and the beautiful.[18]

The bishops seem more comfortable proposing political solutions to complex problems than attempting to clarify for Catholics and other Americans the meaning of virtue and the dependence of the common good on the practice of virtue. The latter, of course, is a very difficult task in the accomplishment of which its bishops need better assistance from Catholic theologians.

When the bishops and theologians do talk about justice, they are often influenced by the new concept of social justice.

> Justice in this new sense, one is tempted to say, is not a virtue at all, in that it has more to do with social structures than with the internal dispositions of the agent. Its proper object is not the right order of the soul but the right order of society as a whole. It shares with early modern liberalism the view that society exists for the protection of certain basic and prepolitical rights, and radicalizes that view by combining it with an emphasis, stemming ultimately from Rousseau, on the need for a greater equalization of social conditions as a means of guaranteeing the exercise of those rights. It thus takes for granted that social reform is at least as important as personal reform and that the just social order depends as much on institutions as on moral character. . . . . Its immediate goal, in short, is to produce happy rather than good human beings.[19]

Commitment to social justice requires having the correct opinions regarding desirable political and social change, mostly through government intervention. It often does not require conversion or the practice of virtue. Having the correct political opinions required by contemporary social justice is wholly compatible with being a prisoner of sin in many aspects of one's life. Being a just person in the biblical sense of justice is not compatible with the practice of any serious sin.

To make his point about the lower standards of behavior introduced by the modern tradition of rights Fortin quotes Edmund Burke: "The little catechism of the rights of men is soon learned; and the inferences are in the passions."[20] Armed with rights, people can defend their interests and make claims on others. The other catechism is not learned and lived so easily. Fortin is not arguing that rights are not important in modern liberal society. Because of the spread of moral relativism, the protection of rights through law may be citizens' most solid bulwark against loss of life, liberty, and property. Nevertheless, thinking about Christian life in terms of rights leads to serious misunderstandings. For example, an important

Christian concept is the dignity of the human person. *Gaudium et Spes* argues that human beings have dignity because God invites them to enter into communion with him. Made in God's image, men and women are capable of knowing and loving their Creator. This having been said, the next comment is usually about rights. Because human beings have dignity, they must be protected by rights—political, civil, and socioeconomic—and the protection of rights requires adequate structures. Thus the dignity of the human person, the foundation stone of Catholic social thought, requires attention first and foremost to structures, especially to public policy. So the argument runs, but it runs too quickly. Vatican II's *Declaration on Religious Liberty* says that people attain true dignity when they seek truth and live their lives in accord with truth.[21] In the measure that people sin and fail to love God and neighbor, they fail to achieve the dignity God intended. The dignity of the human person thus depends primarily on the practice of virtue, especially charity, but also on prudence, justice, temperance, and fortitude. Therefore, it goes without saying that people may have all their rights but still not be inclined to avoid sin and to love God and neighbor.

CHRISTIANITY'S CONTRIBUTION TO CIVIL SOCIETY

The greatest contribution that Christianity makes to civil society, according to Fortin, is to form the character of Christians by teaching them to live the faith with their whole heart, soul and mind. Avery Dulles has emphasized this point on a number of occasions. For example, in one of his annual Lawrence J. McGinley lectures at Fordham University Dulles said:

> The Church, I submit, can make its best contribution to the political order by being itself—by being the community of faith and worship that it was from its earliest days. Where faith is strong, Christians will be honest, loving, merciful, and respectful of the rights of others. They will have a sense of solidarity reaching out to the whole human family.[22]

I note that Pope John Paul II has been trying to persuade the world that "freedom is the measure of [human] dignity and greatness," when it is perfected by solidarity, virtue, and wisdom. Before the General Assembly of the United Nations he said:

> The basic question which we must face is the responsible use of freedom in both personal and social dimensions. . . . . Detached from the truth

about the human person, freedom deteriorates into license in the lives of individuals, and in political life it becomes the caprice of the most powerful and the arrogance of power.[23]

The rationale behind this Christian teaching rests on a simple premise: unless people recognize that the practice of virtue is possible and desirable, most will simply love money and pleasure, as Aristotle observed in his *Nicomachean Ethics*. When people recognize the possibility and desirability of living a virtuous life, they have an incentive to avoid harming others, to stretch themselves to the limit of their capacity, to love God and their neighbors and to do what they can for the common good of the society in which they live. The practice of virtue has the immediate effect of promoting harmony in families and in every other community.

The Christian understanding of virtue takes into account the impossibility of rooting out all the injustices of any particular society because of the perduring effects of original sin. Consequently, Christianity simultaneously encourages individuals to reach for personal perfection, but to have moderate expectations and patience in their work to promote the common good. Christianity, then, can make great demands on human beings without leading them into personal or political fanaticism.

CONCLUSION

It is very difficult to do justice to the richness of Fortin's thought. I have discussed briefly his contribution to the renewal of liberalism and focused on the importance of his works for Christian theology and the life of the Church. Much more should be said about his contribution to the study of political philosophy and the various relations between faith and reason. I would simply note that without some knowledge of political philosophy contemporary Christians will rarely appreciate the connections between faith, virtue and the common good.

I would like to conclude by directing attention to Fortin's hope in the power of liberal education. He shows how this great human good can help us to see more clearly. A passage from Xenophon's *Memorabilia* nicely reveals what Fortin tries to do in all his essays. Xenophon recalls that Socrates used to say:

> As others take pleasure in a good horse, or dog, or bird, so I rather take even greater pleasure in good friends. And if I know anything good, I teach it and introduce my friends to others from whom I think they might derive

some benefit with a view to virtue (*aretē*).

Socrates goes on to say that he and his friends explore the writings of the wise, extract whatever good is to be found, and "consider it a great gain if they become helpful to one another."

·

J. Brian Benestad
University of Scranton
September 11, 1996

NOTES

1. Ernest Fortin, *The Birth of Philosophic Christianity: Studies in Early Christian and Medieval Thought*, edited by J. Brian Benestad (Lanham, MD: Rowman & Littlefield, 1996), lx.

2. St. Augustine, *Confessiones* (*Confessions*) x.23.

3. Alexis de Tocqueville, *Democracy in America*, edited by J. P. Mayer with a new translation by George Lawrence (Garden City, NY: Doubleday and Company, Inc., 1969), vol. 1, part 2, chapter 7, 255-56.

4. James Q. Wilson, "Liberalism, Modernism, and the Good Life." in *Seedbeds of Virtue Sources of Competence, Character and Citizenship*, edited by Mary Ann Glendon and David Blankenhorn (Lanham, MD: Madison Books, 1995), 29.

5. Ernest Fortin, *Human Rights, Virtue, and the Common Good: Untimely Meditations on Religion and Politics*, edited by J. Brian Benestad (Lanham, MD: Rowman & Littlefield, 1996), 148.

6. Machiavelli, *The Prince*, translated and introduced by Harvey C. Mansfield, Jr. (Chicago: University of Chicago Press, 1965), 61.

7. Leo Strauss, *Natural Right and History* (Chicago: University of Chicago Press, 1953), 243.

8. *Human Rights, Virtue, and the Common Good*, 20.

9. Mary Ann Glendon, *Rights Talk: The Impoverishment of Political Discourse* (New York: The Free Press, 1991), x.

10. *Rights Talk*, xi-xii.

11. *Rights Talk*, 4.

12. *Human Rights, Virtue, and the Common Good*, 11.

13. William Galston, *Liberal Purposes: Goods, Virtues and Diversity in the Liberal State* (Cambridge and New York: Cambridge University Press, 1991), 292.

14. *Human Rights, Virtue, and the Common Good*, 63.

15. *Human Rights, Virtue, and the Common Good*, 63-64.

16. Avery Dulles, S.J., *The Challenge of the Catechism* (New York: Fordham University, 1994), 20-21.

17. *Human Rights, Virtue, and the Common Good*, 42.

18. *Human Rights, Virtue, and the Common Good*, 310.

19. *Human Rights, Virtue, and the Common Good*, 273-274.

20. *Human Rights, Virtue, and the Common Good*, 208.

21. Vatican Council II, *Dignitatis Humanae* (*Declaration on Religious Freedom*), no 2.

22. Avery Dulles, S.J., *Religion and the Transformation of Politics* (New York: Fordham University, 1993), 16.

23. Pope John Paul II "Address to the General Assembly of the United Nations, Oct. 5, 1995." *Origins* 25, no. 18 (1995): 297-298.

24. Xenophon, *Memorabilia* I.6.14.

ACKNOWLEDGMENTS

The author and editor wish to take this opportunity to thank all those who had a direct hand in the launching and the realization of this project, among them: Stephen F. Brown, Matthew Lamb, and Patrick J. C. Powers. Daniel J. Mahoney found the right publisher and graciously contributed the Foreword to the second volume. The idea of publishing the three volumes simultaneously and as a set belongs to Jonathan Sisk, the editor-in-chief of Rowman and Littlefield, without whose encouragement and active support the project would probably never have seen the light of day.

Invaluable assistance was provided by the editorial staff of Rowman and Littlefield and, in particular, by Julie Kirsch and Dorothy Bradley, whose expertise and endless patience won both our gratitude and our admiration. We are also grateful to Phillip Wodzinski, a doctoral student in political science at Boston College, whose research talents and proof-reading skills, hitherto unknown to the world, revealed themselves to superb advantage on this occasion. Ann King, the retired secretary of the Department of Theology and Religious Studies at the University ·of Scranton, toiled endlessly and in a completely selfless way on virtually every page of the manuscript, at the risk of being driven out of her mind by its innumerable references to works in at least six different foreign languages, including Greek and Latin. Her dedication was paralleled by that of Shirley Gee, the omnicompetent administrative assistant of the Institute of Medieval Philosophy and Theology at Boston College. Further secretarial help was provided by Marie Gaughan, Patricia Mecadon, and Barbara Quinn, of the University of Scranton. A recent University graduate, Steven Pustay, generously offered assistance to the editor through his work on the endnotes. The staff of the Thomas O'Neill Library, Boston College was always there to help whenever necessary and also deserves special mention. The author is immensely grateful for generous financial help in the form of research fellowships received from the John M. Olin Foundation, the Lynde and Harry Bradley Foundation, the National Endowment for the Humanities, and the Boston College Graduate School.

None of the articles in these three volumes could have been written without constant input from the author's colleagues and daily conversation partners in the departments of Theology, Political Science, and Philosophy at Boston College, along with friends and long-time associates in other colleges and universities here and abroad. They shall not be listed individually for fear that too many names should inadvertently be left out. If they are ever tempted to peruse these books, they will have no trouble identifying their respective contributions.

The editor is especially indebted to his colleagues at the University of Scranton, Father Richard Rousseau, S.J., and Dr. Edward Mathews, as well as to Betsy Moylan, a librarian at the University, for their expert assistance. He thanks his wife, Janet Benestad, and their daughters Katherine and Lizzie, for help with various aspects of the project. He also expresses his appreciation to the University of Scranton for grants to support the writing of his foreword and the typing of numerous articles, and to Dr. Thomas Hogan and Dr. Richard Passon for their support of the whole endeavor.

Chapter 1. "The Regime of Separatism: Theoretical Considerations on the Separation of Church and State" is reprinted from *Modernity and Religion*, edited by R. McInerny, © 1994 University of Notre Dame Press, 145-165. By permission of the University of Notre Dame Press.

Chapter 2. "Human Rights and the Common Good" is reprinted from the *CCICA Annual* 13 (1994): 1-16. By permission of the Catholic Commission on Intellectual and Cultural Affairs.

Chapter 3. "Christian Education and Modern Democracy" in *Liberal Education and American Democracy*, edited by R. Utley, forthcoming. By permission of R. Utley.

Chapter 4. "Is Liberal Democracy Really Christian?" is reprinted from *Free Inquiry* 4, no. 2 (Spring, 1984): 32-35. By permission of *Free Inquiry*.

Chapter 5. "Do We Need Catholic Universities?" is reprinted from *Fellowship of Catholic Scholars Newsletter* 16, no. 3 (June 1993): 2-6. By permission of Fellowship of Catholic Scholars Newsletter.

Chapter 6. "Rome and the Theologians" is reprinted from *Crisis* 9, no. 5 (May 1991): 16-20. By permission of *Crisis*.

Chapter 7. "In the Shadow of the Gallows: Criminal Justice, Its Rationale and Its Limitations" is reprinted from *The World and I* (March 1990): 501-15. By permission of *The World and I*.

Chapter 8. "Augustine, the Arts, and Human Progress" is reprinted from *Theology and Technology: Essays in Christian Analysis & Exegesis*, edited by C. Mitchum and J. Grote, © 1984 University Press of America, Lanham, Maryland, 193-208. By permission of the University Press of America.

Chapter 9. "Science as a Political Problem" is reprinted from *CCICA Annual* 3 (1984): 35-41. By permission of the Catholic Commission on Intellectual and Cultural Affairs.

Chapter 10. "The Bible Made Me Do It: Christianity, Science, and the Environment" is reprinted from *The Review of Politics* 57, no. 2 (Spring 1995): 197-223. By permission of *The Review of Politics*.

Chapter 11. "Otherworldliness and Secularization in Early Christian Thought: A Note on Blumenberg." Previously unpublished.

Chapter 12. "Thoughts on Modernity" is reprinted from *Conditions and Purposes of the Modern University*, © 1965 The Catholic Commission on Intellectual and Cultural Affairs, Washington, D.C., 19-30. By permission of Catholic Commission on Intellectual and Cultural Affairs.

Chapter 13. "Natural Law" is reprinted from *The Encyclopedia of Democracy*, vol. 1, © 1995 Congressional Quarterly Inc., Washington,

D.C., 878-880. By permission of Congressional Quarterly, Inc.

Chapter 14. "Troeltsch and Christendom." Previously unpublished.

Chapter 15. "Gadamer on Strauss" is reprinted from *Interpretation* 12, no. 1 (January 1984): 1-13. By permission of *Interpretation*.

Chapter 16. "Sacred and Inviolable: *Rerum Novarum* and Natural Rights" is reprinted from *Theological Studies* 53 (1992): 203-33. By permission of *Theological Studies*.

Chapter 17. "From *Rerum Novarum* to *Centesimus Annus*: Continuity or Discontinuity?" is reprinted from *Faith and Reason* 17, no. 4 (Winter 1991): 399-412. By permission of *Faith and Reason*.

Chapter 18. "The Saga of Spiritedness: Christian Saints and Pagan Heroes." Previously unpublished.

Chapter 19. "In Defense of Satan: Christian Perspectives on the Problem of Evil" is reprinted from *This World* 25 (Spring 1989): 50-58. By permission of *This World*.

Chapter 20. "Social Activism and the Church's Mission" is reprinted from *Center Journal* 1, no. 3 (Summer 1982): 33-41. The *Center Journal* has discontinued publication.

Chapter 21. "Church Activism in the 1980s: Politics in the Guise of Religion?" is reprinted from *Religion and Politics*, edited by F. Baumann and K. Jensen, © 1989 University Press of Virginia, 33-50. By permission of the University Press of Virginia.

Chapter 22. "Theological Reflections on *The Challenge of Peace: God's Promise and Our Response, A Pastoral Letter on War and Peace* (May 3, 1983)" is reprinted from *Catholicism in Crisis*, vol. 1, no. 8 (July 1983): 9-12. By permission of *Crisis*.

Chapter 23. "Christianity and the Just-War Theory" is reprinted from *Orbis* 27 (1983): 523-35. By permission of *Orbis*.

Chapter 24. "Catholic Social Teaching and the Economy: Criteria for a

Pastoral Letter" is reprinted from *Catholicism in Crisis* 3, no. 2 (January 1985): 41-44. By permission of *Crisis*.

Chapter 25. "The Trouble with Catholic Social Thought" is reprinted from *Boston College Magazine* (Summer 1988): 37-42. By permission of *Boston College Magazine*.

Chapter 26. "Friend and Teacher: Allan Bloom's Obsession with the Mystery of the Soul" is reprinted from *Crisis* 11, no. 1 (January 1993): 38-40. By permission of *Crisis*.

I

CHRISTIANITY AND THE LIBERAL
DEMOCRATIC ETHOS

THE REGIME OF SEPARATISM: THEORETICAL CONSIDERATIONS ON THE SEPARATION OF CHURCH AND STATE

The late Leo Strauss, who did more than any other twentieth-century scholar to restore political philosophy to its place of honor among the human disciplines, is reported to have said that in the last analysis there are two interesting things in life, God and politics, and furthermore that today we have neither. By "God" he meant the realm of speculative wisdom, once thought to be the preserve of the gifted and inquisitive few who are tortured in their flesh by the distinction between truth and untruth and take greater delight in exploring the mysteries of the universe or the human soul than in performing such notable deeds as might win the acclaim of their fellow human beings. By "politics" he meant the realm of practical wisdom, which finds its most perfect expression in what the Greeks called the *polis*, the self-sufficient human association and the one most favorable to the exercise of the noble virtues that characterize life at its human, as distinguished from its divine, best.

These were the fundamental alternatives between which generous souls were summoned to choose, the highest goals to which they could aspire and in terms of which human excellence was defined. The two were in latent and at times open tension with each other insofar as they moved in different directions or pointed to different ends. Civil society depends for its stability and well-being on the attachment of its citizens to a set of laws and a corresponding way of life that are never completely rational, but that attachment is undermined by the philosopher's unswerving dedication to reason. In simple terms, what is good for the philosopher is not necessarily good for the city, and vice versa. Theoretically, the problem was insoluble, but it did admit of a practical solution. Once alerted to the politically dangerous character of his enterprise and therewith to the precariousness of his situation, the philosopher could respect the needs of society and observe a measure of caution in his dealings with it, avoiding in word and deed anything that might unduly disrupt its life. Circumstances permitting, he might even be able to contribute in a remote way to its improvement. The compromise was a sensible one. With a minimum of good will, most people could live with it.

It is nonetheless significant that the political philosopher whom I began by quoting spoke of God and not simply of the theoretical life as the alternative to the life of gentlemanly statesmanship, thereby showing himself more pious in speech than Aristotle had been. The change in terminology draws attention to the fact that the horizon within which the issue was debated by the philosophers of classical antiquity underwent a profound transformation as a result of the emergence of the great monotheistic religions—Judaism, Christianity, and Islam—all three of which called into question the status of philosophy and politics as they had hitherto been understood. By virtue of the numerical preponderance of Christianity, two themes came to dominate the philosophic and theological literature of the West: that of the relationship between Jerusalem and Athens or between divine revelation and philosophic reason, and that of the relationship between Peter and Caesar or between the spiritual power and the temporal power. My main concern is with this second theme, especially as it presents itself within the context of a society such as ours, which holds firmly to the separation of church and state and does not in principle favor any particular religion or religious group. The question is whether the regime of separatism, with which America was the first country to experiment, has benefitted the churches by revitalizing them, purging them of worldly ambition, preserving them from artificial irrationality, and making it possible for them to carry on their work in a true spirit of charity and inwardness, or whether, by depriving them of state support as well as by emancipating them from state control, it has weakened them, jeopardized their integrity, and posed a threat to sound public policy. But first a few remarks about the relationship of Christianity to the political order, which, I hope, will bring the issue into proper focus.

CHRISTIANITY AND THE TEMPORAL ORDER

The problem can best be approached by saying that Christianity differs from both Judaism and Islam on the one hand and from pagan religion on the other by its uniquely apolitical character. Its appearance on the scene marked a turning point in history in that it stood for a principle that claimed to be wholly independent of civil society, and thus sundered the age-old unity of the religious and the political.

Few modern authors have done a better job of elucidating this phenomenon than the great nineteenth-century historian Fustel de Coulanges, in the concluding chapter of his classic study of Greek and Roman political society. Here for the first time was a religion that was not the source of law, did not call for the establishment of a particular regime, refrained from laying down rules for the governance of civil society, and saw no need to prescribe the rites that give shape to the communal life of that society. The God that it proclaimed was not attached to any tribe, people, or race. In contrast to the gods of the *polis*, he did not command hatred between nations and make it a duty to detest foreigners but taught that all human beings were entitled to the same gentle and benevolent treatment. He was a God for whom there were no strangers to profane the temple or taint the sacrifice by their presence. His worshippers were not required to renounce their citizenship in the earthly city but formed a spiritual and sacramental community that transcended national boundaries, held together as it was by faith in Christ and the bonds of brotherly love. To be a member of that community, one had only to render to Caesar what is Caesar's and to God what is God's. A saving doctrine, which promised immortality and in which anyone could be instructed, had replaced the sacred or ancestral law as the high road to happiness.

Though not itself political, the new ideal was fraught with enormous political ramifications. No religion had ever effected such a radical change in the manners and sentiments of people, and never before had God and the state been so clearly set apart from each other. In Fustel de Coulanges's own words,

> The first duty no longer consisted in giving one's time, one's strength, one's life to the state. Politics and war were no longer the whole of man; all the virtues were no longer comprised in patriotism, for the soul no longer had a country. Man felt that he had other obligations besides that of living and dying for the city. Christianity distinguished the private from the public virtues. By giving less honor to the latter, it elevated the former; it placed God, the family, the human individual above country, the neighbor above city.[1]

This is not to imply that henceforth religion and society would have nothing to do with each other but only that half of one's life would be withdrawn from the control of Caesar and directed to the service of God. Human beings would belong to civil society by only one part of themselves, and not the most important one at that. In time, the emperor would surrender his jurisdiction over all religious matters and the title of *pontifex maximus* or "supreme pontiff," which had once been his and his alone, would be passed on to the successor of Peter.

The fact that the New Testament religion was not specifically political in character accounts for both its spiritual dynamism and the peculiar difficulty it has always experienced in adjusting to the temporal sphere. Christianity would probably have gone the way of the radical sects of late antiquity had it not been so well suited to the needs of the time and had its early apologists not been able to make a plausible case for the compatibility of its goals with those of society at large. As history would demonstrate, it had a lot to offer to the society within which it developed. Properly supported, it could be pressed into service to counteract the forces that threatened the dissolution of an inordinately large and unwieldy political structure. The spirit of moderation and law-abidingness that it inculcated was likely to improve people's manners, especially at a time when the traditional sources of morality were showing signs of decay. Laws are effective to the extent that they are accompanied by habits of decency and self-restraint on the part of most citizens; left to themselves, they seldom inspire virtue and are even less capable of containing vice. Yet, it was obvious that neither education nor pagan religion, the two principal agencies on which governments had formerly relied for this purpose, was adequate to the task. A universal and despotic empire is not the most suitable locus of moral education, and the old religion of the city, which had been on the wane for years, was not about to be revived.

Much more could be expected from the new religion, which addressed itself to everyone regardless of language, ethnic background, or local tradition. Christians had long rejected the idea that they constituted a *triton genos*, or separate race, side by side with Jews and Gentiles. Unlike other religious groups, they were not given to living in isolation or withdrawing from society altogether. One found them everywhere, mingling freely with the rest of the population, sharing their customs, their dress, and within broad limits, their way of life. The moral teaching to which they subscribed may have discouraged spiritedness but it enjoined the practice of every other virtue. Once generally accepted, it could be counted on to curb the selfish passions and propagate sentiments of truth, justice, and harmony among people who would regard themselves as common children of the one true God. Thanks to its influence, the dissensions racking the empire were less apt to erupt into bloody strife. In Gibbon's memorable words,

> a prudent magistrate might observe with pleasure and eventually support the progress of a religion which diffused among the people a pure, benevolent, and universal system of ethics, adapted to every duty and every condition of life, recommended as the will and reason of the supreme deity, and enforced by the sanction of eternal rewards and punishments.[2]

The cost on the other side of the political ledger was, all things considered, minimal. Even the depreciation of military valor, which at other moments and under different circumstances would have been a serious liability, had suddenly turned into a distinct asset, favoring the ends to which imperial policy was committed out of self-interest. By the same token, emperors had little to fear from a religion that derived the institution of civil government from the will of God, frowned upon sedition, and discountenanced worldly ambition with as much vigor as it extolled the virtue of obedience to one's divinely sanctioned rulers. If even under a Nero, St. Paul had preached the duty of unconditional submission to the constituted authority (Rom. 13:1-7), how much more willing would Christians be to acquiesce in the rule of a prince who was at the same time a patron and a defender. There were other benefits as well, not the least of which is that it freed rulers from the shackles of ancestral religion and allowed them to govern without having to bend to sacred usages or resort to oracles, auspices, or any other form of divination. The new alliance was clearly advantageous to both parties. By a miraculous convergence, it served the best interests of heaven and earth.

The Constantinian settlement and the subsequent establishment of Christianity as the official religion of the empire brought the era of hostilities to an end and vastly augmented the Church's material resources. Christians were indemnified for the losses incurred during the persecutions, basilicas were built at public expense, new copies of the Bible were procured, bishops were provided with residences and welcomed at the court, and a fixed portion of the provincial revenues was allocated to charity.[3] More important, state power could now be used to suppress religious dissent and insure the triumph of the new faith by liquidating its major rivals. It is not by chance that to this day the geographical limits of Christianity coincide roughly with those of the late Roman Empire, with the exception of Russia, a few small enclaves in the Far East, and the countries of the New World colonized by Europeans. By and large, the Christian world as we know it is a product of the inner transformation of the former mistress of the nations.[4]

What the Church gained in worldly power, however, it tended to lose in spiritual vigor. With increased political protection came a greater dependence on the emperor, who began to play an active role in its affairs, intervening to restore order within it, convoking councils, directing their proceedings, and on occasion imposing dogma. Not surprisingly, from the fourth century onward, the center of gravity of the spiritual life shifted from the larger community to the Egyptian desert and the monastery, to which Christians fled in substantial numbers and whence over the centuries would come virtually all of the great reform movements in the Church.

The same pattern was to repeat itself with interesting variations throughout the Middle Ages, the political history of which is dominated by the creation of the Holy Roman Empire for the express purpose of securing the papacy against the threats to its existence on the part of a heretical Byzantine emperor and his Western allies. I do not wish to rule out considerations of a providential nature, but one has to suppose that without the help of the state, Christianity would not have been able to implant itself in the West and maintain its sway over the minds and hearts of most of its inhabitants. This brings us back to our initial question, which is whether Christianity can be expected to fare equally well under a regime that neither favors nor opposes it but relegates it to the private sphere, on a par with every other religion.

THE SECULAR CITY AND ITS RELIGION

Tocqueville, an acknowledged authority in the matter, was persuaded that the churches would, indeed, prosper under the liberal democratic creed of the separation of church and state. This much was evident from recent European experience, which showed that the real cause of the discredit into which Christianity had fallen lay in its organic ties to a corrupt and oppressive regime.[5] The people had never been hostile to religion as such. The proof was that no sooner had the Revolution accomplished its purpose than a massive return to it began to take shape. Tocqueville himself was impressed by the vibrancy of the churches in America in contrast to those of Europe and attributed that vibrancy to two factors, religious liberty or the relegation of religion to the realm of private choice and the willingness of priests and ministers to steer clear of public office. There was no more desirable solution to the church–state problem than the one for which America had opted. Separatism was the new and divinely ordained rule, and, in a democratic age, it was the only one that had any chance of success. No other arrangement was compatible with the principles of freedom and equality to which the regime was dedicated.

Still, we know that Tocqueville was not a little concerned about the impact of the new regime on the character of the nation. A new type of human being was aborning who could be admired in many ways but was not superior in all of them to the one it was destined to replace. The symptoms of bourgeois mediocrity were everywhere present: in the decline of political oratory, in the impoverishment of America's literary and artistic life, in the overriding preoccupation with material well-being. They were also present in the sermons preached by clergymen. Priests and ministers knew better than to run for public office, but their attention was very much on earthly things. In listening to them, it was impossible to tell "whether the main object of religion is to procure eternal felicity in the next world or prosperity in this."[6] There was no doubt in his mind that Christianity and democracy could live at peace with and support each other, and, anyway, one no longer had any choice in the matter; but the harmony between them had been purchased at the price of an extraordinary accommodation to the spirit of modernity.

Not much has happened since then to cast doubts on the validity of this perspicacious if somewhat troubling analysis. As Tocqueville predicted, religion has not ceased from the land, far from it. There may even be more of it around than there has been for quite some time. According to an extensive survey commissioned by the Graduate School of the City University of New York and conducted among 113,000 people between April 1989, and April 1990, 92.5 percent of Americans describe themselves as religious, and 86.4 percent as Christians. Only 7.5 percent profess no religion at all. The village atheists, who until a few years ago were still advertising their wares in the national papers, have mysteriously disappeared, gone underground, or converted. God has his weekly column in *Time* magazine and *Newsweek* (even if he is not always well served by his editors), a good number of journals of his own, and, lest legislators should fail to take heed, a powerful lobby in Washington.

He also has his regular TV shows, on which he is half expected to make a personal appearance one of these days. If the extraordinary success of a few popular preachers proves anything, it is that he is still very much in business, too much perhaps. The Religious Right has become a major force on the political horizon, and the Religious Left, momentarily thrown off stride by the collapse of Marxist socialism, is not about to admit defeat. In the meantime, cults of all sorts proliferate, some more eccentric or more frightening than others. Even Harvey Cox has taken to trumpeting the rebirth of religion in the midst of a city from which God was supposed to have been banished with the advent of the secular city a quarter of a century ago.[7]

The aim of my facile caricature is not to disparage these outbursts of evangelical or charismatic fervor but to suggest that they remain ambiguous and that all may not be for the best in the state of our divine affairs. If our ubiquitous pollsters were to train their sights on astrology or fortune-telling, they would probably discover that they, too, are on the rise, and for the same reasons. Whereas it is customary to speak of the Middle Ages as the "age of faith," no one would dream of applying that label to our time, if only because the house of religion is more fragmented than ever. Maybe a religious revival is on the way, and there is no reason to doubt the sincerity of many of those who have undertaken to lead it; but that still does not dispense us from raising questions about its true nature.

In a brilliant lecture entitled "The Almost-Chosen People: Why America is Different,"[8] the British historian Paul Johnson noted a few years ago that our country has gone through four great religious "awakenings" in the course of its relatively short history: one in the 1730s, another between 1795 and 1835, a third from 1875 to 1914, and finally the one that we are witnessing at the present moment. If we need to be forcefully reawakened every forty years or so, it must be that the roots of our religious convictions are not as deep or as vigorous as one might think.

Nietzsche had already warned us that the death of God is perfectly consistent with a burgeoning religiosity—he had seen it with his own eyes, even though by the end of the century it was on the verge of exhaustion—and he did not think for a moment that "religion" was finished. What he questioned was its ability to move people and elevate their minds. The morning paper had been substituted for the morning prayer and people were more in touch with one another than ever before. The trouble is that they were not in touch with anything higher than themselves. Their "religion" was a product to be consumed, a form of entertainment among others, a source of comfort for the weak who like to think of themselves as the center of a universe that modern science has effectively destroyed, or an emotional service station destined to fulfill certain irrational needs that it satisfies better than anything else. As one-sided as it may sound, Nietzsche's diagnosis is very much to the point. If, instead of consulting the polls, we were to listen to the sermons that are being preached in most churches—and here I do not presume to speak for any church other than my own—we might easily come to a similar conclusion. One of the accomplishments of Vatican II was to persuade everybody that the Sunday sermon should be called a "homily." Nothing but the name has changed. Before the Council, we had bad sermons; now we have bad homilies. To be sure, this is not the only touchstone, but, as Tocqueville realized, it is an important one, and it has high visibility in its favor.

THE IMPACT OF THE REGIME OF SEPARATISM

In retrospect, it is not easy to determine whether the separation of church and state has done much to reinvigorate religion, make it less worldly, and rid it of its less desirable accretions. The opposite could be true for two main reasons. The first, which is hinted at by Toqueville, is that the disestablishment of the churches renders them more vulnerable to the pressure of public opinion. Deprived of state aid, these churches have been forced to compete for their members and rely on voluntary contributions for their subsistence. They have thus been placed in the position of having to cater to the changing tastes of those whom they serve. Tocqueville observes rightly that the churches in America were not at liberty to contradict the passions that the commitment to the pursuit of material wealth arouses or propound teachings that ran counter to "the prevailing ideas or the permanent interests of the mass of the people." Religion would henceforward owe part of its vitality to the "borrowed support of public opinion," outside of which there was no force capable of sustaining a prolonged resistance.[9] Left to follow its bent, the human spirit tends to "regulate political society and the City of God in uniform fashion."[10] People wedded to worldly prosperity generally feel better when they can combine that prosperity with moral delights, harmonizing as it were heaven and earth. The only solution, then, was to lower the former in order to elevate the latter. By trying to do less one might accomplish more. Under such circumstances, however, the religious spirit was less likely to be tested and ascend to the heights it was able to reach in the souls of the great mystics of former ages, such as Pascal, for whom Tocqueville professed unbounded admiration.

The gravest danger was not that society would become secular but that religion would either go soft or lose sight of its transcendent goal. The early Christians fought bitter battles over the Trinity or the union of the two natures in Christ. Their late twentieth-century counterparts prefer to talk about things that have rather more to do with the needs of the body. It is a sure "sign of the times," to use one of their favorite expressions, that the Catholic bishops of this country have poured huge sums of time, energy, and money into the study of nuclear warfare and the American economy, that is, self-preservation and comfortable self-preservation, at the risk of giving the impression that they regard these matters as more important than the ones that pertain to the good of the soul. It would have astonished our predecessors to be told by religious authorities that mere life is the greatest good and the one that must be protected at all cost. True, there are other items on bishops' pastoral agenda, but they are not necessarily more spiritual in nature. One cannot wait to hear what these same bishops will finally decide to say about other "timely" issues, such as priestly celibacy or the role of women in the church, again in response to pressures from below. If even the Roman Catholic Church, with its notorious reputation for authoritarianism, is so easily influenced by the voices of the world, it is hard to imagine that things might be different elsewhere.

The objection to this argument is that, far from being swayed by worldly voices, our church leaders are ahead of their flocks, that their function is precisely to form public opinion by interpreting the "signs of the times," and that they are being "prophetic" when they espouse political views for which ordinary churchgoers have thus far shown little sympathy. Until recently, the liberals in the American Catholic hierarchy were thought to outnumber the conservatives by a ratio of roughly four to one. How, then, can one accuse them of being followers rather than leaders? I note only that the generation that recently began to move into positions of authority received their training during the turbulent sixties and tend to be out of step with the rest of the country. If so, one can expect a major shift before long. The laity will again speak, with deeds if not with words. The straws are in the wind. When funds are withheld, the mind is wonderfully concentrated and not infrequently changes. In the name of "accountability," vast organizations suddenly discover that the time has come for a sweeping overhaul, such as the one announced some time ago by the National Council of Churches. God may be dead, but in no way must he be allowed to go broke.

The second reason, which is a corollary of the first, is that, while the separation of church and state favors religion by guaranteeing its free exercise, it also saps its vitality by turning it into a matter of private choice. In view of the sad history of religious strife among Christian groups during the early modern period, one might well look favorably upon an arrangement that disestablishes religion once and for all, and America prides itself on having successfully implemented just such an arrangement; but it was able to do so only by officially treating all religions as equal, leaving it to each individual to make up his own mind about their relative merits. Tocqueville predicted that, because human beings are innately religious, this privatization would be more of a boon than a handicap to religion and was therefore something to be welcomed rather than resisted. The truth of the observation seems to be borne out by recent history, which shows that the Western countries where the Church retains its established status—Great Britain and Scandanavia may be the prime examples here—are not as a rule noted for the intensity of their spiritual life.

At first glance, it makes sense to say that people will be more devoted to a church or a religion that they have chosen of their own free will. Yet we know from experience that the combination of freedom and attachment, or of individuality and commitment, behind which lies the Nietzschean dream of a "free spirit" coupled with a "tethered heart,"[11] is not as easily achieved as our social theorists seem to believe. For all his outrageous statements, Nietzsche had a clearer perception of what it would take to make such an unlikely combination work. The condition of its possibility, as he saw it, was "creativity," and since there was so little creativity left in the modern world—his contemporaries were highly "civilized" but they had no "culture"—it was imperative that it be restored. What human beings needed for this purpose was not more freedom, of which they already had too much, but will, discipline, and a tyranny the like of which the world had never seen. The prospects, horrifying as they were, had to be faced squarely, lest the process of degeneration upon which the West was embarked be allowed to run its course.

Although nobody in his right mind would wish to experiment with such a remedy, it remains to be seen whether our own liberal and more gentle regime can perform the extraordinary feat of intensifying our commitments without diminishing our freedom. If the state is indifferent to religion or, at the very least, to the distinction between religions, chances are that most of its citizens will be indifferent to it as well. Better in a way that the government should be straightforwardly antagonistic toward Christianity, for such antagonism usually has the effect of strengthening the resolve of believers, as it did during the early centuries and still does in some parts of the world today.

Therein lies the difficulty to which from the start liberalism was exposed. Liberal democracy is unique among regimes in that it does not seek to define the goals of human existence or produce a specific type of human being. Its object is rather to provide a neutral framework within which each individual is allowed to choose his own goal and find his own way to it. But by according the same respect to all religions, it implicitly denies that any of them has an intrinsic claim to this respect. To that extent, it inevitably works against religion, for few people are likely to acquiesce in the stringent moral demands made on them by their religion unless they believe in it, heart and soul.

Differently stated, if the vaunted pluralism of our liberal democratic system is to have any meaning, it must exclude its opposite; for, like every other "ism," it too is a monism. Its basic premise, asserted absolutely, recoils upon itself. This leaves us with a neutrality that is more apparent than real. Contrary to its stated aim, liberal democracy does breed a specific type of human being, one that is defined by an unprecedented openness to all human possibilities. What this leads to most of the time is neither Nietzschean creativity, nor a noble dedication to some pregiven ideal, nor a deeper religious life, nor a rich and diversified society, but easygoing indifference and mindless conformism.

Partly because of the steady influx of immigrants to this country, many of whom came from strong religious backgrounds, the corrosive effects of the regime of separation remained hidden from sight for a long time. They have since become visible to the naked eye and are reflected in the way people have taken to speaking about themselves. One "happens" to be a Jew or a Christian, a Protestant or a Catholic, but one could just as easily be something quite different, a "secular humanist," say, or maybe nothing at all. Or else one is "into" religion, almost in the same way that one is "into" painting or drugs. Religion is a "value" among others, and values, we are told, are subjective. One has to be committed to them, for otherwise a properly human life is impossible, but one cannot give any reasons for them. Only intellectual laziness prevents us from taking other religions seriously, and that laziness is itself engendered by the conviction that no religion, not even our own, deserves to be taken with ultimate seriousness.[12]

THE ORIGINS AND TRIUMPH OF SEPARATISM

It goes or should go without saying that the liberal position just outlined is alien to the Hebrew Scriptures as well as to the New Testament, both of which look upon the way of life that they enjoin as the one true way and thus implicitly condemn modern religious pluralism. How the liberal view came into being is a question over which scholars are divided, some holding that it is a product of the later Christian tradition or some form thereof, such as the left wing of the Protestant Reformation, and others that it emerged by way of a radical break with the whole of the premodern Christian tradition. According to the first group, the seventeenth-century conflict out of which it arose was a civil war, that is, a war between two Christian groups, the moderates and the fanatics, or "enthusiasts," as they were then called. According to the second, it was a war between two intellectually different continents, one made up of believers and the other of rationalists or nonbelievers. For obvious reasons, the rationalists, if such they were, had much to gain by creating the impression that their defense of religious freedom and toleration was grounded in the Bible. Whether this was their own personal conviction is unlikely. I, for one, fail to see how anybody can come away from Spinoza's *Theologico-Political Treatise* or Locke's *Letter on Toleration* thinking that both books are nothing but a fresh version of what the Bible and Christianity had always taught.

Be that as it may, the point to note is that prior to that time no religious group, Christian or otherwise, ever subscribed to a "theory" of toleration. Toleration, while often encouraged, was taken to be a matter of practical policy rather than of universal principle. Each case had to be judged on its own merits. Religious minorities could be tolerated, but they could also be repressed if their presence or their conduct was thought to pose a threat to the common good of society. Admittedly, the solution left the door open to a wide variety of abuses, but, given the fluidity of human affairs, no better one was available. Besides, it had its redeeming features. Judiciously applied, it had the twofold advantage of making it possible to hold to the superiority of one's religion without insulting the religious convictions of others; for there is a real sense in which only the religiously committed person is capable of understanding the commitment of the religious person and thus of developing a genuine appreciation for what is most profound in him. The same cannot be said of the dogma of the separation of church and state, which wittingly or unwittingly demotes all religions by placing them on the same footing.

For better or for worse, the only organized resistance to this dogma on the part of a major church body came from Roman Catholicism. From the Restoration of 1815 to the early 1960s, the problem of the Catholic Church's relation to modern liberal society was typically couched in terms of the distinction between "thesis" and "hypothesis." The "thesis" represented the ideal situation, often nostalgically identified with medieval Christendom, which demanded that the whole of the public life be informed by Christian principles. Since, as the saying went, "error has no rights," civil governments were not permitted to adopt a neutral stance toward religion. They had to recognize the truth of the Catholic faith, enact laws that were favorable to the Church, support it financially, and allow themselves to be judged by it. The "hypothesis" referred to the concessions that the Church had been led to make by recognizing the neutrality of the state in matters of religion, not because such neutrality was desirable in itself, but in the interest of peace or out of necessity. I hasten to add that when the thesis-hypothesis distinction was first invoked, little was known concretely about the political realities of the Middle Ages, which Catholic scholars were then in the process of rediscovering and of which they formed a somewhat idealized picture.

This unfashionable approach was publicly abandoned only with Vatican II and the *Declaration on Religious Freedom,* one of the most controversial of the sixteen documents promulgated by the Council and the one that signals perhaps its most striking departure from previous Church teaching. The *Declaration* opens with a forceful endorsement of the "right of religious freedom," which is said to have its "foundation in the very dignity of the human person as it is known through the revealed Word of God and by reason itself."[13] It protests against the abridgments of this right in various parts of the world and demands that it be enshrined everywhere as a "civil right."[14] It forthrightly acknowledges that "a harmony exists between the freedom of the Church and the religious freedom which is to be recognized as the right of all human beings and communities and sanctioned by constitutional law."[15] In so doing, it tacitly does away with the so-called "double standard" of which it could be accused so long as it claimed for itself a privilege that it denied to others.

Finally, it accompanies these statements with a timely reminder that "the disciple is bound by a grave obligation toward Christ his master ever more adequately to understand the truth received from him, faithfully to proclaim it, and vigorously to defend it."[16]

It is too early to say how much this more liberal stance and posture will have contributed to the religious renewal that the Council sought to promote. The record to date is not reassuring. How one reads it depends on whether one regards the dissensions that have since racked the Catholic Church as a sign of growing maturity or as the harbinger of a deepening crisis that has barely begun to play itself out. All that can safely be asserted is that for the most part the Catholic intelligentsia has responded more positively to the Council's statements that "human beings are bound to obey their own consciences" and are to be "guided by their own judgment"[17] than to the injunctions that limit the exercise of that judgment.

THE MORALIZATION OF CHRISTIANITY

This brings me to my final point, which is directly related to the preceding one and which has to do with what Robert Nisbet referred to not long ago as the "decline of the transcendental Christianity and its replacement by ever softer religions and ever harder systems of social metaphysics."[18] The decline is already prefigured in a number of early nineteenth-century attempts to counter the Enlightenment attack on revealed religion by means of an argument that stresses the usefulness rather than the truth of Christianity. Reversing Pascal's procedure, some of the better known apologists of that generation, beginning with Chateaubriand, engaged in an all-out attempt to prove, not that Christianity was good because it came from God, but that because it was good it had to come from God. This much was clear from a survey of the innumerable benefits—science, the arts, democracy, and the abolition of slavery, to mention only a few—that had supposedly accrued to society as a result of its adoption.

The end product of these efforts was a Christianity emptied of a large portion of its dogmatic content and animated to an unprecedented degree by thisworldly concerns. The popularity of the social gospel at the turn of the century is only one local manifestation of the same syndrome. Years before, Saint-Simon had written in his influential work, *The New Christianity*, that "the whole of society ought to strive toward the amelioration of the moral and physical existence of the poorest class and organized in the way best suited to the attainment of that end."[19] A new civil religion was in the making, inspired more by Enlightenment humanitarianism than by the Bible and concentrating the bulk of its energies on matters pertaining to social welfare or the restructuring of society along egalitarian lines. To quote Nisbet once again, "it is difficult not to wonder how much the moral zeal evident in some quarters today is the consequence, not of any genuine reassertion of Christian or transcendental belief, but of continued Christian fragmentation and of immersion still deeper in the waters of secularism."[20]

In light of this situation, the thought with which I began, namely, that today we have neither God nor politics, takes on a sharper meaning. Our religious leaders have indeed recovered their moral zeal, but while in the past this zeal was mainly directed toward the reform of individuals, today it finds its chief outlet in the clamor for social reform. Bad institutions rather than human frailty or malice are blamed for the evils that plague human existence. All sin, we are given to understand, is social sin, and it will not disappear until the whole of society is restructured in accordance with the norms of universal justice. It follows that the just society is no longer an object of deliberation on the part of wise and just rulers. It is everywhere the same and everywhere possible. When all is said and done, there is only one legitimate regime, whose outlines are known in advance and for the establishment of which everyone is morally obliged to work.

As a result, politics, in the old sense of the word, loses its primary significance and is reduced to the science or the art of public administration. It used to be that there were two noble alternatives to the political life, revealed religion and philosophy. Now there is a third, which is specifically moral in character and which is the more attractive as it addresses itself to all decent human beings, whether they be religiously inclined or not. All of our mainline churches and most of the others now endorse it as the one that best accords with the demands of the gospel. The irony is that in their eagerness to build up the earth, they may have hastened the destruction of heaven. James Turner is probably right when he says that "the church played a major role in softening up belief" and that "having made God more and more like man—intellectually, morally, emotionally—the shapers of religion made it feasible to abandon God, to believe simply in man."[21] My only query is whether the suggestion goes far enough, for it presupposes that the churches themselves had already been softened by modern secular thought and its political offspring, the separation of church and state. On that score, it matters little that in a few Western countries the Church should still enjoy state support, since the supporting state does not appear to be any less secular in its outlook.

Some of our theologians are obviously of the opinion that secularization is a phenomenon over which one should rejoice rather than grieve since, lo and behold, this is what Christianity was all about in the first place. Even if this were the case, however, and there are good reasons to think that it is not, we should still be faced with a problem, for the moral ideal in question does not appear to be anchored in any religious or meta-physical view of life capable of imparting real vitality to it. That problem was brought home to us with dazzling clarity by Nietzsche, who had reflected more deeply than any of his contemporaries on the implications of godlessness and come to the conclusion that a fatal contradiction lay at the heart of the modern theological enterprise: it thought that Christian morality, which it wished to preserve, was independent of Christian dogma, which it rejected. This, in Nietzsche's mind, was an absurdity. It amounted to nothing less than dismissing the architect while trying to keep the building or getting rid of the lawgiver while claiming the protection of the law. Christian belief and Christian morality belong together, to such a degree that the latter cannot long survive the death of the former.

> When one gives up the Christian faith, one pulls the right to Christian morality out from under one's feet . . . . Christianity is a system, a whole view of things thought out together. By breaking one main concept out of it, the faith in God, one breaks the whole: nothing necessary remains in one's hands. Christianity presupposes that man does not know, cannot know, what is good for him, what evil: he believes in God, who alone knows it. Christian morality is a command; its origin is transcendent; it is beyond all criticism, all right to criticism; it has truth only if God is the truth—it stands and falls with faith in God.[22]

A century later, the situation remains pretty much the same. Ethical theories vie endlessly with one another and the competition is encouraged in the name of "pluralism," Isaiah Berlin's euphemism for what bolder thinkers did not hesitate to call "relativism." None has managed to win the support of a majority of our contemporaries and all of them can be seen as last-ditch attempts to restore some meaning to human life on the basis of our modern scientific or materialistic conception of the universe. This is true of the older theories of Kant and Hegel, and it is also true of our more recent theories, most of which are scarcely more than revised or syncretistic versions of their predecessors. No one has been able to bridge the gulf between the "is" and the "ought" or to maintain the "ought" in spite of the "is," and few seriously think that in some crucial instances the "is" might be an "ought." The results speak for themselves, and we are daily treated to the spectacle of a mass of people who talk as if the "ought" were the only thing that matters and act as if there were nothing but the "is." In Allan Bloom's terminology, "commerce" prevails over "culture" and a more or less intelligent selfishness determines the course of our impoverished lives.[23]

Will religion be able to reassert itself and regain its lost ascendancy, with or without the aid of the state? The possibility cannot be ruled out a priori, but, humanly speaking, the prospects are not good. Not being a prophet, I do not know what the future holds in store for us, and, not being a pessimist, I am not about to put the worst possible face on the situation. The consensus among historians is that science won the nineteenth century. Religion appears to be winning the twentieth, but doubts persist about the quality of that religion. No one who knows anything about the disastrous political entanglements of the medieval church wants to return to that model of church-state relations, and no one who has given any thought to the spiritual confusion, not to say bankruptcy, of modern life thinks that the separation of church and state is always and everywhere the best possible solution to the problem. Contrary to what Hegel, Marx, and their followers would have us believe, there are some problems in life that defy any universally valid solution and call for the kind of prudential judgment of which only a well-informed and well-intentioned person is capable.

By reason of a tradition that is coeval with its founding, America is not about to renege on its commitment to the principle of separation, to which, significantly, both present-day liberals and present-day conservatives profess their undivided allegiance. The lesson that we learn from Tocqueville is that this principle will yield its choicest fruits if its application is accompanied by an awareness of, and a concomitant attempt to mitigate, some of its less desirable features.

NOTES

1. Fustel de Coulanges, *The Ancient City: A Study of the Religion, Laws, and Institutions of Greece and Rome* (Garden City: Doubleday, 1955), 395.

2. Edward Gibbon, *The Decline and Fall of the Roman Empire*, vol. I (New York: Modern Library, n.d.), 639.

3. The allocation was considered "so large that even when cut to a third at its restoration after the suspension under Julian's pagan revival, it was reckoned generous." H. Chadwick, *The Early Church* (Baltimore: Penguin, 1967), 128.

4. See on this subject the penetrating remarks by Paul Valéry in the famous essay entitled "La crise de l'esprit," in *Variété, Oeuvres*, Bibliothèque de la Pleiade, vol. I (Paris: Gallimard, 1957), 1007-1014.

5. Alexis de Tocqueville, *Democracy in America*, edited by J. P. Mayer with new translation by George Lawrence (Garden City, New York: Doubleday and Co., 1969), vol. I, part 2, chapter 9.

6. *Dem. in Amer.*, vol. 2, part 2, chapter 10.

7. See, for example, Harvey Cox's *Religion in the Secular City: Toward a Postmodern Theology* (New York: Simon and Schuster, 1984).

8. Published in booklet form by the Rockford Institute, Rockford, Illinois.

9. *Dem. in Amer.*, vol. 2, part 1, chapter 5.

10. Ibid., vol. 1, part 2, chapter 9.

11. Friedrich Nietzsche, *Beyond Good and Evil*, translated by Walter Kaufmann (New York: Vintage Books, 1966), aphorism 87.

12. With a purist's dismay, I notice that even the Roman Catholic liturgy has adopted the new language, with no apparent awareness of its ideological implications. The collect for the Monday of the fifth week of Easter begs God to help us "seek the values that will bring us eternal joy in this changing world."

13. Vatican Council II, *Dignitatis humanae* (*Declaration on Religious Freedom*) no. 2.

14. Ibid., no. 6.

15. Ibid., no. 13.

16. Ibid., no. 14.

17. Ibid., no. 11.

18. Robert Nisbet, *The New York Times Book Review*, 28 April 1985, 17.

19. Henri de Saint-Simon, *Nouveau christianisme: dialogue entre un conservateur et un novateur* [*The New Christianity*] (Paris: Bossange père, 1825).

20. Robert Nisbet, *The New York Times Book Review*, 28 April 1985, 17.

21. J. Turner, *Without God, Without Creed: The Origins of Unbelief in America* (Baltimore: Johns Hopkins University Press, 1985), quoted in Nisbet, *The New York Times Book Review*, 28 April 1985, 17.

22. Friedrich Nietzsche, *Twilight of the Idols*, "Skirmishes of an Untimely Man," no. 5. The same point is developed at great length in Nietzsche's early essay, *David Strauss, the Confessor and Writer*, published as part of his *Untimely Meditations*, translated by R. J. Hollingdale (New York: Cambridge University Press, 1993).

23. A. Bloom, "Commerce and 'Culture'" *This World* 3 (Fall, 1982): 5-20.

2

HUMAN RIGHTS AND THE COMMON GOOD

Not long ago, an American teenager living in Singapore was caught spray-painting a few cars as a prank and assessed the usual penalty for this kind of crime: "caning." The American government promptly intervened to have the sentence commuted, claiming that such a harsh punishment violated the youth's "rights." In a gesture of good will, Singapore officials agreed to reduce the number of strokes, but nevertheless insisted on going ahead with them. Their argument, which they based on Confucius, was that social order takes precedence over human rights. The incident was widely interpreted by the American press as a "clash of cultures" between East and West, the former emphasizing societal stability and the latter individual freedom. One does not have to go all the way to Singapore to find examples of this tension between the two poles of human existence. We have plenty of them at home.

For all the benefits that it provides and for which we can be immensely grateful, modern liberal democracy does not of itself produce a strong attachment to the common good of society and was not calculated to do so. This means that anyone in search of proper models of public-spiritedness will have to look elsewhere for them, as did our American forefathers, who sought them in classical antiquity, often naming their cities after such famous ancient places as Athens, Rome, Syracuse, Troy,

Ithaca, Utica, and the like.

From its inception in fifth-century Greece and for the next 2,000 years or more, scholarly debate concerning moral matters focused on natural right or natural law; today it focuses almost exclusively on individual or subjective rights. Nowhere in the older tradition does one run across anything like a theory of natural rights. I am referring to rights that inhere in individual human beings qua human beings and independently of their membership in the larger society to which they belong, as distinguished from civil rights or rights that have their source in some duly enacted law. The passage from natural law to natural rights and later (once "nature" had fallen into disrepute) to "human" rights represents a major shift, indeed, *the* paradigm shift in our understanding of justice and moral phenomena generally. Prior to that time, the emphasis was on virtue and duty, that is to say, on what human beings owe to other human beings or to society at large rather than on what they can claim from them. This is surely the case with the Bible, which invites us to think in the first instance of others rather than of ourselves (we do not need to be reminded to think of ourselves) and which does not promulgate a Bill of Rights, but the Ten Commandments, a "Bill of Duties," as it were. But this is also the case with all of premodern literature, classical as well as Christian, whose foremost representatives—Plato, Aristotle, Cicero, Ambrose, Augustine, and Thomas Aquinas, to mention a few—wrote treatises or dialogues on natural right in the singular, on moral virtue, on laws, or on duties. It never occurred to any of them to publish a book entitled *The Rights of Man* or to issue such documents as the *Declaration of the Rights of Man and the Citizen* or the *Universal Declaration of Human Rights*.

To be sure, the expression "natural rights" does appear sporadically in the older texts, but never as part of a theory that claims independent status for them. In one of his anti-Pelagian tracts, St. Augustine alludes in passing to the "natural rights of propagation," *iura naturalia propaginis*,[1] in connection with original sin, the sin transmitted by way of generation from parent to child, without explaining the expression or showing any awareness of its novelty. A little known contemporary of Augustine, Primasius of Hadrumetum, speaks in a metaphorical sense of the "natural rights of places," *iura naturalia locorum*,[2] apropos of the antlers that burst forth from the heads of certain animals and keep on growing, seemingly without end. None of these cursory statements amounts to anything like a bona fide natural rights doctrine of the kind that would later be propounded by Hobbes, Locke, Pufendorf, and a host of other early-modern thinkers.[3]

Altogether different is the view that comes to the fore in the seventeenth century, where the equation is reversed and where rights become the fundamental moral phenomenon, the source rather than the result of such natural laws as will enable people to live comfortably and at peace with one another. In his classic treatment of this subject, Hobbes begins precisely by laying down a basic right, the right of self-preservation, which he defines in terms of freedom and from which he proceeds to infer the self-enforcing laws, nineteen in all, by which human beings are required to abide lest they should jeopardize that freedom.[4]

Professor Brian Tierney unwittingly furnishes us with a striking example of the shift in perspective that marks the transition from one world to the other, namely, the classic question of whether a criminal who has been convicted of an offense punishable by death and is in jail awaiting his execution can flee if the opportunity presents itself—if, for example, the jailer has fallen asleep and inadvertently left the jail door unlocked.[5] The answer was yes, but with the proviso that the convict was not to kill or maim the jailer, something that would have constituted a second punishable offense. What Tierney fails to note, and it is the key point, is that, when we come to Hobbes and Locke, this crucial proviso has been lifted on the ground that the right of self-preservation is strictly inalienable and may legitimately be exercised against anybody, including one's executioner.

The standard objection to my argument regarding the disparity between the premodern and modern understandings of rights is that, even though our premodern forebears made little use of the language of natural rights, they understood the reality to which it refers. Hence, to oppose rights and duties is to set up a false dichotomy between two complementary rather than antithetical approaches to the subject of morality. Fly from San Francisco to London via Chicago or via the North Pole and you arrive at the same destination. On any given day one route may be preferable to the other, as would be the case if, say, the air traffic controllers were on strike in the Midwest; but otherwise the choice between them is a matter of indifference. As correlatives, rights and duties imply each other. If I have a duty to do something, I must have the right to do it. For whatever reason, the premoderns did not speak explicitly of natural rights, but this is not proof that they would have rejected them had the problem been put to them in those terms.

At first hearing, the objection sounds unimpeachable, but it does not get to the heart of the matter inasmuch as it fails to address the question of the priority of duties over rights or vice versa. Did Socrates have the right to defend himself or was it his duty to do so? Which of the two is the primary moral counter and takes precedence over the other in the event of a conflict between them? All indications point to the fact that in the premodern view the duty came first. This accounts among other things for the natural inclination to sacrifice oneself for the whole whenever necessary and to do so because one instinctively perceives that the good of the whole—the "common good," as it used to be called—is one's own good; for such a good would not be truly common were it not at the same time the "proper" good (albeit not the "private" good) of the individuals who constitute the whole.

In the final analysis, we are confronted with two vastly different conceptions of morality, one that looks at it from the point of view of what a given action does to the person who performs it, and the other from the point of view of what it does to the recipient. The problem had not escaped the great Hercule Poirot, who, when asked one day by that impossible woman, Mrs. Ariadne Oliver, whether he did not think that some people "ought" to be murdered, had the good sense to reply: "Quite possibly, madam, but you do not comprehend. It is not the victim who concerns me so much; it is the effect on the character of the slayer."[6] This is not to say that what happens to the victim is unimportant but only that it is not the primary consideration and the one through which the nature of the moral act reveals itself most profoundly.

The same holds, of course, for good deeds, which likewise have an effect on the character of the doer and benefit him as much as they do the recipient of the good deed.[7] We find this thought expressed in any number of places, among them the fable of the old man and the three youths who made fun of him when they saw him struggling to plant trees that he would never live to see; to which the old man, who had his children and grandchildren in mind, replied: "Are the wise forbidden to work for the pleasure of others? This is itself a fruit that I enjoy today."[8]

The point is not at all far-fetched. It was brought home to me in an unexpected way a few years ago by an incident that occurred in a small restaurant on Beacon Hill to which I had gone for lunch with a Jesuit colleague. Sitting two tables away from us was Ted Williams, whom my friend recognized and to whom he waved discreetly. Williams acknowledged the greeting with a smile and we went on with our meal. When it came time to pay, the waiter announced that our tab had been picked up by someone else in the room. At this point, my friend had no choice but to go over to Williams's table and recite the usual platitudes: that, although we appreciated the gesture, it was not necessary, etc., etc. Williams listened patiently and then replied in the simplest, most unpretentious way: "Father, please don't deny me this small pleasure. I have so few of them left in life."

Similar remarks could be made about numerous biblical texts, among them the parable of the Good Samaritan, which, according to the interpretation that I find most plausible, inculcates the lesson that the Good Samaritan is the one who is indebted to the man to whose rescue he comes for the opportunity to serve him and not conversely.

To generalize on the basis of these homely examples, human beings are most closely bound to others not because they hope to receive *from* them but because they find their deepest satisfactions in doing something *for* them. This is the prototypical premodern understanding of one's relationship to one's fellow human beings. It is emphatically not the understanding that informs the original modern rights approach, which denies the natural sociality of human beings and views them instead as atoms that are complete in themselves and hence not essentially dependent on others for the achievement of their perfection. Not being ordered to any preexisting end or ends, these free-floating individuals are at liberty to choose their own ends, along with the means by which they may be attained. As was recently asserted by the Supreme Court (in *Planned Parenthood* v. *Casey*), "At the heart of liberty is the right to define one's own concept of existence, of meaning, of the universe and of the mystery of human life."[9] The only just society is the one that grants to each individual as much freedom as is compatible with the freedom of every other individual. It has nothing to say about the good life and is not concerned with the promotion of virtue. Its sole function is to insure the safety of its members and provide both for their comfort and, as we now see everywhere, the satisfaction of their vanity.

I do not wish to imply that the advocates of this new scheme were themselves monsters of inhumane self-centeredness. Theirs was no ordinary selfishness or hedonism, a dog-eat-dog outlook in which everyone runs the risk of being devoured. It was a political hedonism, something entirely new on the intellectual horizon of the West. Behind it lay the laudable desire to put an end to the evils that had always plagued society and that had become particularly acute in the wake of the wars of religion that were then ravaging Europe. Its great advantage was that it did not depend for its success on a painful conversion from a concern for worldly goods to a concern for the good of the soul. Pursuing one's selfish interest, it was decided, was the best way to serve others. Properly managed, private vices could lead to public benefits. It thus became a moral duty to encourage people to think of themselves rather than of others, for by so doing one necessarily contributes to the good of the whole. The entrepreneur who is out to enrich himself and himself alone benefits the whole of society by creating lucrative jobs for the rest of its members. In the end, everybody is materially better off. The scheme was a clever one indeed, for by reconciling selfishness with altruism, it enabled everyone to reap the rewards of virtue without going to the trouble of acquiring it.

The question is whether the narcissistic atomization of individuals can coexist with genuine community—whether any society is likely to endure, let alone prosper, without a shared notion of the good life. The problem has been with us for a long time. It was raised as far back as the middle of the eighteenth century by Rousseau, the first modern thinker to criticize modernity and call attention to the deficiencies of the society to which it was giving rise: its lack of nobility or elevation, its pettiness, its manifest disunity, its magnificent boringness, and the multiple alienations that it inevitably produced. Modernity's typical product was the bourgeois, as Rousseau called him and as we have been calling him ever since: The man who has been taught to live for himself in the midst of people for whom he does not care but in whom, for his own sake, he is obliged to feign interest. The citizens of the new society are not held together by the love of a common good in which everyone can share. Rather they are attached to society by bonds of self-interest alone, have nothing to die for and, by the same token, nothing to live for. As Rousseau put it famously: "We have physicists, geometers, chemists, astronomers, poets, musicians, painters; we no longer have citizens."[10] All of the virtues on which society normally relies for its well-being, such as civic-mindedness, patriotism, religion, military valor, and self-restraint, have been weakened beyond recognition.

I do not exaggerate when I say that the whole of modern thought since then has been a series of heroic attempts to reconstruct a world of human meaning and value on the basis of Rousseau's and our own purely mechanistic understanding of the universe. This is obviously not the place to enter into a discussion of these attempts, whether it be Kant's bloodless categorical imperative, a direct descendant of Rousseau's "general will," for which it sought to provide a philosophic justification; Hegel's entrusting to History with a capital "H" the carrying-out of a similar task; Nietzsche's and Heidegger's appeals to creativity as a means of overcoming the predatory nihilism of the age; or, to begin with, Rousseau's own project, the aim of which was to recreate the Platonic soul by means of a complex process of sublimation whereby all the higher things in life are made to originate in the impulses of our lower nature and specifically our sexual impulses.

Not coincidentally, it is to Rousseau, *the* seminal writer of the late modern period, that we owe the first formulation of the nonmercenary notion of rights that has come to prevail in our times. I know of no finer statement of the basic issue than the passage in Rousseau's *Emile* in which Emile and his tutor, Jean-Jacques, fail to show up for an announced visit to Sophie—the person to whom Emile is to be married—without informing her of their change of plans. Afraid that something terrible has happened, Sophie is beside herself with anxiety, until she learns that Emile and his tutor are safe and will soon be arriving, at which point her anxiety turns to rage at the affront she has suffered. The matter is finally cleared up when Emile explains that he and his companion were on their way to Sophie's house when they stumbled upon a man lying in the forest with a broken leg and carried him to his home. At the sight of her crippled husband, the man's wife promptly went into labor and had to be helped by Emile and Jean-Jacques. This forced them to spend the night where they were, unable to send word to Sophie of their whereabouts. Emile ends his account by declaring: "Sophie, you are the arbiter of my fate. You know it well. You can make me die of pain. But do not hope to make me forget the rights of humanity. They are more sacred to me than yours. I will never give them up for you."[11] Moved by this declaration, Sophie replies: "Emile, take this hand. It is yours. Be my husband and master when you wish. I will try to merit this honor."[12] In his commentary on this passage, Allan Bloom writes: "Emile has won his fair maiden. Dedication to human rights has taken the place of slaying dragons or wicked knights as the deed that makes him irresistible to his beloved."[13]

In view of the fragility of the foundations on which they rest, it is hardly surprising that none of the previously mentioned solutions to the problem of modernity should have won the endorsement of a majority of our educated contemporaries; and that is why the search goes on. The currently most popular alternative is the one that goes under the name of communitarianism, a label that captures at least part of what was once meant by the "common good." I note in this connection that *The Ethics of Authenticity*, the latest book by Charles Taylor, one of the best known representatives of the new movement, bears in its original edition the title of *The Malaise of Modernity*, which could easily have served as the subtitle of Rousseau's landmark *Discourse on the Arts and Sciences*, the first but by no means the only full-scale attack on modernity from the side of modernity to which we have been treated over the years.

If I allude to contemporary communitarianism (according to some newspaper reports, the philosophy to which President Clinton and his wife subscribe), it is because it is symptomatic of our current predicament. My reservation about it is that it is still too much committed to finding within modernity itself the intellectual and moral resources needed to overcome the limitations of modernity. One does not transcend the "ego" by expanding it into a "we" through the incorporation of other "egos" into it; for the "we" of modern thought is not a community. For this, a crucial further step is required: the realization that the "we" is more fundamental than the "I" and, hence, not derivable from it. It is something with which we start and not something with which we end.

We find this fundamental "we" (assuming that we want to continue the artificial practice of using pronouns as nouns) not in modern but in premodern thought, whose approach to these matters is at once more natural and more attractive to people who have not been brainwashed into believing that modern science and the philosophy that comes out of it are the sole arbiters of our intellectual and moral tastes.

I, for one, am always pleasantly surprised to see how much more en- thusiastically college students respond to Plato and Aristotle than to Kant, Hegel, or for that matter, Nietzsche, despite his enormous appeal to young minds. I have just finished going through Aristotle's *Ethics* with a group of freshmen and received the greatest compliment of my teaching career from one of them—not an "A" student, mind you—who said one day as we were walking out of class, and here I apologize for his language, which is more colorful than mine: "You know, I eat this shit right up!"

There is much to be said for the fact that the most significant intel- lectual development of our time is the recovery of classical thought and the reopening of the life-and-death struggle in which it was once engaged with modern thought for the minds and hearts of our seventeenth-century predecessors and their descendants, ourselves included. When I started teaching many years ago, the ongoing debate among academics was be- tween three forms of modern thought: Kantianism, Utilitarianism, and either logical positivism or its offshoot, linguistic analysis. That debate is not dead by any means, but it has gradually been taking a back seat to a much livelier debate, the one between modern thought as whole and a premodern thought that was supposed to have been laid to rest once and for all at the dawn of the modern era. In my mind, this is the most important debate of our time.

Fortunately, we do not have to go far to find living examples of the premodern mentality. For starters, I would recommend a small work with the charmingly ironic title of *Paterno by the Book*, the autobiography of the famed head football coach of the Penn State Nittany Lions, who is almost as sharp on the subject of virtue and the common good as my old friend, Agatha Christie. It leads right back to the classics and, in par- ticular, to Virgil's masterpiece, the *Aeneid*, a book to which Paterno was introduced by one of his prep school teachers and by which he learned to play the serious game of life long before making a name for himself as a football coach. "Aeneas," he says,

> is not a grandstanding superstar. He is, above all, a Trojan and a Roman. His first commitment is not to himself but to others. He is bugged constantly by the reminder, the *fatum*, "you must be a man for others." He lives his life not for "me" and "I" but for "us" and "we." Aeneas is the ultimate team man.
>
> A hero of Aeneas's kind doesn't wear his name on the back of his uniform. He doesn't wear "Nittany Lions" on his helmet to claim star credit for his touchdowns and tackles that were enabled by everybody else doing his job. For Virgil's kind of hero, the score belongs to the team.[14]

Part of the book deals with an important decision that Paterno was summoned to make when, toward the end of the 1972 season, he was offered the job of coach and general manager of Billy Sullivan's old Boston Patriots (now the New England Patriots) at an annual salary forty times as great as the one he was making at Penn State—$1,400,000 versus a measly $35,000, not counting such perks as a summer home on Cape Cod and the chance to move in the highest professional circles. The offer was, as they say, irresistible and he accepted it, only to reconsider it during the course of a sleepless night and finally turn it down because "it was not right for him." His personal destiny or calling, his *fatum*, as Virgil would have said, was to remain where he was. Here again, the decisive factor was not what he could do for Penn State by staying there—one cannot predict the outcome of one's efforts—but what leaving the school would do to him.[15]

There would be a lot less talk about community everywhere today if we all had a better idea of what a true community is like. How we go about recovering our lost sense of community I do not know. The emptiness of so much of what is currently being said on the topic makes me think that the proper place to begin is where I ended, namely, by listening to those who have experienced it in their own lives and demonstrated not only in speech but in deed that they understand both what it demands of us and what it promises us in return.

NOTES

1. St. Augustine, *Contra secundam Juliani responsionem opus imperfectum* (*Unfinished Work Against Julian's Second Response*) 6.22.

2. Primasius of Hadrumetum, *Commentary on the Apocalypse* 2.5.

3. Cf. E. L. Fortin, "On the Presumed Medieval Origin of Individual Rights" in *Classical Christianity and the Political Order: Reflections on the Theologico-Political Problem*, edited by J. Brian Benestad (Lanham, MD: Rowman and Littlefield, 1996), chapter 12.

4. Cf. *Leviathan*, chapter 14.

5. Cf. "Natural Rights in the Thirteenth Century: A *Quaestio* of Henry of Ghent," *Speculum* 67 (1992): 58-68.

6. Agatha Christie, *Cards on the Table* (New York: Dodd, Mead and Company, 1963, © 1936), 162.

7. The difference is that the good deed always benefits the doer. It does not always benefit the recipient, for we cannot foresee all the consequences of our actions.

8. LaFontaine, *Fables*, XI.8.

9. *Planned Parenthood v. Casey* 60 LW 4800 (1992).

10. Jean-Jacques Rousseau, *The First and Second Discourses*, edited with an introduction and notes by Roger D. Masters (New York: St. Martin's Press, 1964), 441.

11. Jean-Jacques Rousseau, *Emile or On Education*, introduction, translation and notes by Allan Bloom (New York: Basic Books, 1979), 441.

12. Ibid.

13. A. Bloom, *Love and Friendship* (New York: Simon and Schuster, 1993), 132.

14. Joe Paterno with Bernard Asbell, *Paterno by the Book* (New York: Berkley Publishing Company, 1991), 45-6.

15. Cf. Joe Paterno, *Paterno by the Book*, 3-14. I am grateful to my good friend and former student, Prof. J. Brian Benestad, of the University of Scranton, for calling Paterno's book to my attention.

3

CHRISTIAN EDUCATION AND MODERN DEMOCRACY

I begin with the Nietzschean observation that the flood of books, pamphlets, and reports on education with which we are currently being inundated is as much a symptom of the decay of the contemporary educational scene as it is of its renewed vitality. These various publications, most of them critical of what is going on in colleges and universities across the land, are not necessarily without merit; but even the better ones, the ones that are more than just exercises in cultural criticism, outbursts of crankiness, or nostalgic pleas for a return to an often idealized past, have done little to relieve the malaise that afflicts us. In any event, we should have less of a need for them if we had a better idea of what genuine education is all about.

Interestingly enough, our ancient forebears, who were the first to teach us the importance of education, did not leave us a single book by that title. The closest we come to a *Peri Paideias* in classical antiquity is Xenophon's *Education of Cyrus—the Cyropaideia*—which contains barely more than six lines on the topic with which it purports to deal. This is not to say that the Greeks were not concerned with the proper formation of their children, far from it. For them, education was and remained a prize possession, the key to one's existence, "the first among noble things for the best human beings," as Plato calls it in the *Laws*.[1] On it, they thought, depended the fate of nations and empires. This much is evident from the *Cyropaedeia*, which suggests by the irony of its title the overwhelming significance of Cyrus's education or, more precisely, his lack thereof. I know of few more moving testimonies to the prestige that education enjoyed in the Greek world than an incident to which Thucydides refers in the *History of the Peloponnesian War*,[2] and which has to do with a contingent of Thracian soldiers hired to participate in the Sicilian expedition. The mercenaries arrived too late at the port of embarkation, and, unable to sail with the Athenian army, were forced to return to their country. En route, they fell upon the peaceful town of Mycalessus, which they proceeded to ransack, mercilessly killing off much of the population. The reader inevitably wonders why Thucydides has interrupted his narrative at this point in order to dwell on an episode that bears no obvious relation to the rest of the story and appears almost insignificant by comparison with the epic grandeur of the events that we are about to witness. We find part of the answer in the seemingly casual remark that Mycalessus, though small in size, had a large school. It was a Greek city, a place where the cult of the beautiful was held in high honor. Education, and not the human nature that is common to all of us, is what made the Greeks superior to the barbarians by whom they were surrounded and sometimes threatened. Still, it was not something to be treated by itself or apart from the various intellectual disciplines that supply its content. It required no special justification and made perfect sense as the means that fitted human beings for the pursuit of the highest goals to which they could aspire, those of humanity and noble citizenship. There was no need for separate treatises on it. Classical literature had no other theme. The whole of it was devoted to the subject of education.

The same is not true of our own society, which, no matter how we look at it, is not the natural home of liberal education. It is not the Ancients but the Moderns, beginning with Locke's *Some Thoughts Concerning Education*, who write books on education; and if they have been compelled to produce them in ever larger numbers, it is for the precise reason that education itself has become problematic in our time. No doubt, liberal democracy grants to its citizens the freedom to set their own goals and seek the education that is best suited to them. Liberal democracy is often lavish in its praise of education, subsidizes it to a degree that would have been unthinkable in the past, strives to make it available to more people than ever before, and, because liberal democracy does not give us much in the way of moral guidance, needs it more than any other regime. The American founders in particular went out of their way to show the compatibility of education with popular government and, whenever necessary, to defend it against its detractors. They knew from experience that the case for the superiority of their own regime could only be made by liberally educated persons who were thoroughly acquainted with the possible alternatives to it. As is clear from the epitaph that he wrote for his tomb, Thomas Jefferson was as proud of having established the University of Virginia as he was of having authored the Declaration of Independence. Like most of his associates, he realized that there was more to be feared from the ignorance of the people than from the elitist or anti-democratic tendencies inherent in any kind of high learning.[3]

Much as these and other modern writers may have admired classical education, the fact remains that the principles underlying the regime to which they were committed offer little rational support for it and in a deeper sense actually run counter to it. I shall argue that this, more than anything, is what accounts for the limited impact of all recent appeals for the reform of our educational programs and institutions. It is no accident that the two philosophers from whom we have the most to learn about the status of education in the modern world, Rousseau and Nietzsche—and I shall have more to say about them later—also happen to be the greatest critics of liberal democracy, one from the Left and the other from the Right. I shall argue further that the Christian churches, which for reasons of their own have always had a stake in this matter, have themselves capitulated to the new trend and contributed in their own way to the demise of liberal education.

From the beginning, modern liberal society, by which I mean the kind of society that is based more or less explicitly on the principles first elaborated by such early modern giants as Hobbes, Locke, and Spinoza, was bent on making the development of a sound public life less dependent on liberal education. Its proponents took issue with premodern political thought on the ground that the latter made impossibly high demands on human nature and was thus unable to guarantee the actualization of the kind of society that it regarded as most conducive to human well-being. Classical political thought took its bearings from the twofold end of virtue and wisdom to which human beings are ordered by nature. Education was meant to facilitate the attainment of those ends by clarifying them, providing a rationale for them, displaying them in their attractiveness, and showing concretely all that was to be gained by striving for them. Its aim was to help young people acquire by dint of human exertion the moral and intellectual virtues to which nature inclines us but does not itself supply inasmuch as their specific form tends to vary from one historical context to another.

This lofty program had later been taken over and further enlarged by the writers of Christian antiquity. It would be difficult to exaggerate the magnitude of the effort that went into this prodigious enterprise, which entailed nothing less than a reinterpretation of the whole of classical literature with a view to bringing its ideals into greater harmony with the requirements of the gospel.[4] In the process education lost none of its importance. As a learned religion, Christianity relied on it as much as did the pagan society in the midst of which it had first developed. The only difference was that henceforth it would have as its goal not only the formation of the human being and the citizen but that of the Christian or the citizen of the city of God as well.[5] To be sure, there were risks on both sides: the pagan elite could not help wondering what would happen to the Homeric hero once his accomplishments were measured by the standard of the Sermon on the Mount, and the Christian educator had to think twice before exposing his young charges to the glittering examples of noble pride and self-reliance with which classical literature abounded. Still, the risks were worth taking. As time went on, a mutually acceptable solution was worked out which allowed for the continued use of pagan literature in the schools, albeit under the vigilance of its new Christian masters.

The flaw that the early modern political theorists detected in such an education is that, by making the health of society contingent on the acquisition of virtue, it compelled one to speak of an ideal that is seldom if ever attained in real life. Both the pagans and the Christians knew, of course, that most people are neither wise nor particularly moral, but there was always the possibility and the hope that by living under good laws they might become better. The problem is that these laws are themselves framed or enacted by human beings. One was caught in a circle from which there appeared to be no escape: good government makes for good human beings and good human beings make for good government. According to the modern argument, the only solution was to devise a scheme whose successful implementation depended less on moral goodness and the exercise of prudence than on the establishment of such political institutions as did not contradict the selfish desires by which the vast majority of people are actuated. The new goal may have been less exalted but it had the advantage of being more easily attainable. It did not require that one undergo a painful conversion from the premoral or amoral concern for worldly goods to a moral concern for the good of the soul. Plato and his followers had taught that the passions were more or less amenable to reason and could therefore be guided by it. Prudence was the charioteer leading the horses by which the chariot was drawn.[6] Keeping the same image, one could say that for Hobbes and his successors, the recalcitrant horses could be given free rein on the sole condition that they be forced to remain within the barriers that calculating reason had erected for them. Simply put, people could remain as they are as long as the institutions under which they lived were as they ought to be. There was no danger that the new society would enter into conflict with the passions since it was structured in such a way as to be in accord with them from the start. Under such circumstances, one could dispense with the kind of education that the ancient authors had sought to promote and even frown upon it as a positive hindrance to the achievement of the legitimate, but rather more modest, goals of self-preservation and comfortable self-preservation that everyone was now encouraged to pursue. Strictly speaking, the only education needed was the one that reminded people of their right to run their lives as they themselves saw fit, that is to say, of what down deep they most want to hear anyway.

Enter Rousseau, the father of modern educational theory and the author of what is far and away the most influential book on it, *Emile or On Education*. Modern liberal society promised security and freedom, along with the means whereby they could be secured, namely, the ever greater prosperity that would accrue to us from our rapidly growing mastery of nature and the rational exploitation of the world's natural resources. It nonetheless suffered from one crucial deficiency: it lacked dignity or elevation. Its ethos succeeded to a remarkable degree in curbing the excesses of moral fanaticism and religious intolerance, but by the same token it was not especially favorable to the development of public-spiritedness or the deepening of one's spiritual life. Henceforth, the people held up as models to be emulated would not be the saint, the prudent statesman, or the noble warrior; they were the entrepreneur, the merchant, or the banker who was convinced that in the long run crime does not pay and that honesty is still the best policy. Instrumental virtue, which tended to equate moral goodness with utility, was all that was required to make the system work. Anything beyond that was a matter of personal choice, allowable in the name of freedom and only insofar as it did not interfere with anyone else's freedom. If the new society excelled at anything, it was at teaching individuals to live for themselves in the midst of others. In a word, it was hopelessly pedestrian or bourgeois. What it most begged for was an infusion of something akin to the disinterested virtue that had characterized the lives of the ancient republics at their best, and this is what Rousseau tried to give it.[7] But he did so in a novel way—not by going back to the Ancients, however much he may have admired them for their civic-mindedness, but by recreating the structure of the human soul on the basis of our modern scientific or nonteleological understanding of nature. All of the qualities that had once lent nobility and distinction to human life were to be reinstated and could be reinstated as long as a way was found to make them grow out of the impulses of our lower nature. None of these qualities had an independent status. In one fashion or another, they were all rooted in the sexual drive, of which they were pure and simple modifications. Even noble love, that most delicate and refined of human sentiments, had no other foundation. It was not a natural but an

artificial feeling, one that was ultimately grounded in selfishness.[8]
The different stages of this elaborate technique are outlined in great detail in the *Emile*, the first part of which deals with the formation of Emile as an individual and the second part with his formation as a citizen. The process eventually acquired a name. It was called "sublimation," the implication being that a genuine conversion or a real turning around of the soul is neither necessary nor possible.[9] The so-called "higher" things for which people sometimes yearn have no existence in and of themselves; they are human constructs, which come into being via the repression of one's natural instincts rather than their redirection toward more suitable objects. One has to look to oneself rather than to nature for them. Ours is a disenchanted, demystified, and strictly mechanistic universe. It offers nothing lovely or exquisite to which one might be attracted and for the sake of which one might be willing to forgo the satisfaction of one's bodily appetites. The one point to be added is that in Rousseau the notion of sublimation still retains something of its lofty connotation, the one implied in the word "sublime" from which it is derived. This, needless to say, will not always be the case with Rousseau's innumerable nineteenth- and twentieth-century followers.

The argument took a new twist in the nineteenth century, which marks another crucial turning point in the intellectual history of the West and in the course of which the seeds of much of present-day religious, political, and philosophic thought were sown. Without any doubt, the seesaw battle that had pitted the representatives of the old order against their modern adversaries for upwards of two centuries continued unabated and, indeed, was enlivened by the spectacular comeback that religion had made in the aftermath of the French Revolution, thanks in part to Rousseau's earlier call for the restoration of civil religion. The most interesting developments nevertheless lie elsewhere, in the novel attempt that was then made by philosophers and liberal theologians alike to mediate the Enlightenment conflict between Christianity and modernity.

One name stands out among all others in this vast and startlingly new project, that of Hegel, the first philosopher to develop a systematic argument for the unbroken continuity of the Western tradition from its inception in ancient Greece down to our time. Instead of opposing premodern and modern thought, as had previously been the custom and as Rousseau himself had done, it became fashionable to stress the link that bound them one to the other. The two worlds, it was discovered, were not the immiscible opposites or the eternal irreconcilables for which they had been mistaken. One was a child of the other. The tensions, if not always the open conflict, that had marred their relationship since the dawn of the modern era, were simply due to a lack of perspective on both sides. Seen from within, the modern world was nothing but an unfolding of certain premises that had been latently at work in the historical process from the beginning. What we moderns had finally come up with was exactly what our predecessors had been after all along. Having recognized this, the two groups had only to lay down their arms. The combat could cease for want of combatants. The perfectly rational society, the society that had finally grasped the principles of justice and knew how to put them into practice, was about to emerge, and it found its natural ally in the Christian religion, which could itself be understood in a fully rational way and of which it was but a secularized version to begin with. At long last, human beings would be spared the trouble of agonizing over important decisions or of having to face any problem to which the solution was not already available. They could even do without philosophy, which, by Hegel's own admission, always came too late to be of any use to them.[10] For the first time ever, one had reached an understanding of human life which was complete in every respect and which for that reason left no room for ethics, did away with the need for character formation, removed prudence from the realm of action, and made education superfluous except perhaps as a kind of popular initiation into the insights of Hegelian thought.

The reaction was as swift as it was vehement. It burst into the open with Kierkegaard shortly after Hegel's death and, some years later, with Nietzsche, neither of whom could reconcile himself to the idea that history was at an end; that all notable achievements belonged to the past; that the whole had lost its elusiveness; that the human soul would henceforward be deprived of any depth or inwardness; that we moderns no longer had any real choices to make or anything important to look forward to, and that all that remained was to reap the benefits of what had been wrought at the price of untold labor and suffering by those who had gone before us. With Kierkegaard and Nietzsche, philosophy recovers something of its original impulse in that it is again intended to guide human action and illumine our choices. The whole of it becomes, as it had been in former times, an educational enterprise. In retrospect, it seems fair to say that of these two authors, Nietzsche is the one who in the long run was destined to have the most profound impact on our thinking. The paradox is that he unintentionally did more than any of his immediate predecessors to sap the vitality of modern education.

The premise from which Nietzsche started is that everything in the modern era is a conspiracy against human greatness. This holds for Hegelian philosophy, which relegates all greatness to the past and invites us to look upon ourselves as late-comers in a world grown old; and it also holds for post-Hegelian historicism, which denies that history had come to an end and replaces Hegel's absolute moment with the notion that all truths and values are relative. In Nietzsche's eyes, the danger to which such a view gives rise is not that, once persuaded that nothing is eternally true or false, human beings will feel free to embark upon reckless courses of action, but that they will lack the incentive to do anything at all. Human greatness hinges on one's willingness to devote oneself heart and soul to the pursuit of a noble cause. It presupposes that one is convinced of the righteousness of that cause and prepared to make whatever sacrifices it may impose on him. Yet it is doubtful whether the individual who has glimpsed the ephemeral character, and hence the ultimate vanity, of all such causes will be motivated to sacrifice himself for any of them. Human life can only thrive within a limited horizon. It requires for its well-being a protective atmosphere that restricts one's vision and blinds one to the injustice implied in the choice of any goal. By disclosing all specific horizons as mere horizons, historical relativism destroys them and thereby poisons the wellsprings of human activity. It teaches not only that all ideals are perishable but, inasmuch as life is inseparable from injustice, that they "deserve" to perish. Its truth is a "deadly truth,"[11] breeding inertia or apathy and leading by degrees to the blessed narcosis of a life dedicated to the pursuit of base, selfish, or purely private ends. With nothing higher, whether it be politics or religion, to occupy their minds and only themselves to live for, human beings inevitably end up by taking egoism as their god. This typically modern phenomenon is again well illustrated by Flaubert's heroine, Emma Bovary, who commits adultery twice in the course of the novel, once with Rudolph after attending the county fair and listening to the inane speeches delivered on that occasion by local politicians, and once with Leon after visiting the cathedral at Rouen, where she found nothing that could give to her life the meaning for which she was so desperately looking.

For Nietzsche, the main culprit in all of this was Christianity or, to be more precise, the "Christian movement," of which the "democratic movement," with its rampant egalitarianism, its sentimental humanitarianism, its irrepressible craving for sameness, and its instinctive abhorrence of any higher form of life, was the unconscious heir.[12] The whole of modern political thought was fundamentally Christian in its inspiration; but so also was modern philosophic and scientific thought, however much some of its representatives may have tried to deny it.[13] It is not uninteresting to observe that the thematic discussion of Hegel in chapters seven and eight of the *On the Advantage and Disadvantage of History for Life* is flanked on both sides by an analysis of the deleterious influence of Christianity on the moral tone of society, as if to suggest that Hegelian philosophy was Christian at its very core. All modern philosophers, Nietzsche notes elsewhere, are "Schleiermachers," that is to say, crypto-theologians, or taking Schleiermacher's name in its etymological sense, "spinners of veils." As the latter-day propagator of the slave morality first invented by the Jews, Christianity was the sworn enemy of human greatness and the single most powerful obstacle to its restoration. Hence Nietzsche's overriding concern with it even in those works, such as *The Birth of Tragedy*, in which it is not explicitly mentioned.[14]

By comparison, Rousseau had little to say on the subject of Christianity, which he hardly felt the need to criticize, save for the purpose of defending himself against the censures of the ecclesiastical establishment. His own quarrel was not with the transcendent Christianity of the Middle Ages, which he had no desire to resurrect by attacking it;[15] it was with early modern liberalism and its devastating effect on the quality of human life, its prosaic down-to-earthness, and its utter lack of nobility or grandeur. Like most of his contemporaries, he realized that most people cannot live without religion, and he was well aware that one does not create a religion out of whole cloth. Discredited as it was in intellectual circles, Christianity was here to stay. The best one could hope for was to render it as harmless as possible by retranslating it into something like the nebulous, soothing, and politically useful myth that he places in the mouth of his Savoyard Vicar.[16]

Aside from the brilliant synopsis of all the arguments that could be marshalled against it in the name of modern political or scientific reason,[17] one does not learn much that is new about Christianity by reading Rousseau. One does learn a great deal about it by reading Nietzsche, who saw more clearly than anyone else the depths of its roots, the extent to which the human soul had been transformed by it, and all that it would take to overcome it. Nietzsche had the greatest admiration for Pascal, who, as a mystic and a scientist rolled into one, had endured, within his own tormented soul, the war that faith and modern science, those two mortal enemies, had begun to wage on each other.[18] The tragedy was that no one was left to appreciate the poignancy of that conflict or the price that Pascal had to pay for the immolation of the scientific impulse that animated him. As was clear from the remarks that Voltaire devotes to him in his *Philosophical Letters*,[19] nobody could understand Pascal anymore, let alone admire him. The whole modern development since that time had been nothing but a concerted attempt to unbend the bow and relax the tension that was now found to be unbearable.[20]

The whole thrust of Nietzsche's argument was to save human beings from themselves by alerting them to the peril that they faced. Accordingly, his own philosophy culminates in an appeal for creativity, self-transcendence, the intensification or enhancement of life beyond itself. The notion of human greatness had become all but totally eclipsed, but it could be recovered. True, there were no preexisting goals on which one could set one's sights, but insofar as it presented us with a new and unprecedented challenge, the situation was not without remedy. Now that God was dead, one could aim higher than ever before. One could become one's own creator. There were no fixed limits to what could be accomplished through a combination of absolute freedom on the one hand and total commitment to one's freely chosen ideal on the other.[21]

One cannot begin to describe the impact that this new program was to have on twentieth-century thought. To an astonishing degree, everybody today wittingly or unwittingly thinks in the shadow of Nietzsche, and that includes not only theologians, philosophers, and political theorists, but sociologists, psychologists, artists, and men of letters as well. The phenomenon is unparalleled in the history of Western thought. No other philosopher with the possible exception of Rousseau has done more to shape our vision of things and transform the intellectual climate of our age. I do not wish to underestimate the importance of such other great or influential thinkers as Kant, Hegel, and Marx, but no one has been equally successful in infiltrating the thought of even his staunchest adversaries and in establishing the *primum quoad nos* or "first for us" from which in one way or another everyone else would start.

How well Nietzsche has been understood is another matter. Most people have responded enthusiastically to his call for freedom and commitment; very few have heeded his admonitions regarding the rigorous training and the harsh discipline required for their proper exercise. For Nietzsche, a superhuman effort was needed to keep them together and prevent them from destroying each other. They were both predicated on the death of God, which was a catastrophe, an intimation of the terribleness of life without any compelling religious or metaphysical horizon. The death of God marked the onset of a new "world night" and meant that humanity would henceforth be "straying as through an infinite nothing."[22] The consequences were as dire as they were unpredictable. Only by taking the strongest measures to reverse a trend that had been gaining strength for centuries would it be possible to set modern man on the right course. One had to think through the whole tradition of Western thought from beginning to end with a view to surmounting it and equipping one-self for the new possibilities that had recently been opened up. The task was enormous. It necessitated nothing less than the recovery of the highest experiences of the past, those of the pagan Socrates and of the Christian Pascal, reliving them within ourselves, and, on the basis of what had transpired in the meantime, going beyond them. No one was yet ready for such a task and no one knew to what it could lead, but, given the depths to which the modern herd animal had sunk, it was imperative that it be undertaken. The future of the human race was in the balance.

By the time the message reached the level of popular thought, it had been adulterated almost beyond recognition and all sense of urgency had gone out of it. The proclamation of the death of God was greeted not just with equanimity but with a sigh of relief, as it is, for example, in Harvey

Cox's *The Secular City*, the most widely read theological work of the 1960s. The basic premise remains the same:

> All things are relative. Everything depends on how you look at it. . . . Secular man must live with the realization that the rules which guide his ethical life will seem just as outmoded to his descendants as some of his ancestors' practices now appear to him . . . How, in such a situation, is it possible to avoid a dizzy descent into pure anarchic relativism? How can secularization, if it results in the deconsecration and consequent relativization of values, lead in the end to anything but nihilism?[23]

Cox's questions are the ones that haunted Nietzsche, but they have suddenly become far less alarming. The relativization of all values, we are now being told, need not plunge us into either individual or collective solipsism. It is not a mortal danger threatening all of us but a positive gain, leading to nothing more than the recognition that "since everyone's perspective is limited and conditioned, no one has the right to inflict his values on anyone else."[24] As for the nihilism against which Nietzsche had warned, it was merely the "adolescent phase of the relativization of values,"[25] the one we had to go through in order to be freed from the dangerous precritical illusion that our values are ultimate. That phase was happily behind us: "Paradoxically, the mutual discovery by all parties concerned that their various value systems are in fact relative puts them all into the same boat. It supplies a common experience and outlook which can become the basis for a new social consensus."[26]

As might have been expected even then, that consensus has not materialized. If anything, it appears to be more elusive than ever. Partly for that reason, the past twenty years or so have witnessed a massive resurgence of the old Kantian human rights doctrine over against both the easygoing relativism extolled in Cox's book and the utilitarianism that had long been prevalent in Anglo-American academic circles. Bad or obsolete institutions rather than human frailty or malice are to be accounted as the root cause of the evils from which human beings continue to suffer. Rousseau, who had incurred the wrath of the ecclesiastical authorities for denying the existence of original sin, was right after all. Since human beings are naturally good or inclined to benevolence, all sin is necessarily social sin, and it will not cease from the land until the structures of society as we know it are altered and in some instances overthrown in accordance with the norms of universal justice. Contrary to what was formerly thought to be the case, the just society is everywhere the same and everywhere possible. When all is said and done, there is only one legitimate regime, which is known in advance and for the establishment of which everyone is morally obliged to work. To that extent, it can be said to transcend the realm of the political and does not constitute a proper object of deliberation on the part of wise rulers, who are no longer presumed to have any choice in the matter. Politics in the old sense thus loses its primary importance and is reduced to the science or the art of public administration.

The larger question is whether the currently fashionable human rights theory is anchored in a religious or metaphysical view of the universe that guarantees its power and vouches for its plausibility. That question was raised by Nietzsche, who had reflected more deeply than most of his contemporaries on the practical implications of the liberal theology of his day and, as we saw in chapter one, denounced as futile its attempt to preserve Christian morality while jettisoning Christian dogma. Such is the flagrant inconsistency of which in an early essay he accuses the illustrious David Strauss, the author of the first modern *Life of Jesus*. Strauss, a convinced Darwinian, failed to see that his theoretical premises led by an inexorable logic, not to the ethic of love and self-abnegation that befits a being created in the image of God, but to a Hobbesian war of each against all.[27] Such also is the inconsistency that he found at work in the writings of the English moralists à la George Eliot and the "flathead" John Stuart Mill, who were likewise "rid of the Christian God" but clung all the more firmly to his moral commands.[28]

There is some comfort to be taken in the thought that God is not dead for everybody and that the majority of our own contemporaries seem to have lost their taste for the practice of deicide. But for all the difference that he makes to our lives, he might just as well be. Sad to say, the theoretical foundations of the rights doctrine remain shaky and largely unexplored.[29] The problem is not new by any means, as we know from the debates that pitted Kant against Bentham in the eighteenth century and Mill against Kant in the nineteenth. It was never fully resolved at that time and there is little reason to think that it has been resolved since then. Kantian philosophy had no completely satisfactory reply to Mill's question regarding those foundations and left it at speaking of the "facticity of freedom," an answer that, Mill thought, simply begged the question.

One is tempted to go a step further and suggest that no modern philosopher has succeeded in demonstrating the validity of his position or in fulfilling the promises held out at the outset of his investigations. In 1781, when the first edition of the *Critique of Pure Reason* appeared, Kant expressed the conviction that by the end of the century philosophy as such would be completed and that all skeptical questions concerning the status of morality and its requirements would have finally been laid to rest.[30] That was not to be, or so one gathers, not only from the attacks mounted against him by later philosophers, beginning with Hegel, but from the series of short and somewhat inconclusive pieces that he wrote toward the end of his life, all of which show him wrestling with the loose ends that his three great *Critiques* had left untied. Nietzsche did not live to write the major work that was meant to nail down the brilliant, if troubling, insights contained in his published works. We shall never know whether he himself could have brought the fantastic project upon which he had embarked to a successful completion. The aging Husserl declined an invitation to attend a concert with his students, alleging that he needed the time to finish his philosophy as a "rigorous science." He never did. Heidegger died without producing the long-awaited sequel to his epoch-making *Being and Time*, despite the fact that, unlike Nietzsche, he was given a full fifty years in which to do it. Finally, so as not to prolong this enumeration, there is not a shred of evidence that at the time of his death Merleau-Ponty had written a single line of a work to be entitled *La Pensée Fondamentale*, which was supposed to answer all the questions raised by his *Phenomenology of Perception*, the book that had launched his career. If anything, the most widely shared opinion is that there is in fact no such thing as a presuppositionless science—that not only mathematical systems, as Gödel tried to show, but all philosophical and scientific positions have their roots outside of themselves and are thus *essentially* incomplete.

In the meantime the controversy goes on, pitting "deontologists" against "teleologists"[31] and both groups against Marxists, social science positivists, and historical relativists ("pluralists," as they are now more politely called). No single school managed to impose itself, and, if recent experience is any indication of things to come, none is likely to at least for the foreseeable future. Behind the gigantic efforts expended in defense of these various theories lurks a subtle anguish provoked by the realization that the mechanistic universe in which modern science has taught us to believe leaves our intellectual and moral "oughts" without any solid foundation in nature. Michael Sandel puts the matter nicely when he writes:

> Bound up with the notion of an independent self is a vision of the moral universe this self must inhabit. Unlike classical Greek and medieval Christian conceptions, the universe of the deontological ethic is a place devoid of inherent meaning, a world 'disenchanted' in Max Weber's phrase, a world without an objective moral order. Only in a universe empty of *telos*, such as seventeenth-century science and philosophy affirmed, is it possible to conceive a subject apart from and prior to its purposes and ends. Only a world ungoverned by a purposive order leaves principles of justice open to human construction and conceptions of the good to individual choice. In this the depth of opposition between deontological liberalism and teleological world views most fully appears.[32]

The predicament described by Sandel is unfortunately not as innocuous as some of us would like to think. Human beings who do not see themselves as inherently directed to the goals of knowledge and virtue will generally be inclined to view the demands of their corporeal natures as more "real" than any of the higher things to which they might otherwise be drawn. The language of business, commerce, or any of the other pursuits that conduce to greater material well-being will make more sense to them than that of the disinterested activities of the mind.[33] Education becomes a luxury of sorts, a part of what since Kant we have been naming "culture," that artificial supplement destined to relieve the ennui of an everyday life which serves the body and its needs admirably well but does so little to accommodate the longings of the soul.[34]

These seem to be the more profound reasons for which the proposals that are being made for the improvement of our current educational policies and practices, well-intentioned and pertinent as they may be, are bound to fall on deaf ears. Liberal education does not fit naturally into our modern thought patterns and must struggle to adjust itself to them. Lest I should leave you dangling in a nihilistic vacuum, however, let me end with a few brief remarks about what I take to be a good education. My answer to that question is ridiculously simple and old-fashioned. It amounts to little more than saying that true education consists in reading old books together with one's friends.[35] By old books I mean not only the ones written by the authors of classical antiquity but those books which have withstood the test of time and which recommend themselves to us by the subtlety of their thought, the grace of their style, and their insight into the fundamental human problems. Identifying these books is easy enough. We know roughly what they are, and besides, there are more of them than we shall ever be able to study with the proper care. Furthermore, all or most of them are now available in cheap or relatively cheap editions. The real problem is that we have forgotten how to read them and lack the incentive to do so. In true Nietzschean fashion, we are all imbued with the idea that they should all be approached as documents of the age that produced them and whose prejudices they reflect. Any thought that they might have something of ultimate importance to teach us about ourselves and the way we should live is out of the question. This is true even of the Bible, which a surprisingly large number of our educated contemporaries now study for the purpose of rewriting it, the assumption being that we have as much to teach the sacred authors as they have to teach us.

From that point of view, the situation does not seem to have improved very much over the past thirty years. If anything, it has gone from bad to worse. In a book entitled *The Organization Man*, which everybody in academia was reading in the late fifties, William H. Whyte talks about a young suburban couple who aroused the suspicions of their neighbors when, shortly after their arrival, it was discovered that they had a copy of Plato's dialogues on their bookshelf. I took that to be a good sign: people who were suspicious of Plato still had a vague sense that reading him might have an effect on us, that one can be moved by such books and become a different kind of person as a result of having steeped oneself in them. I doubt whether their children feel the same way today. The new wisdom is that it is perfectly safe to read Plato or anybody else for that matter if that is your personal preference, because it no longer occurs to us that a book might change our lives. After all, Plato dealt with "values," and values, everybody is now convinced, are subjective. I am not suggesting for the time being that these old books are better than the newer ones, although a strong case could be made for that view,[36] but only that we shall never be able to see ourselves as we are unless we become aware of what we could be and might have been had our circumstances been different.

It is amazing to observe the degree to which these new philosophic and scientific theories, once the preserve of an intellectual elite, have come to permeate the thinking and the language of our society. One Saturday morning not long ago, while in a hotel room waiting for the start of a conference, I happened to turn on the TV, only to discover that at that hour most of the programs are cartoons geared to juvenile audiences. In one of them two little girls were shown standing next to a hedge out of which there suddenly leapt a huge animal. One of the girls exclaimed: "Gee, that's scary"; to which the other replied in a tone of reproach: "'Scary' is a value judgment!" If even the youngest generation must now learn to use social-scientific jargon to express its most natural reactions, one can just about imagine the condition the rest of us are in.

Hopeless as the situation may appear to be to some people, it nevertheless has a relative advantage over other, more stable situations. Insofar as it is characterized by the shaking of all traditions and cultural horizons, it allows for a reconsideration of the fundamental human alternatives in ways that would have been impossible at other moments in our history. The sense of disintegration that so many of our thoughtful contemporaries have begun to experience is itself an invitation to undertake a fresh or nontraditional assessment of these traditions. This is precisely where Nietzsche can be of immense help to us. He himself could find no better way of replying to the accusation that he had tried to kill Socrates than by saying that, whereas others had mummified him, he, Nietzsche, had brought him back to life. The effort paid off. For the first time in close to two centuries, it became possible to view the development of Western thought as a unified whole and reexamine the basic choices with which it confronts us. For this we shall never be grateful enough. When asked whether he would have liked to live at another moment in history, a distinguished twentieth-century political theorist, not known for his partiality to the modern age, is reported to have answered that he still preferred our time if for no other reason than that it had afforded him the privilege of reading Nietzsche.

Nietzsche's ultimate purpose in going back to Socrates was not to rehabilitate him; it was to take issue with the entire tradition of Western thought on what, he was convinced, was its highest level. Yet Socrates remained a "problem" for him, so much so that at the very end of his active life, he was still not sure of having won his contest with him.[37] He did, however, succeed in creating a new interest in him and that interest has been with us ever since. I have no idea of how many of us would be reading the Platonic dialogues today if Nietzsche had not so forcefully drawn our attention to them, but my guess is that the number, which is perhaps not very large, would be smaller by a wide margin.

There is no telling what direction our intellectual world and the myriad factors that impinge upon it might yet take. Cultural cycles do not conform to the immutable patterns of nature; and, besides, our culture critics, their critics, and the critics of their critics do not all read the signs of the times in the same way. Which one will eventually prove to be correct is something that can only be demonstrated in the event. As one of them has suggested, however, it may well be that the present chaos is a parody of the transcendence for which our trendy theologians are groping and for which our worldly scientists have likewise begun to thirst.[38] Much as we may deplore the parody, we can still glimpse in its distorted features the grandeur of the lost original to which it points. The extremity into which we have fallen could indeed be our greatest opportunity.

NOTES

1. Plato, *Laws* 644b1.

2. Thucydides, *History of the Peloponnesian War* VII.29.

3. On the educational ideals of the Founders, see, most recently, E. F. Miller, "On the American Founders' Defense of Liberal Education," *The Review of Politics* 46 (1984): 65-90. Also see, E. Brann, *Paradoxes of Education in a Republic* (Chicago: University of Chicago Press, 1979).

4. For some indication of the manner in which this was accomplished, cf. E. L. Fortin, "Christianity and Hellenism in Basil the Great's Address *Ad Adulescentes*," and "Basil the Great and the Choice of Hercules: a Note on the Christianization of a Pagan Myth," in *The Birth of Philosophic Christianity: Studies in Early Christian and Medieval Thought*, edited by J. Brian Benestad (Lanham, MD: Rowman and Littlefield, 1996), 137-168.

5. The classic statement on this subject in antiquity is St. Augustine, *On Christian Doctrine*, book II, where all of the arts and sciences are enlisted as aids to the understanding of the Scriptures.

6. Plato, *Phaedrus* 246a-b; 253a-256e.

7. In a remarkable note appended to the "Profession of Faith of the Savoyard Vicar," Rousseau explains that even fanaticism is preferable to the "secret egoism" engendered by the liberal theories of his day: "Bayle has proved very well that fanaticism is more pernicious than atheism, and that is incontestable. But what he did not take care to say, and which is no less true, is that fanaticism, although sanguinary and cruel, is nevertheless a grand and strong passion which elevates the heart of man, makes him despise death, and gives him a prodigious energy that need only be better directed to produce the most sublime virtues. On the other hand, irreligion—and the reasoning and philosophic spirit in general—causes attachment to life, makes souls effeminate and degraded, concentrates all the passions in the baseness of private interest, in the abjectness of the human *I* and thus quietly saps the true foundations of every society . . . . Thus fanaticism, although more deadly in its immediate effects than what is today called the philosophic spirit, is much less so in its consequences. Moreover, it is easy to put fair maxims on display in books; but the question is whether these maxims really are well connected with the doctrine, whether they flow from it necessarily; and that is what has not appeared clear until now. It still remains to be known whether philosophy, if it were at its ease and on the throne, would have a good command over vainglory, interest, ambition, and the petty passions of man, and whether it would practice that gentle humanity it lauds to us in its writings." J.-J. Rousseau, *Emile or On Education*, introduction, translation, and notes by Allan Bloom (New York: Basic Books, 1979), 312n.

8. Cf. *Emile*, book IV, 329: "And what is true love itself if it is not chimera, lie, and illusion? We love the image we make for ourselves far more than we love the object to which we apply it. If we saw what we love exactly as it is, there would be no more love on earth." See also, under Rousseau's influence, Turgenev, *Fathers and Sons*, (New York: Modern Library, 1950), ch. 25, 207 and ch. 26, 219.

9. The term "sublimation" in this new sense appears to be a creation of Nietzsche (e.g., *Beyond Good and Evil*, no. 189), from whom it was inherited by Freud. For the prehistory of the term, cf. W. Kaufmann, *Nietzsche: Philosopher, Psychologist, Antichrist*, 4th ed. (Princeton: Princeton University Press, 1974), 218-19.

10. Hegel's best known statement to this effect is the one that occurs at the very end of the Preface to the *Philosophy of Right*: "When philosophy paints its grey in grey, then has a shape of life grown old. By philosophy's grey in grey it cannot be rejuvenated but only understood. The owl of Minerva spreads its wings only with the falling of the dusk." Translated by T. M. Knox (London, 1952), 13. Hegel's "grey in grey" is an allusion to a passage in Goethe's *Faust*, I.iv, 509-14, in which Mephistopheles says: "Grey are all theories and green alone is life's golden tree." Nietzsche, who criticizes Hegel on this count, remarks: "Historical culture is indeed

a kind of inborn grey-hairedness, and those who bear its mark from childhood must instinctively believe in the *old age of mankind*: to age, however, there pertains an appropriate senile occupation, that of looking back, of reckoning up, of closing accounts, of seeking consolation through remembering what has been, in short, historical culture." Friedrich Nietzsche, *On the Uses and Disadvantages of History for Life*, in *Untimely Meditations*, translated by R. J. Hollingdale (New York: Cambridge University Press, 1983), 101.

11. Friedrich Nietzsche, *On the Advantage and Disadvantage of History for Life*, translated by Peter Preuss (Indianapolis, IN and Cambridge: Hackett Publishing Co., 1980), 38-39.

12. Friedrich Nietzsche, *Beyond Good and Evil* (New York: Vintage Books, 1966), 115-18.

13. Friedrich Nietzsche, *On the Genealogy of Morals*, third essay, aphs. 24 and 25. The same idea was later taken up by A. N. Whitehead, who writes in *Science and the Modern World* (New York: Macmillan, 1931), 19: "Faith in the possibility of science, generated antecedently to the development of scientific theory, is an unconscious derivative from medieval theology." This view has since become a commonplace in the literature on the history of science.

14. The observation is Nietzsche's, who, in the "Attempt at Self-Criticism" that prefaces the second edition of *The Birth of Tragedy* (1886), speaks of the "careful and hostile silence with which Christianity is treated throughout the whole book."

15. Cf. Rousseau, *Discourse on the Arts and Sciences*, first part, init.

16. Cf. J.-J. Rousseau, *Emile or On Education*, introduction, translation and notes by Allan Bloom (New York: Basic Books, 1979), 266-313.

17. Rousseau, *Emile*, book IV.294-308.

18. See, for example, *Beyond Good and Evil*, aph. 45.

19. Voltaire, *Lettres philosophiques*, twenty-fifth letter.

20. Cf. *Beyond Good and Evil*, preface and aph. 205.

21. See, among other texts, *Thus Spoke Zarathustra*, first part, "On the Gift-Giving Virtue," 3; second part, "Upon the Blessed Isles," and "On Self-Overcoming."

22. *The Gay Science*, book 3, aph. 125.

23. Harvey Cox, *The Secular City* (New York: Macmillan, 1965), 27.

24. Ibid.

25. Ibid.

26. Ibid., 30.

27. Friedrich Nietzsche, *David Strauss, the Confessor and the Writer*, 7, in *Untimely Meditations*, translated by R. J. Hollingdale (New York: Cambridge University Press, 1993), 30.

28. Friedrich Nietzsche, *Twilight of Idols*, "Skirmishes of an Untimely Man," 5. Cf. *Will to Power*, aph. 18: "Those who have abandoned God cling that much more firmly to their faith in morality." George Eliot, for whom Nietzsche obviously had

little use, had translated Strauss's *Life of Jesus* into English in 1846. See also in support of Nietzsche's point Leslie Stephen's journal entry, dated Jan. 26, 1865 (cited by James Turner, *Without God, Without Creed: The Origins of Unbelief in America* [Baltimore and Lanham: Johns Hopkins University Press, 1985], 203): "I now believe in nothing, to put it shortly; but I do not the less believe in morality."

29. See on this matter the observation by R. Tuck, *Natural Rights Theories: Their Origin and Development* (Cambridge: Cambridge University Press, 1979), 2.

30. Cf. Kant, *Critique of Pure Reason*, end.

31. The new labels were introduced by C. N. Broad, *Five Types of Ethical Theory* (London and New York: Routledge and Kegan Paul, 1930), 206f., and were proposed as a more adequate substitute for Sidgwick's threefold classification of ethical theories into intuitionism, egoistic hedonism, and utilitarianism. Broad's "teleologists" correspond roughly to the old utilitarians. His "deontologists" were in the main the advocates of Kantian morality in one or another of its forms.

32. M. Sandel, *Liberalism and the Limits of Justice* (New York: Cambridge University Press, 1982), 175.

33. Nietzsche notes with dismay that the vocabulary of the marketplace was already being employed to characterize the work performed by scholars: "I regret the need to make use of the jargon of the slave-owner and employer of labour to describe things which in themselves ought to be thought of as free of utility and raised above the necessities of life; but the words 'factory', 'labour market', 'supply', 'making profitable', and whatever auxiliary verbs egoism now employs, come unbidden to the lips when one wishes to describe the most recent generation of men of learning." Friedrich Nietzsche, *On the Uses and Disadvantages of History for Life*, 7, translated by Hollingdale, 99.

34. See the brilliant analysis of this phenomenon by A. Bloom, "Commerce and Culture," *This World* 3 (Fall 1982): 5-20.

35. The definition is borrowed from Xenophon, *Memorabilia* 1.7.14.

36. Even Locke and J. S. Mill, whom no one would accuse of being biased in favor of the Ancients, were convinced that education in the classics was not only important but indispensable to the formation of modern gentlemen. Cf. Locke, *Some Thoughts Concerning Education*, nos. 93-94, 164f.; 186. J. S. Mill, *Autobiography*, in James and John Stuart Mill, *On Education*, edited by F. A. Cavanagh (Cambridge, 1931), 151-157. Also Tocqueville, *Democracy in America*, vol. 2, part 1, chapter 15.

37. The second section of *The Twilight of Idols*, the last of Nietzsche's work to be published by himself, is entitled precisely "The Problem of Socrates."

38. Cf. R. Unger, *Knowledge and Politics* (New York: Free Press, 1975, 1984), 27.

4

IS LIBERAL DEMOCRACY
REALLY CHRISTIAN?

The complex problem of the relationship between Christianity and civil society has not been solved but only further complicated by the rise of modern liberal democracy and its more recent competitor, social democracy or democratic socialism. There are good reasons for preferring democracy to other possible and possibly legitimate regimes, and I doubt whether, in our Western countries at least, a plausible case could be made for any alternative form of governance. This is not to say that democracy as we know it is self-evidently Christian or more Christian than its now discarded predecessors. If anything, the opposite would seem to be true.

The widely held view that democracy was summoned into existence or decisively shaped by Christianity is of relatively recent vintage and is open to serious question. From the first, liberal democracy was marked by a latent but profound hostility to the Christian faith. Its earliest and perhaps still greatest theorists did not draw their inspiration from the works of their Christian mentors, for which they had little use even if they occasionally quoted from them, and they knew better than to try to ground their principles in the religious tradition, which offered little support for them. One probes their theory of society in vain for anything that cannot be explained without recourse to that tradition. Even the notion of the state of nature, for all its theological antecedents, is more of a denial than an adaptation of the biblical teaching. The state to which it refers was meant to replace the Garden of Eden and reflects a totally different understanding of the origins of human life. It is not an ideal state to which one aspires to return but a state of utter misery from which one seeks to escape. Human beginnings were not perfect but harsh and threatening. One had no choice but to leave them as far behind as possible. This could best be done, not by invoking divine grace, but by entering into a civil society, a society founded on the desire for self-preservation, which no one is likely to forget under any circumstances, rather than on principles of justice and morality, which people are too often tempted to disregard when these principles come into conflict with their private interests.

The scheme was a clever one and all the more so as it was no longer necessary to acquire any of the moral or Christian virtues in order to reap their benefits. Justice played no role whatever in it and, contrary to what is sometimes asserted, occupied no place in the literature on it. It is interesting to note that the references to John Locke under the heading of "Justice" in Mortimer Adler's *Syntopicon* have nothing to do with that virtue and were inserted along with the others on the misleading assumption that a writer of such preeminent stature had to have a theory of justice.[1] (I am told by persons close to the project that the graduate assistant initially charged with supplying the mandatory references was fired when she failed to come up with any.) The "spirit of capitalism," as it was later called, would more than compensate for the loss of virtue by providing material blessings far in excess of anything that had previously been thought possible or even desirable. The greatest benefactors of humanity, the true heroes of the coming age, were not the dedicated parish priests or the saintly Christians toiling selflessly for the benefit of others, but the new captains of commerce and industry, who by enriching themselves would enrich everyone else as well. As Locke wrote, day laborers in England would henceforth enjoy a standard of living that was denied to kings elsewhere and ten acres of well-tilled land in Devonshire would yield more than a thousand acres in the wild and uncultivated wastes of America.

Nor was this all, for in the process humanity would at long last be rid of the scourges of war and religious fanaticism. The trick was to divert everyone's energies into more productive channels. People who were busy making money might take the duties of religion less seriously, but by the same token they were less apt to go off on crusades.

This being the case, it is hardly surprising that for so long—centuries in fact—liberal democracy was viewed with suspicion and only grudgingly accepted, if accepted at all, by thoughtful Christians. As far as I know, the first attempts to effect a reconciliation between modern democracy and

Christianity date back to the early years of the nineteenth century and form part of the different outlook on religion that was typical of the writers of that generation. Christianity, which throughout most of the eighteenth century had been blamed for the evils of contemporary society, was suddenly hailed as the source of all that was supposed to be good in it, science and democracy included. The modern world was indebted to it for everything—*le monde moderne lui doit tout,* as Chateaubriand proudly announced in the introduction to the first part of *The Genius of Christianity,*[2] one of the most influential books of the century and the fountainhead of so many of the ideas later to be expounded with all sorts of new twists by Tocqueville, Nietzsche, and countless others who likewise came to the conclusion that the origins of modern political and scientific thought were to be sought in the ancient or the medieval Christian tradition. (I shall not mention Hegel, who restated the problem with a philosophical depth to which Chateaubriand could not and did not aspire.) Far from opposing Christianity and modernity, it became fashionable to proclaim their fundamental agreement and stress the links that bound them together.

The fly in the ointment is that, quite apart from the question of historical accuracy, the new argument was a pure and simple inversion of the old one. It pulled the bag inside out, so to speak, but did nothing to alter the terms in which the problem was posed. Accordingly, its leading proponents were largely unaware of the extent to which they adopted the perspective of their erstwhile adversaries. They, too, had come to look upon Christianity as a political or cultural phenomenon, and they proceeded to defend it on those grounds. In retrospect, their account of it is barely more than a mirror image of the one they rejected. At no point does one sense that a real breakthrough had occurred and that the issue had been raised to a level on which it could be, if not completely settled, at least more profitably debated. The basic question, which was rarely addressed, is whether one does full justice to the Christian faith by politicizing it, secularizing it, or reducing it to a moralism and a culturalism that are foreign to its native impulse.

Christianity may be politically useful in some ways and politically dangerous in others, but the least that can be said is that it was not originally conceived as a political religion. However valuable it may be as a complement to the political life, Christianity does not offer itself as a replacement for it or as a prescription for its improvement. Anyone who reads the New Testament attentively soon discovers that it has virtually nothing to say about the organization of civil society, its proper functioning, or the manner in which power ought to be distributed. It does not tell us *who* should rule, which is the political question par excellence, but only *how,* in general, human beings, whether rulers or subjects, should conduct themselves, which is something else altogether. The New Testament's guiding theme is not justice but love, and love as a political principle is at best a fuzzy concept. Neither does it promulgate any universal laws to which we could appeal in order to solve our problems. The commandments that it issues are not laws of nature, possessing an intrinsic intelligibility that would give us an inkling of how they might be applied to concrete situations; they are expressions of the will of a personal and loving God who expects—nay, demands—the same kind of response from his creatures.

The only maxim proper to the gospel that might be construed as a principle of justice is the Golden Rule: Do unto others as you would have them do unto you. This is as good a rule of thumb as any—when in doubt, place yourself on the receiving end of the action—but it does not specify any content and has about as much practical value as the recommendation to be "as sweet as pie" in our dealings with others. Besides, it is not without its ambiguity, as Augustine was careful to point out. Suppose that I am in love with your wife and you are in love with mine. Does that mean that a swap is permissible as long as we both consent to it? The old bishop hardly thought so.

If the history of the Western tradition demonstrates anything, it is the utter futility of any and all attempts to derive a coherent political doctrine from the pages of the New Testament alone. Any such attempt is bound to end in madness for the simple reason that there is no way of harmonizing the conflicting aspirations to which the love enjoined upon us is likely to give rise. One can read the New Testament and decide to go into the wilderness, as Antony did, or set out to conquer the world for Christ, as Don Quixote and the Spaniards whose spirit he embodied tried to do. The early Christian writers knew this only too well, and that is why they turned either to the Hebrew Scriptures or to Greek philosophy for the practical guidance that the gospel failed to provide.

In brief, Christian agapism is neutral in regard to all legitimate regimes and gives as little or as much encouragement to any one of them as it does to the others. None of its teachings translates directly into programs of political action that would automatically bear its stamp of approval. God does not vouch for the worth or success of our political initiatives but is apparently willing to let us try our hand at any number of them. On the basis of reflection and experience we are to decide which regimes and which programs are best suited to the needs of the hour or, at least, to the possibilities of the hour.

The only way to determine whether a particular regime is acceptable from a Christian point of view is to examine its goals and see to what degree they are compatible with those of the New Testament. This is not an easy task for a variety of reasons, not the least important of which is that liberal democracy is far from being as clear as it once was (and as its opponents still are) about its aims and purposes. The ones with which it is most commonly associated are relief of man's estate on the one hand and freedom on the other. While one should be grateful for both of these precious commodities, one cannot help feeling uneasy about the uses to which they are being put. There are some prominent people, John Paul II among them, who think that the Western democracies are in danger of being devoured by their consumerism (if that metaphor is not too offensive) and who rightly wonder whether the spiritual interests of the human race are always best served by the type of regime that is most conducive to the production of wealth and material goods.

As for our much vaunted freedom, it seems to have become practically indistinguishable from a permissive egalitarianism that guarantees everyone the right to think and choose as he pleases. Democracy has always tolerated and even encouraged differences of opinion, but these were formerly looked upon as an incentive to seek the truth. Today they are interpreted as a sign of the futility of that quest. The new wisdom is that there is no wisdom. Since all absolutes are ruled out, only freedom is left—absolute freedom, of course. As the main character in a recent play says so well, in defense of suicide no less: "Personal choice is the only thing one owns."

The obvious objection to this critique is that liberal democracy has acquired a new radiance over the years and that, whatever one may think of its tainted origins, its decency and moral integrity are now amply attested to by its dedication to the cause of human rights. This alone would suffice to invest it with a religious aura of sorts and recommend its endorsement by Christians everywhere. The hitch in this case is that the "rights" approach to the problem of justice is not necessarily the one that best fits the philosophy of the New Testament. The sacred writers might have had some sympathy for the language of "duties" but the language of "rights" probably would not have made much sense to them. Their own emphasis was on what one owes to others rather than on what one can claim from them.

Differently stated, the New Testament is more preoccupied with what happens to the perpetrator of an unjust or unkind deed than with the plight of his victim. The nuance, and it may be more than just a nuance, is not insignificant. For better or for worse, the New Testament is completely silent on the subject of rights and of what is now called "social justice," (to be discussed in chapter 21) most often championed by people who have given little thought to the philosophical implications of that popular nineteenth-century notion.

Problems of a different and less specifically religious nature come to the fore when we turn our attention to some of the other difficulties inherent in liberal democracy. The prospect of a large group of reasonable beings freely subjecting themselves to the rule of law is attractive in many ways and has about it an air of universality that renders it particularly appealing to Christians. Like Christianity, liberal democracy takes the whole world as its stage, but it sees that world in the context of production and consumption rather than in the context of love. Whether this narrow and highly unerotic view of life is good enough to satisfy most people is another matter.

Human beings do not live by rational calculation alone, and all recent efforts to temper the harshness of economic society by such misty sentiments as humanity, compassion, and philanthropy, or else to embellish it with the aid of such lofty but artificial notions as culture and aesthetics—not to mention the other playthings of a bored and boring bourgeois world—have been at most only partially successful. Compassion is not the same thing as charity and hardly qualifies as a substitute for it. As its name indicates, it is not a virtue but a passion—more likable than a lot of other passions perhaps, but not any less self-regarding. And modern art that is so completely divorced from the rest of life as ours has become will not do the trick, either. Its end result is not a new integration of our fragmented and disoriented selves but the emergence of the aesthetic state as a state capable of determining one's existential attitude, alongside other possible states, such as the scientific state or the political state. One is merely concerned with bringing about a harmony of the feelings on the basis of which one's sentiment of life and one's view of the world is formed. Where that leaves us, nobody knows or even bothers to ask. Culture is taken for granted as a requirement of human progress, but the crucial question of the goal of that progress receives no answer. All of this, as Heidegger notes, is still only a facade, behind which lurks the metaphysical despair so perspicaciously unmasked and so powerfully analyzed by Nietzsche. Little wonder that our society should be dividing more and more of its time between the office or the assembly line and the couch. The scheme is not totally bad as long as the economists make enough money to pay for our psychiatrists, but it does leave something to be desired.

The overall situation is neither reassuring nor particularly edifying. Its consequences may be observed on any university campus these days, where, next to a few "idealists" and a few "cynics," one encounters an ever-growing number of students whose attitude can only be described as one of indifference or skepticism. My experience is that it is usually among this third group that the best and most gifted minds are to be found. Unfortunately, they are seldom given an intellectual fare capable of satisfying their natural curiosity and of generating some clarity about fundamental issues, which happen to be the only ones they really care about. The graduate counterparts, although more sophisticated, do not fare much better. From them one can expect to hear all about "values," "life-styles," "authenticity," "creativity," "culture," "fulfillment," "consciousness raising," and a host of other abstractions by means of which philosophers have been trying for over a century to restore some meaning to human life on the basis of the modern scientific understanding of the universe. The whole structure is amazingly fragile, as the students themselves soon realize when they are made to see the foundation on which it rests.

The purpose of these remarks, as you may have guessed, is not to denigrate liberal democracy—enough people are doing that as it is—but to suggest that all is not well with it and that Christians ought to be as eager to improve it as they sometimes are to praise it. Since the Christian God does not have a political platform, let alone a set of recipes for its implementation, the burden is on us to devise our own solutions to the problems at hand and come up with ways of making them work better. Christianity has made its peace with stranger regimes in the past. It would be astonishing if, without forfeiting its identity, it were not able to enter into a marriage of convenience or, short of that, some kind of liaison with this one as well. Happily, it does not have to baptize it in order to do so. Having lived through centuries of sacred kingship, almost to its undoing, the last thing the Western world needs is a few more centuries of sacred democracy.

If I have insisted so much on the transcendent nature of the Christian faith, it is only because it is the point that is most easily misunderstood by contemporary religious and political thinkers. By "transcendent" I do not mean "otherworldly," as is so often said since Nietzsche, but "apolitical" or, better still, "transpolitical." This transpolitical character is precisely what enables Christianity to become effective in the midst of the changing configurations and innumerable contingencies of human exis-tence; this is also, and for the same reason, what makes it impossible to give any final or universally valid answer to the vexed question of its rela-tionship to the political order.

The churches of America have never been more visible than they are at the present moment and have never had a greater opportunity to contribute to the enhancement of the spiritual life of the nation. The irony is that they are more confused than ever about their own mission and goals. The situation is not one that can be remedied overnight, but, if a start is to be made anywhere, I would venture to say that it is with our universities, theological schools, and seminaries. These are the prime incu-bators of the political fantasies that are currently being propagated under the banner of the gospel by church leaders, journalists, and religious edu-cators on all levels from the highest to the lowest. The task will take at least two generations, for, as John Stuart Mill reminded us long ago, the teachers themselves have to be trained before they can train others. One would like to think that the Christian tradition still has enough strength and vitality to renew itself from within. Instead of aping the worst features of modernity, its present-day spokesmen could take a leaf from that important segment of the nonreligious academic community that in recent years has brought to light the hidden roots of our modern consciousness and taught us how the wisdom of the past can be made to serve the needs of the present.

NOTES

1. Mortimer Adler, *The Great Ideas: A Syntopicon of Great Books of the Western World*, 2 volumes (Chicago: Encyclopedia Britannica, 1952).

2. Francis-René de Chateaubriand, *The Genius of Christianity* (Baltimore, MD and New York: J. Murphy Company, n.d.).

5

DO WE NEED CATHOLIC UNIVERSITIES?

For better or for worse, the defining moment in the history of American Catholic higher education came with the publication of John Tracy Ellis's celebrated article, "American Catholics and the Intellectual Life," in the October, 1955, issue of *Thought*. I say "for better or for worse" because, in retrospect, it is not clear that the changes for which Ellis was calling are the ones we most need at the present moment.

Ellis's thesis was that Catholic universities lagged far behind their secular counterparts and had a lot of catching up to do if they were to establish themselves as major players on the intellectual scene. There were good reasons for this. None of our universities was very old, the majority of their students were of immigrant peasant stock among whom higher education was a rarity, and subtly applied quotas had long kept most Catholics out of the best non-Catholic universities. Formidable obstacles indeed, against which, despite its rapid growth, the Catholic community had yet to make much headway.

To say that Ellis's article touched a nerve would be an egregious understatement. For months, it was a prime topic of conversation at faculty gatherings on Catholic campuses everywhere. Yet there was something oddly anachronistic about it. Its statistics were gleaned from studies that often dated back a few years and reflected a situation very different from the one that began to emerge shortly after the end of World War II. By the mid-fifties the picture was still not very rosy, but neither was it as bleak as Ellis's bookish account would have us believe. Even the minuscule, all but invisible college at which I had just been hired to teach was winning its share of national fellowships and having a surprisingly large number of its graduates admitted to the Yales and Harvards for their advanced degrees.

There was another conspicuous flaw in Ellis's article, this one having to do with the yardstick by which the author had chosen to measure the performance of Catholic universities. These were deemed inferior because they failed to live up not to their own standards but to the external standards of the modern research university, to which the Humboldt reform had given birth in Germany a century earlier and which the great American universities had been frantically striving to emulate since the beginning of this century. They were less well endowed, lacked adequate research facilities, received less support from outside sources, produced fewer Ph.D.'s, and contributed little to the advancement of science. The article said next to nothing about the kind of education that Catholic schools were attempting to provide versus that of the secular institutions to which they were being compared.

If I mention Ellis's article, it is not because it is solely responsible for the subsequent evolution of Catholic higher education but because it implicitly drew attention to a problem that had been festering beneath the surface of Catholic academic life for half a century or more, viz., the long-deferred but inevitable confrontation between Catholicism and the modern world. The irony in the story, as has often been noted, is that Ellis's stinging indictment came at a time when the secular universities that so impressed him were about to enter a period of crisis from which they have been trying to recover ever since.

In support of his general thesis, Ellis cited a remark once made to a group of Catholic educators by the legendary Robert Maynard Hutchins to the effect that Catholics had "imitated the worst features of secular education and ignored most of the good ones."[1] Ellis's use of this statement shows only how little he had understood it. Hutchins had no intention of holding up secular education as a model to be imitated; he was merely warning his Catholic audience that they were in danger of selling their own heritage for a mess of pottage. As president of one of America's premier research universities—the home of Enrico Fermi, James Franck, and I do not know how many other Nobel laureates—Hutchins himself recognized the importance of scientific research and supported it wholeheartedly. His own project was nonetheless quite different. It harked back to an earlier age and evinced a deep concern about the fractionalization of knowledge that the new university endorsed as the condition of scientific progress. Whatever else it might be, Hutchins thought, a university was first and foremost a place where the old Socratic question of the good life could be addressed squarely and subjected to rigorous scrutiny in the light of what the greatest minds had said about it. This explains his interest in the books that deal thematically with it and are best able to illuminate it, the Great Books, as they are now pompously called. It also explains his respect for the twin disciplines of philosophy and theology, once the centerpieces of the university curriculum and, as synoptic disciplines, the only ones capable of showing how the parts of that curriculum are related to one another and to the whole.

Hutchins's vision had a noticeable impact on a good number of Catholic colleges in this country, particularly the smaller ones. It showed them how they might be able to broaden themselves in a manner consistent with their own principles. It confirmed them in their attachment to Thomas Aquinas and his philosophic mentor, Aristotle. Above all, it helped them shake off the remnants of an archaism that the neo-Thomistic movement had inherited from its romantic origins and of which it had never fully succeeded in divesting itself. Hutchins's subterraneous influence still persists in a few of these schools. I, for one, like nothing more than to see their graduates show up in my classes once in a while. They are generally well ahead of the pack and have a lot less "unlearning" to do before they can settle into their new programs.

The same cannot be said of our larger and wealthier universities, many of which have been busy aping the features of secular education that most troubled the reform-minded Hutchins. By and large, philosophy and theology have been stripped of their status as architectonic disciplines and survive, if they survive at all, as parts of a democratic arrangement within which the quest for first principles and the unity of knowledge are dismissed not only as irrelevant but as inimical to modern egalitarian ideals. The Great Books have not fared much better. A syllabus from a well-known school that I happened to glance at recently did not list a single book that was more than three years old, even though the course was billed as a general introduction to moral theology. Needless to say, what passes for a great book today is a far cry from what Hutchins had in mind. The "canon," as it is derisively called and about which we have been hearing so much lately, has been expanded in such a way as to include a host of newcomers whose merit lies mainly in their being just that—newcomers.

Equally characteristic of the new trend is the "objective," "detached," or "scientific" spirit in which theology and philosophy are approached. The existential questions are either bypassed altogether or relegated to some undefinable realm of feeling, to be dealt with elsewhere, preferably in a nonacademic setting. Value-neutrality is the order of the day and the touchstone of one's integrity as a scholar. As often as not, students are told to "make up their own minds" about all important questions. Any attempt to teach in the old sense of the word is branded a form of indoctrination and denounced on those grounds.

It would nevertheless be unfair to say that our new university is not governed by any overarching principle. Consciously or unconsciously, it subscribes to at least one such principle, viz., the relativity of all human horizons of meaning and value. The point was brought home to me not long ago by an administrator who proudly announced that his college's newly established "substance-free" dorm had been a total success. The fifty-odd students who opted for it had been fully "accepted" by the rest of the community. No one had tried to make fun of them. In a fit of perverse curiosity, I inquired whether anybody had expressed admiration for them. Receiving no answer, I reluctantly came to the conclusion that these broad-minded students were well on their way to becoming dogmatic relativists. They "accepted" their substance-free fellows in the same way that they will soon be "accepting" the campus gay-lesbian club, if they have not done so already. The "orthodoxy" to which they are committed is that one can do whatever one chooses provided no one else is hurt by it. Have sex any way you like, but don't smoke while doing it. A strange orthodoxy, no doubt, but like all orthodoxies nowadays, it binds everyone under pain of ostracism, the harshest penalty that can be inflicted on young students.

The foregoing account may sound like a caricature of what is going on in our Catholic universities, but the frequency with which the pattern is repeated suggests otherwise. Two newer and seemingly innocuous developments are there to remind us that all is not well in the state of our educational affairs.

One is the emphasis on advisement, now a veritable campus industry requiring an enormous investment of time, energy, and money on the part of both administration and faculty. Advisement had hitherto been closely linked to what went on in the classroom, where courses dealing with questions of more than purely academic interest were taught and where healthy relationships between professor and student were forged as a matter of course. Informal contacts outside the classroom were taken for granted and considered an indispensable part of student life. Much of this has gone by the board. What we have instead are formal advisement sessions in which a group of students who are total strangers to one another are assigned to individual faculty members who have never met them, know nothing about them, and spend no more than a few minutes with each of them once or twice a year at registration time. The whole system is about as effective as a Band-Aid on a wooden leg. Maybe we need it, but to say that we do is to admit that real advisement is in trouble, something of an endangered species for the protection of which extraordinary means must be used.

The second development is a fixation on what is now euphemistically called "cultural diversity," along with the decision to institutionalize it by making it a "core" requirement—as if cultural diversity were a discipline on a par with other bona fide academic disciplines, as if it could be "taught," or as if a one-semester course in Mahayana Buddhism was likely to effect a significant change in anybody. What students mostly find in such courses is what they bring to them, namely, their own unconscious and typically Western prejudices. However far and wide our modern-day Columbuses may travel in search of new continents, they never seem to discover anything but Genoa. I much prefer to see the few students who are capable of it immerse themselves in some great author and learn to see the world through *his* eyes for a change. This and only this will rid them of their provincialism and give them an intimation of the differences that separate human beings. As matters stand, very few of them ever live with a good book long enough to find out what a true alterative to their own way of thinking might be and thus learn something of importance about themselves.

Why is it that the bulk of the education provided at such great cost to both graduates and undergraduates by our best schools is so often perceived by the students themselves as anemic and antiseptic to the nth degree? The reason, I suspect, is that most faculty members are themselves products of the modern research university and imbued with its peculiar ethos. Being very bright for the most part, they realize they will be promoted or otherwise rewarded on the basis of their scholarly achievements and are more likely to identify with their professional guild than with the school that pays their salaries. They are academic "free agents," ready to move whenever a better offer comes along. Their courses correspond less to the needs of the students than to their own academic interests, and in all but the rarest of cases the subject matter is approached from a religiously and metaphysically neutral standpoint that supposedly guarantees their objectivity. They could teach the same courses anywhere to almost anybody. Little of what they have to say touches the hearts of the students or is apt to have the slightest effect on the way they live.

Do not misunderstand me. The Catholic Church needs theological and historical research to sustain its intellectual life, and the collapse of so many of the great European institutes where this research used to be carried out is a calamity of the first magnitude, one of the tragedies of twentieth-century Catholic life. All I am saying is that this type of scholarly work is not a substitute for the education that many dedicated Catholic students are demanding and to which they are entitled. One wonders what the Catholic individuals and foundations who have been contributing handsomely to the support of this work would think if they knew how their money is being spent.

Catholic institutions are materially better off today than they were when Ellis dropped his bomb, and they have vastly superior intellectual resources at their disposal. Of even greater significance to their life and work than the spectacular advances of modern science is the fresh under-standing that we now have of classical philosophy, made available to us by the most brilliant minds of our century, and therewith of the creative use to which that philosophy was put by the luminaries of the Christian tradition. These are the issues that today's Catholic universities would do well to explore, instead of consuming themselves in endless efforts to circumvent the regulations Rome would like to impose on them or in stillborn disputes between "liberals" and "conservatives" that remind me of nothing so much as the silly quarrel between the apothecary Homais and the abbé Bournisien in Flaubert's *Madame Bovary*, the worst of the new locked in bitter combat with the worst of the old. The opportunity is there, perhaps for the first time in the modern period. It remains to be seen whether enough Catholic educators will recognize it and find within themselves the courage and the imagination needed to exploit it.

NOTE

1. John Tracy Ellis, "American Catholics and the Intellectual Life," *Thought* 30, no. 18 (1955): 375.

6

ROME AND THE THEOLOGIANS

The nagging tensions inherent in any attempt to reconcile the demands of academic freedom with a firm commitment to the Church's official teachings have again been forcefully brought home to us in an important document issued in June of 1990 by the Congregation for the Doctrine of the Faith under the title, *Instruction on the Ecclesial Vocation of the Theologian.*[1] Coming as it did on the heels of the Vatican's investigation of such well-known figures as Edward Schillebeeckx, Leonardo Boff, Hans Küng, and Charles Curran, the last two of whom were eventually stripped of their mission to teach as Catholic theologians, the new document was bound to make headlines, especially in a country where most Catholic universities operate under public charters and subscribe to the principles of academic freedom in vogue in secular universities.

The reaction of the liberal press was predictable. *Time* magazine called the *Instruction* a thunderbolt, "the toughest and most sweeping pronouncement Rome has made in modern times on the limits of intellectual freedom in the Roman Catholic Church."[2] *Commonweal* followed suit, ignoring completely the broad ecclesial context within which the problem of dissent is taken up. It labeled the *Instruction*'s overall message "disheartening" and read it as a "sweeping rejection of what constitutes much of the framework of human consciousness in the twentieth century," thereby implying that this framework should have the same normative value for the Catholic Church as it supposedly does for everyone else in our time. The text, it added, was the latest salvo in an ongoing offensive against dissidents, and it warned that there was more to come—perhaps a new encyclical or apostolic letter proclaiming the infallibility of the norms on which the Church bases its stand on artificial birth control and other matters pertaining to sexual morality. A "major disaster" was in the making. Catholic intellectual life would be discredited and "for great numbers of Catholics of all ranks, the choice would be to fall silent, dissemble, or take their leave of the Catholic Church."[3]

For weeks, the *National Catholic Reporter* was its usual strident self. Its Vatican affairs expert, Peter Hebblethwaite, did not have a kind word to say about the *Instruction*: its content was amiss, its argument flawed or nonexistent, and its spirit anything but Christian. To make matters worse, there is "a lot of anger" in it, more so, it would seem, than in Hebblethwaite's own article, if that is possible. One can sense it in "the venom with which theological positions are caricatured, the better to be denounced," and in the allegation that the theologians are trying to displace or replace the bishops as the Church's official teachers.[4] In the meantime, letters to the editor poured in, the vast majority of them bitterly critical of the Vatican and comparing its tactics to those of the Nazi regime or Stalin's Great Terror.

Non-Catholics were equally quick to jump into the fray. In a *New York Times* op-ed piece that verges on the hysterical, Harvey Cox spoke of the *Instruction*'s "chilling" effect in outer as well as inner Catholic space. The self-censorship for which it calls was sure to "weaken the presently robust level of Catholic biblical and theological scholarship, thus endangering the field of religious studies as a whole."[5] More than ever, theologians will have to be on their guard, never knowing when a phone call might come from some chancery watchman in response to a casual remark made on a TV panel or in a magazine article. To forestall that dreaded call, they will have to abandon their "unflinching candor" and become dissemblers.

Cox saw a good deal of irony in all of this. Catholic theologians will now "have to cope with some of the same numbing prohibitions that Pope John Paul II helped abolish in Eastern Europe and the Soviet Union."[6] Whereas they are forbidden to use the mass media to make their views known beyond the narrow circles in which they move, their ecclesiastical superiors, the American bishops, have no qualms about hiring a public relations firm to manage their campaign against abortion. Cox likewise faults the *Instruction* for its failure to note that today's heretic is often tomorrow's saint. The books of Thomas Aquinas, Roman Catholicism's official theologian, were once "publicly burned in Paris by church authorities," or so Cox would have us believe. In our own day, theological luminaries who had long been out of favor with the Vatican—Congar, De Lubac, Daniélou, and von Balthasar among them—were spectacularly rehabilitated and in some cases elevated to the rank of cardinal.

True, says Cox, the document recognizes that dissent is sometimes legitimate, but it will not permit theologians who have serious difficulties with the Church's teaching to take the matter up with anyone save the responsible authority. Theological discourse is thus reduced to a private enterprise, to be carried on "under the seal of the confessional." This, for Cox, is "elitism at its worst," one of the great crimes of our time. The laity is shut out altogether and religion is left in the sole hands of . . . . theologians!

But why should Cox, a Protestant who teaches at Harvard, a university not known for its close ties to the Vatican and well beyond the reach of its thunderbolts, be so exercised about the internal affairs of the Catholic Church? The answer, we are told, is that everybody stands to lose by the new ban. Even when they go unheeded, gag rules have a stifling effect on life, both for the gagged and for those who are trying to talk with them. Just as theology is too important to be left to theologians, so Catholic theology is too important to be left to Catholics. No, the ban is not a purely internal matter. It affects everyone who cares about the truth: "Ask not for whom the phone ringeth. It ringeth for us all."[7]

The shrillness of these reactions stands in sharp contrast to the tone of the Vatican document, which is hardly inquisitorial or reminiscent of what went on in Germany and the Soviet Union under Hitler and Stalin. The *Instruction* does not claim to do more than restate for our time the principles by which the Church has long been guided in dealing with the complex issue of theological dissent. Its position is not noticeably different from the one taken by Vatican II and already advocated by Thomas Aquinas, who states flatly that there are times when a subordinate has not only the right but the duty to criticize an ecclesiastical superior, as long as he does so with internal respect, external deference, and discretion.[8] It was understood that this respect is not limited to the Church's solemn pronouncements but extends to its so-called "ordinary magisterium" as well.

What the *Instruction* rules out is not dissent as such, unless one insists on using the term in a pejorative sense, but organized and public dissent, which nowadays frequently takes the form of an appeal to the media for the purpose of mobilizing popular sentiment in favor of opinions at odds with the authoritative teachings of the Church. But this is nothing new, either. Pope Paul VI had already said as much and for the same reasons. What sets Christianity apart from the religions to which it is most closely related, Judaism and Islam, is that it first comes to sight as a nonpolitical religion or, in St. Paul's words, a "life-giving doctrine." It teaches that henceforth one would be justified by "faith" rather than by obedience to a God-given "law." In its view, orthodoxy is more important than orthopraxy and what one holds as a believer takes precedence over any of the political or legal arrangements by which human beings are wont to order their temporal lives. Accordingly, no other religious tradition has ever placed a greater premium on purity of doctrine. This explains the need for a unique teaching office, known since the nineteenth century as the "magisterium" and made up of the Pope and the bishops, who speak for the Church as a whole in matters of faith and morals.

Hebblethwaite is probably right when he says that few theologians have any desire to set themselves up as a separate magisterium, on a par with and at times in opposition to the hierarchy; but if, by going over the heads of the bishops and taking their case to the general public through press conferences, TV appearances, interviews, widely circulated petitions, and full-page ads in our national dailies, they regularly succeed in undermining the hierarchy's doctrinal authority, the effect is much the same.

No one denies that theologians have an irreplaceable contribution to make to the life of the Church. Part of their task is precisely to enlighten the magisterium and, by so doing, to help everyone else arrive at a clearer or firmer grasp of the divinely revealed truth. Vatican II showed us all that can be accomplished when bishops and theologians collaborate in genuinely collegial fashion. This does not alter the fact that the theologian's authority is essentially different from that of the magisterium. It is an "epistemic" authority, based on knowledge alone, as distinguished from the effective or "deontic" authority of the person to whom the governance of the community has been entrusted.

What this means concretely is not that there is never any room for discussion about points of doctrine but only that, when complete agreement cannot be reached, the hierarchy retains the right of final decision. The exercise of that right, an act of the practical intellect (as the old Scholastics would have said), requires that the present and long-range good of the whole community be taken into account. As such, it calls for a prudential rather than a purely speculative judgment on the part of the one who makes the decision. If the originality of the *Instruction* lies anywhere, it is in the forthright acknowledgment that magisterial decisions of this sort are not necessarily the last word on the subject and need to be interpreted in the light of the historical situation that is being addressed. Church doctrine is expected to remain the same in all essentials, but pastoral considerations dictate that the modalities of its presentation vary in accordance with changing times and circumstances. A prime example of this kind of adaptation is Vatican II's reformulation of certain well-known nineteenth- and early twentieth-century declarations concerning religious freedom whose restrictive language was deemed inappropriate to our time. The interesting question, and it is not a question on which we need to dwell here, is whether the new formula does not in effect represent a substantial departure from traditional Church teaching.

None of this is to imply that Church authorities have always demonstrated the proper wisdom in their efforts to implement the foregoing principles. Rome itself has admitted that mistakes were made in the past, that those who acted in its name were occasionally inspired by motives that were more political than religious, and that its procedures still leave something to be desired. The other side of the story is that most popular accounts of the Church's dealings with dissenters and heretics fail to rise above the level of Enlightenment propaganda. At a distance of several centuries, it is difficult to arrive at a clear picture of the issues involved in any particular case. The most frequently cited examples are those of Giordano Bruno and Galileo, who were dragged before the Inquisition among other things for upholding the Copernican theory. Yet there is mounting evidence to suggest that more was at stake in this matter than the defense of an astronomical system whose nonhypothetical character would not be established to the satisfaction of the scientific community for at least another century. Some modern thinkers, such as Spinoza, descried in the new theory further proof of the spurious character of the would-be miraculous events narrated in the Bible. If the earth revolved around the sun and not vice versa, Joshua could not have stopped the sun in its tracks. Another blow was being struck at the Bible, whose credibility was thought to rest in part on the historical truth of such events. The argument was the more telling as medieval theology regarded Joshua's miracle as the greatest of all possible miracles, one greater even than the resurrection of a dead person. Other segments of the intellectual elite were attracted to the Copernican theory because it symbolically assigned to natural reason, represented in the Western tradition by the sun or the Greek god Apollo, the place reserved for Christ at the center of the universe. It was part and parcel of a powerful movement the object of which was again to challenge the authority of divine revelation. This alone was more than enough to arouse the curiosity of a group of understandably suspicious inquisitors.

Even so, the Church appears to have been slow to take action against philosophers or scientists as long as the discussion of such matters was confined to learned circles. Galileo did not come under ecclesiastical scrutiny until his *Dialogue on the Two Chief Systems of the World*, a book written for nonspecialists, was published. By its own admission, the famous Condemnation of 1277 alluded to by Cox was prompted by a concern for the spiritual welfare of the nonacademic community, among whom the censured theses, most of them deeply hostile to the faith, had begun to spread.

For the record, it should be noted that Thomas Aquinas's books were not burned by order of the bishop of Paris, Etienne Tempier. Tempier's list of 219 "errors" does include a few propositions that Thomas, who had been dead for three years, would probably have endorsed, but this hardly amounts to a bonfire of the kind that Cox seems to have in mind. Luther is the one who urged his disciples to burn their copies of Thomas's works, something they apparently refused to do for reasons over which it is tempting to muse. Tempier's main target was not Thomas but Thomas's adversaries, Siger of Brabant and his cohorts in the arts faculty, whose radical Aristotelianism threatened to sap the foundations of Christendom.

As for the most common objections to any restriction on freedom of speech—that it smacks of totalitarianism, stifles creativity, and breeds hypocrisy—they are neither new nor very profound. The lesson to be drawn from history is that unlimited freedom of speech is or can be as much of a threat to freedom as its opposite. Our own century has seen how rapidly societies that grant the same privileges to the enemies of freedom as to its friends can lose their freedom. Not entirely by accident did Weimar, Germany, the freest society the world had ever seen, become the breeding ground of the most hateful tyranny known to humankind.

Only if one accepts the now largely discredited view that human progress is inevitable and as it were a necessity of nature, as did Spinoza and John Stuart Mill, the two staunchest advocates of free speech in our tradition, is it possible to maintain that absolute freedom of expression always redounds to the benefit of the society that sanctions it. The glory of Christianity is to have introduced the spirit of free inquiry, the necessary condition of intellectual progress, into the religious community by making the study of philosophy a mandatory part of its curriculum of theological studies. Its genius is to have realized that this spirit would die as quickly as it did elsewhere if it were to be cultivated without any regard for the needs of society at large.

We forget too easily that throughout most of history freedom of speech as we now understand it was an unknown commodity, not only because governments did not tolerate it, but because of the restraints that writers who had something of importance to say saw fit to impose upon themselves. St. Augustine went so far as to argue that one "commits a grave sin," *graviter peccat*, by revealing certain truths to people who are not prepared for them and liable to misuse them.[9] The "creativity" of these writers does not appear to have been thereby stifled. Many if not most of the great classics of world literature owe their depth and subtlety to the severe constraints under which they were written. None of their authors believed in the possibility of popular enlightenment and most of them cautioned against it. This did not make them any less eager to share their thoughts with others. They solved the problem by having recourse to a dual language the purpose of which was to conceal certain truths from the general public while allowing them to transpire for those to whom it could be of greatest benefit.

Cardinal Newman, who made a special study of that dual language in the works of the Church Fathers (and got into a lot of trouble with his former coreligionists for defending it, as he tells us in his *Apologia*), referred to it as the "economy" or the "reserve."[10] Unlike his nineteenth-century critics, he did not think it would automatically turn people into dissemblers and hypocrites. There is a vast difference between keeping certain opinions to oneself out of a noble regard for the needs or sensibilities of others on the one hand, and doing it out of self-interest or cowardice on the other. In the former case, one exercises prudence, which is a virtue; in the latter, one indulges in hypocrisy, which is a vice. Foreign as it may be to our modern taste, the pedagogy had much to recommend it since it served equally well the interests of the truth and those of the political or religious community.

Not much has been gained by abandoning it. Tocqueville and John Stuart Mill both make the paradoxical observation that there is less freedom of thought in our modern democratic societies than in any of the old aristocratic societies. People today may be politically freer than any of their predecessors, but they are that much more subject to what the popular press has since named the hegemony of "political correctness." Tocqueville called it the "tyranny of the majority," adding that the American regime had done a better job of preventing the circulation of ideas that ran counter to majority opinion than even the Spanish Inquisition. "Literary genius," he said, "cannot exist without freedom of the spirit, and there is no freedom of the spirit in America."[11] The remark may be truer than most of us care to admit.

The odd feature of the contest that has been pitting the Vatican against our dissident theologians is that both sides seek to justify their positions by appealing to the authority of Vatican II, which each side accuses the other of trying to highjack. This has been possible only because of the Council's ambiguous stand on many of the issues that divide the two groups. A decision was made at the outset that in its deliberations the Council would pursue two different courses of action or move along two different lines. It was to be an effort at *ressourcement*, that is, a return to the premodern sources of Christian life and thought—principally the Bible, the Church Fathers, and the liturgy—and it was to be an updating of the Church's message or, to use Pope John XXIII's term, an *aggiornamento*, by which was meant an attempt to catch up to or come to terms with the modern world. The trouble is that few people bothered to ask how these two lines of attack were supposed to converge. It was taken for granted that they would complement each other and the crucial question of the

relationship of modernity to premodernity was never addressed thematically. As a result, little serious attention was given to the possibility of
a tension, not to speak of an outright conflict, between these two vastly
different modes of thought.

The tension or the conflict, whichever it may be, is particularly evident in two of the Council's most important documents, the *Declaration
on Religious Freedom* and the *Constitution on the Church in the Modern
World*, both of them promulgated in 1965. Among the many examples that
one could cite is the *Declaration*'s teaching on individual freedom, which
stresses the fact that "human beings are bound to obey their own consciences" and are to be "guided by their own judgment" and later proceeds
to remind them that they are not as free as they may have been led to
believe, for the disciple is "bound by a grave obligation toward Christ his
master ever more adequately to understand the truth received from him,
faithfully to proclaim it, and vigorously to defend it."[12] Not surprisingly,
a sizable portion of the Catholic population seems to have responded with
greater enthusiasm to the first of these two statements than to the second.
Over the centuries, the Church had, whenever possible, encouraged the
laudable practice of toleration. In a sudden about-face, it now seemed
ready to endorse, however ambivalently, the typically modern principle of
toleration and its corollary, complete freedom of religion. It is unfortunate
that few of the Council Fathers had any deep knowledge of modern
thought or much time to acquaint themselves with it, for the Council,
already in its fourth year, was rapidly coming to a close. A significant part
of its agenda was thus left unfinished and the stage was set for the free
fall into which theologians in increasingly large numbers would be caught.

The consequences were not long in manifesting themselves. They became visible to the naked eye in the splintering of Catholic theology
during the years that followed. Like the proverbial hen that has laid duck's
eggs, the once "liberal" theologians whose views the Council had finally
embraced found themselves gazing in amazement, not to say horror, at
what their erstwhile disciples were saying and doing and spent the last
years of their lives bemoaning the drift of post-Vatican II theological
thought. One would like to believe that the dissensions that have been
racking the Church ever since are a sign of growing maturity, but they can
just as easily be seen as the harbingers of a deepening crisis the full
dimensions of which have yet to be revealed.

The fact of the matter is that, for all its penetration, the *Instruction*
adopts the same ambiguous stance as Vatican II toward modern thought,
which it courts in one way and repudiates in another. It sticks to the old
Roman habit of trading in "isms"—liberalism, relativism, pluralism, positivism, and the like—without probing their different meanings or seeking
to uncover the principles that underlie them. It evinces no keen awareness

of the reasons that led to the break with premodern thought, the force of the arguments in favor of modern thought, and the possible answers to these arguments. Above all, it does not explain how the legitimate insights of modernity might be divorced from their original context and inserted into a more adequate framework. In consequence, it lacks the freshness that one looks for in a document of this sort, which cannot be content with merely reaffirming the Church's basic teachings but must strive to instill new life into them by means of an invigorating dialogue with their most thoughtful critics. Quite apart from its substantive merits, the *Instruction* sounds cranky. It thus unwittingly lends credence to the view that Rome is engaged in a concerted attempt to repeal the modern age and roll back the Council.

Much outstanding work has been done by the greatest minds of our century to bring to light the true character of modernity, clarify the nature of its break with premodernity, and propose new and better ways of relating them to each other. With rare exceptions, theologians, who have a vital stake in the matter, have yet to familiarize themselves with this work and learn how to use it both to overcome the limitations of pre-Vatican II theology and curb the excesses of post-Vatican II theology. Only when this task is accomplished will Catholic theology be able to recover its lost equilibrium and turn the stillborn debates in which it has been consuming itself into instruments of genuine intellectual progress.

NOTES

1. *Origins*, vol. 20, no. 8, 5 July, 1990, 117, 119-26.

2. "Drawing the Line on Dissent," *Time*, 9 July, 1990, 62.

3. "Dangerous Opinions," *Commonweal* 117, n. 44 (10 August 1990), 435.

4. Peter Hebblethwaite, "Alternative of Cowed Silence: Strange Way to Defend Faith," *National Catholic Reporter*, 13 July 1990, 15, 28.

5. Harvey Cox, "The Vatican Needs a Dose of Glasnost," *The New York Times*, 17 August 1990, A29.

6. Ibid.

7. Ibid.

8. St. Thomas Aquinas, *Summa theologiae* II-II, qu. 33, a. 4.

9. St. Augustine, *Contra academicos* (*Against the Academics*) III.17.38.

10. Cardinal Newman, *Apologia pro vita sua* (New York: Random House, 1950).

11. Alexis de Tocqueville, *Democracy in America*, edited by J. P. Mayer with a new translation by George Lawrence (Garden City, New York: Doubleday and Company, 1969), vol. 1, part 2, chapter 7, 256. The French text reads: "Si l'Amérique n'a pas encore eu de grands écrivains, nous ne devons pas en chercher ailleurs les raisons: il n'existe pas de génie littéraire sans liberté d'esprit, et il n'y a pas de liberté

d'esprit en Amérique (cf. *De la Démocratie en Amérique*, vol, I [Paris: J. Vrin, 1990]), 200.

12. Vatican Council II, *Dignitatis humanae* (*Declaration on Religious Freedom*), no. 11 and no. 14.

7

IN THE SHADOW OF THE GALLOWS: CRIME AND PUNISHMENT IN MODERN SOCIETY

Aficionados of Gilbert and Sullivan operettas are not likely to forget the sonorous proclamation by means of which the Mikado of Japan informs the citizens of Titipu that his "object most sublime" is to "let the punishment fit the crime." His Imperial Majesty was not always in the same equitable frame of mind, for he had previously threatened to strip the local Lord High Executioner of all power unless he put someone to death within a month.

Both dispositions, one more "humane" than the other, point in different ways to the problematic character of human justice and its corollary, punishment. Assessing guilt and fixing the appropriate penalty is not an easy task, and we know from long experience that a craving for retribution can lead to crimes that greatly exceed the ones for which the guilty party is being made to pay. As far back as the early chapters of Genesis, Lamech, whose lust for vengeance knew no bounds, could sing to his wives: "I have slain a man for wounding me, a young man for striking me. If Cain is avenged sevenfold, truly Lamech seventy-sevenfold" (Gen. 4:23).

Much as we may regret it, there is no completely satisfactory solution to the problem. Good government is government by laws rather than by the will of the ruler, which leaves too much to chance and lends itself to too many abuses. But laws are effective to the extent to which they are backed up by force. Persuasion alone, though preferable and certainly more noble, has never sufficed as an instrument of just rule. It is no accident that *dike*, the Greek word designating the goddess of Justice, also means punishment, trial, or atonement. Laws differ from general rules of conduct, recommendations, admonitions, sound advice, or plain old good ideas precisely in that their observance is enjoined under pain of sanction. A law that has no teeth is not a law in the strict sense. It may be that every so often laws are allowed to be broken with impunity, but this implies only that, though still on the books, they have fallen into desuetude and thus effectively ceased to be laws.

Part of the trouble is that strict obedience to the law does not guarantee that justice will prevail in each instance. Lawgivers, who must legislate not only for whole societies but for the future as well as for the present, cannot take into account all of the circumstances that surround human actions. There will always be moments when, for some unforeseen and perhaps unforeseeable reason, even the best law must be set aside in the interest of justice itself. Aristotle and the long line of legal theorists who followed him tried to circumvent the difficulty by appealing to a special virtue called "equity" or "fairness," which allows one to disregard the law whenever compliance with it would run athwart the intention of the legislator.

Thus understood, however, equity is still only a correction of legal justice. It assumes the necessity and the binding force of the law and merely strives to compensate as best it can for its possible deficiencies. We run into heavier waters when, instead of taking the legal and penal system for granted, we pause to examine its theoretical foundations. Two closely related issues have traditionally dominated the philosophic agenda in regard to this matter: the degree of responsibility that human beings bear for their crimes, and the reasons that ultimately justify the punishment of criminals.

HUMAN FREEDOM AND RESPONSIBILITY: THE CAUSES OF CRIME

It goes without saying that only guilty persons deserve to be punished, and, furthermore, that guilt is imputed only to those who are presumed to have acted of their own free will; for no one can reasonably be held accountable for acts over which he or she had no control. Deeds perpetrated by insane persons or performed under duress, for example, clearly fall outside the pale of morality, as does any damage caused unintentionally or in a purely accidental way, even though in the public interest the law may require that the victim be indemnified. Contrary to modern Western practice, the ancient world likewise prescribed religious expiations for inadvertent slayings and similar mishaps, both out of consideration for the feelings of the relatives of the slain person and as a means of assuaging any remorse that the slayer might experience in spite of his innocence.

The pertinent question is how, in cases other than the ones just mentioned, one decides whether and to what extent the agent was free at the time of the deed. We shall find the beginnings of an answer to that question if, taking our cue from the philosophers of classical antiquity, we look briefly at the most general or most common causes of crime. These may be said to be three in number: passion or spiritedness, pleasure and the means of procuring it—such as money—and ignorance.

There is, to start with, an obvious difference between crimes of passion and crimes committed in cold blood or for the sole purpose of obtaining the material goods that one craves. A human being who, in a fit of rage or indignation, kills or wounds a fellow human being by whom he has been wronged will usually be treated with greater leniency than one who has spent weeks or months plotting a murder or a heist and thus proves that he is acting with malice aforethought. The reason is that the angry person is presumed to have been at least partially blinded by the violence of the emotion to which he fell prey. Although guilty of a crime, he is not himself a criminal insofar as he does not question the authority of the laws and will most often regret his deed once the passion has subsided. His moral character is markedly different from that of the true criminal, who, having cast his lot with injustice, is not apt to repent and will have few qualms about engaging in the same kind of behavior should the opportunity again present itself.

This is not to suggest that crimes of passion necessarily cause less harm than the others. If anything, the opposite is frequently the case. Spiritedness, which can inspire the noblest of deeds, such as the defense of justice or of one's country, also happens to be the stuff of which fanatics are made. It is by nature a self-righteous passion, one that feeds on itself and is the more dangerous as it draws its strength from the conviction that the object pursued is morally good or just. Anyone dedicated heart and soul to a particular cause is prone to exaggerate the justice and nobility of that cause. He must find it worthy of the sacrifices that it demands of him, and he proves to himself that it is by opposing as unjust anyone who would attack it or refuse his allegiance to it. The consequences can be disastrous. One shudders at the number of atrocities that have been committed in the name of justice over the centuries. Hence the peculiar dilemma that confronts civil society, which for its own defense must encourage spiritedness on the part of its citizens and at the same time punish the crimes to which it is liable to give rise.

As for the third cause, ignorance, it directs our attention to a different set of problems, the nature of which is perhaps best illustrated by Socrates's famous dictum to the effect that no one does evil voluntarily. On that premise, all crimes, including the ones just mentioned, would be reducible to crimes of ignorance. Human beings could never be blamed for their misdeeds, and society would have no choice but to decriminalize evil by promulgating something like a "no-fault" murder or theft policy.

Simple common sense should suffice to convince us of the impracticality of any such policy. As Saint Augustine, echoing Cicero, explains in Book V of the *City of God,* if human beings are not accountable for their actions, laws are enacted in vain, exhortation becomes futile, praise and blame are revealed to be meaningless, and there ceases to be any justice whatever in the apportionment of rewards for the good and punishments for the wicked; in short, the whole economy of human life is subverted. Not surprisingly, Plato himself seems to have had some reservations about this thesis insofar as, in Book IX of the *Laws,* he goes out of his way to distinguish crimes of ignorance from crimes of passion or malice.

Civil society has a rough and ready way of dealing with the problem: It insists that the laws be obeyed and does not allow anyone to plead ignorance of them. Still, there is a sense in which it is true to say that all crime is rooted in ignorance. If, as Socrates maintains, to act morally is to follow one's reason, evil must be seen as a form of madness or a miscalculation of some sort. The wrongdoer would then be the one who is led astray by false opinions about human life and, not knowing any better, ends up hurting himself as much as he hurts anyone else. What his conduct betrays above all is a lack of education or, as Plato called it, a lack of experience in things beautiful, *apeirokalia.* The matter will prove to be of some consequence when we turn in a moment to the reasons that stand behind the use of punishment.

A less obvious but no less interesting implication of Socrates's dictum comes to the fore when we recall that the laws enforced by society are relative to the regime and share in its imperfection. Some laws are clearly better than others, and any law—to the extent that it constitutes a barrier to purely arbitrary rule—is better than no law at all; but, just as there is no such thing as a perfectly just regime, save perhaps in theory, so there is no such thing as a perfectly just law or code of laws. A democratic society will enact laws that favor the many, and an aristocratic society, laws that favor the noble few. What one enjoins as praiseworthy the other condemns as blameworthy, and vice versa. In all cases, preference is given to one part of the society, which only claims to rule in the interest of the whole. Bluntly stated, laws are made for the benefit of those who write them. All of them are in some way open to rational criticism. Pascal said it more concisely and more eloquently than anyone else when he wrote: "True on this side of the Pyrenees, false on the other."[1]

This leads to the interesting possibility that some crimes may be due not to ignorance but to a particular individual's more perfect knowledge of the demands of justice. Society itself acknowledges that possibility, however dimly, when it distinguishes between political criminals, who act in the name of a higher principle, and common criminals, who have nothing but their own private interests in view. In order to protect itself, society may feel the need to punish the former as well as the latter, but not without showing them greater respect than is normally accorded to common criminals. One thinks straight away of the paradigmatic case of Socrates, whose sense of justice was vastly superior to that of Athens, however guilty he may have been in the eyes of its leading citizens. Cities, after all, do not philosophize, and even if they did, they would still be compelled to operate within a horizon that is severely limited by the need to combine wisdom with the consent of the multitude.

PUNISHMENT AND ITS RATIONALE

With these ideas in mind, we may be in a better position to tackle the larger question of the reasons that justify the recourse to punishment. Three words sum up the thinking of the philosophic tradition on this score: rehabilitation, deterrence, and retribution.

Plato, whose *Laws* contains the first and still one of the most penetrating discussions of this subject, was of the opinion that punishment had as its chief goal the rehabilitation or betterment of the offender. Once crime is defined as a mindless act, the best treatment is bound to be the one that seeks to make the criminal aware of the irrationality of his ways. One deals with him as a parent would deal with an unruly child, applying coercion whenever necessary to bring about the desired reform. This is not to rule out persuasion, which obviously has its proper place in the process, but only to indicate that with some people punishment, or the threat thereof, is the only effective remedy. The hope is that through this kind of training the delinquent will develop a taste for virtue and begin to take pleasure in its pursuit. Such is the goal for which, ideally, society and its law-enforcement agencies should strive. Implicit in this view is the thought that if the wrongdoer can be rehabilitated by means of persuasion and rational arguments alone, one might dispense with punishment altogether, demanding only that reparations be made for any damage that may have been caused.

Unfortunately, not all criminals are curable. To make matters worse, society cannot afford the luxury of waiting to see whether they are or not. Its first duty is to protect itself and its citizens against those who would prey on them. Necessity, if nothing else, compels society to resort to punishment both for the purpose of preventing the incorrigible evildoer from indulging in further crimes and of discouraging others from emulating him. Deterrence replaces rehabilitation as the overriding motive, and, in the case of the convict who is sentenced to death, the only motive, since the nature of the punishment is such as to preclude any possibility of reform.

Much has been written of late to show that, as a deterrent, punishment is counterproductive or only marginally effective; that it merely embitters the wrongdoer against society or further corrupts him by forcing him to associate with other prisoners, some of them possibly more corrupt than he is; that anyone attracted to a life of crime is unlikely to be dissuaded by the force of the law or the severity of its sanctions; and that the psychology of the criminal mind is such that nothing, not even the fear of death, is capable of making an impression on it.

Whether this is in fact the case is by no means certain. Reliable statistics are practically impossible to come by, since anyone who values his reputation will think twice before volunteering the information that the image of the electric chair is all that kept him from indulging his criminal desires. The only valid method of inquiry would appear to be the one that tries to ascertain whether, for example, the incidence of major crime is higher in the states that have abolished capital punishment than in the ones that have not; but in view of the many social, cultural, and economic variables that have to be taken into account, the method is anything but foolproof. The contradictory, or at best inconclusive, results at which such studies tend to arrive lead to the suspicion that the outcome has been largely predetermined by the assumptions that guided the research. Anyone who is against capital punishment on moral or other grounds, it seems, will deny on the basis of the available data that it functions as a deterrent, and anyone who has no such qualms about it will use the same data to prove the opposite.

Most earlier writers on the subject were content to rely on common observation, which suggests that the prospect of the scaffold, the "terror of evildoers" (Spinoza), is indeed a powerful inhibitor of crime. In *The Journal of a Voyage to Lisbon*, Henry Fielding tells the story of a convicted felon who complained that it was very harsh to hang a poor man for stealing a horse; to which the judge replied in substance: "You are not to be hanged, sir, ... for stealing a horse, but you are to be hanged that horses may not be stolen."[2] The judge's logic was by no means flawless, for one can only be punished for a crime that one has committed, but it brings out an aspect of the problem to which the thief had not given sufficient thought. Fielding's sentiment was apparently shared by Samuel Johnson, who is famous among other things for having said that nothing concentrates the mind more wonderfully than the thought of swinging from the gallows. In a similar vein, Jean-Jacques Rousseau argued in *Emile* that even the most desperate man—the one "least in command of his senses"—will instantaneously overcome his temptation to commit murder or adultery if he sees "the apparatus of torture and is sure of perishing on it in torments a quarter of an hour later."[3]

As was mentioned earlier, the only other major argument in favor of punishment is the one that looks upon it as a form of retribution, that is to say, as a means of restoring the moral order that the criminal deed has called into question. Involved in this argument is an issue which, although less amenable to rational analysis, has nevertheless been brought home to us in a forceful way by the great crimes of our century, crimes so heinous and of such magnitude that they affront mankind's deepest moral sensibilities and cry out for justice: genocide, the slaughter of millions of innocent peasants in the name of some abstract notion of justice or progress, and the brutal repression of whole populations guilty only of clinging to the way of life bequeathed to them by their ancestors.

The Nuremberg War Crimes Trials, along with the unprecedented prosecution of captured Nazi leaders by the International Military Tribunal for "crimes against humanity," is only the most spectacular manifestation to date of a syndrome deeply rooted in human nature itself. In cases such as these, rehabilitation and deterrence hardly come into play, since the chances of achieving either result are so remote as to be virtually nonexistent. Were these trials and the sentences to which they led motivated by an unconscious or unavowed desire for retaliation, or are they justifiable on rational grounds? To formulate the question in more general terms, is there such a thing as a moral order that human beings are bound to uphold within the limits of possibility, and which requires that criminals be brought to justice regardless of whether or not society at large has anything to gain by prosecuting them? Before venturing any answer to that thorny question, it may be to our advantage to acquaint ourselves at least in a superficial way with some profound changes that the classical theory underwent in the course of the modern period.

THE MODERN SOLUTIONS

From the sixteenth century onward, that theory—and the view of justice on which it is predicated—was challenged by a group of thinkers, beginning with Machiavelli, who questioned its effectiveness and sought to replace it with a supposedly better one. The argument ran roughly as follows: Classical political thought had failed because it demanded too much of human nature. It stressed the need for moral virtue and made happiness contingent on one's ability to attain it. By so doing, it was compelled to speak of an ideal that is seldom, if ever, seen among human beings, most of whom habitually prefer their selfish interests to the common good of society. More could be accomplished by going to what Machiavelli called the "effectual truth of the matter," taking human beings as they are rather than as they ought to be.

The new program would soon be taken up and further refined by a host of prominent thinkers—Hobbes, Locke, and Spinoza among them—who agreed with Machiavelli that everyone would be better off if moderate ideals and absolute expectations were substituted for the absolute ideals and moderate expectations of the ancients. The trick was to devise a scheme whereby "private" vices could be channeled in such a way as to conduce not indeed to public virtue, but to "public benefits." Human beings would no longer be required to undergo a painful conversion from a premoral concern with worldly goods to a concern for virtue or the good of the soul. They could remain as they are as long as the institutions under which they lived were as they ought to be.

The writer in whom these ideas, as they relate specifically to punishment, finally crystallized is the eighteenth-century Italian jurist and criminologist, Cesare Beccaria, whose short but enormously popular treatise, *On Crimes and Punishments,* published in 1764, when the author was only twenty-six years old, soon led to a substantial reform of the penal system in Europe and elsewhere. (Under Beccaria's influence, Michigan and Maine became the first of our states to eliminate capital punishment in the 1830s.)

The thrust of Beccaria's argument reflects that of modern thought as a whole in that it makes less stringent demands on people but insists that they be met. One of its novel features is that it leaves no room for pardon or clemency. The system calls instead for mild but inexorable laws and stipulates that punishments, which had often been unduly harsh and arbitrary, are to be administered with the utmost regularity and uniformity. These punishments aimed at nothing higher than to discourage crime and prevent recidivism among those who had already indulged in it. Gone were the refinements that had characterized the penal theory of the premodern world. Punishment was to be defended on purely utilitarian grounds. It had only one motive, deterrence. Simple guidelines and a more efficient apparatus by which they could be implemented would secure the only kind of justice that an enlightened society needed for its proper functioning.

Equally influential in the long run were the ideas of Rousseau, who maintained (against Hobbes) that human beings were innately good and laid the blame for their corruption on society. If society is solely responsible for the wickedness of its citizens, it, and not they, must be changed. To Rousseau, above all, may be traced the sentimental attitude toward criminals that pervades so much of the literature of the nineteenth century and finds its superlative expression in Hugo's *Les Misérables,* a novel that places all guilt squarely on the shoulders of society and whose true heroes are the convicts, the fallen women, the abandoned children, and the wayward victims of poverty and oppression for whom it expresses boundless pity. It should be added that Rousseau himself was too much of a political thinker to advocate the suppression of punishment, for which he thought there would always be a need, lest the wicked should profit from the probity of the just or their own injustice.

No one grasped the implications of the modern drift toward leniency more clearly than Nietzsche, who viewed society's growing unwillingness to punish as a sure sign of its degeneration. For centuries, the West as he saw it had been caught in a downward spiral from which there seemed to be no escape. All of modernity, with its levelling tendencies, rampant egalitarianism, and soft humanitarianism, was at bottom a vast conspiracy against any and all manifestations of greatness. The future belonged to the "last man," the man entirely without longing or aspiration, whose hegemony was daily becoming more apparent. To be sure, life would go on, but it would be a diminished and impoverished life, dedicated only to the satisfaction of one's bodily needs, the pursuit of comfort rather than excellence, and—lest boredom should set in—the cultivation of an infinite variety of decadent art forms. The real culprits in the drama were Christianity and its political expression, liberal democracy—which had been waging a war unto death against every superior type of human being. The time had come to reverse the trend and set humanity on a different course.

According to Nietzsche, the way to do this was to restore spiritedness to the place of honor that it had once occupied in human life. Biblical morality, grounded as it was in the spirit of revenge and suited only to the mentality of slaves, would have to yield to a morality situated "beyond good and evil." New forms of greatness needed to be invented through the unfettered exercise of human creativity. The death of God, which Nietzsche boldly proclaimed, meant that there were no longer any objective principles of right and wrong by which one could be guided. Life operated "essentially through injury, assault, exploitation, and destruction and could not be thought of at all without this character."[4] It had nothing rational about it. Accordingly, Nietzsche denied that punishment served any deliberate purpose and, in the *Genealogy of Morals,* dismissed as naive all attempts to find such a purpose. The link between justice and punishment was severed altogether. From a divinely ordained moral order to which everybody is obliged to conform one had passed into a world in which punishment is nothing but a euphemism for the suffering and destruction entailed by any genuinely creative act. A crisis of major proportions was in the making, to which the horrifying events of our own century have since borne ample witness.

THE DARK SIDE OF HUMAN JUSTICE

Whatever one may think of these developments, one has to admit that the dispensation of justice has always been shrouded in considerable obscurity. Criminals do not always get caught, far from it; and if so much publicity is given to those who do, it is probably because of the widespread feeling that more often than not the big ones get away. Conversely, we occasionally hear of human beings who spent years in jail or were put to death for crimes of which they were later discovered to have been innocent.

Even more alarming is the amount of invisible crime that goes on in society. Most crimes, we now realize, are never reported, and others are so well concealed that nobody is aware of their having taken place. We shall never know, save in the most speculative way, how many rapes and child-abuse cases have occurred over a given period of time, how many murders have been made to look like natural deaths, how many scandals have been hushed up, how many scams, swindles, or insider-trading deals have been pulled off, how many wills have been falsified, how many widows have been cheated out of their incomes by unscrupulous administrators, or how many billions of dollars government and industry lose each year through fraudulent schemes or dishonest accounting procedures. The IRS itself, for all its sophisticated methods of investigation, has no way of tracking down all tax-evaders, as it tacitly admitted not long ago when it ran a series of ads urging citizens to consult their consciences when filing their annual returns. Nobody would be talking of a war on crime if criminals were not prospering in a measure that staggers the imagination.

One likewise wonders how strictly justice is enforced in those cases where both the crime and its perpetrator are known. Too many of our convicted felons receive only suspended sentences, and, even though the death penalty has been restored in a majority of the states since the Supreme Court declared it constitutional in 1976, less than one condemned murderer in a thousand is executed.

The situation, needless to say, is aggravated by a number of factors over which society has yet to gain control. One thinks of such matters as the overcrowding of jails, the proliferation of drug-related crimes, the growing complexity of modern life, the general permissiveness of liberal society, the breakdown of the family, and the technological advances that make it easier for crooks to escape detection.

The paradox in all of this is that the law, which is understandably hard on perjurers, is itself the biggest liar of all insofar as it states flatly that *all* criminals will be punished. In truth, only *known* criminals are or can be punished, something that society cannot publicly acknowledge lest by so doing it should appear to be teaching not that crime must be avoided, but that the criminal should not get caught.

The picture, though less scary, is not much rosier when we look at criminal justice from the standpoint of the one who is charged with carrying it out. No one admires, let alone envies, the executioner, who performs a necessary but nonetheless degrading task. To a delicate soul the spilling of human blood will always smack of murder. Luther, who was sensitive to the problem, used it as an example to illustrate the "two kingdoms" theory by which he sought to reconcile the lofty ideals of the gospel with the harsh requirements of the political life. His point was that a Christian could indeed discharge the function of hangman, not qua Christian, but rather in his capacity as a member of the civil society to which God had assigned him.

This being the case, it is easy to see why so many people would want to replace capital punishment with life imprisonment, as has already happened in a number of countries. The question is whether society can afford to do so without prejudice to other pressing obligations. At the present moment, it costs the government—that is, the taxpayer—a minimum of $25,000 to keep a prisoner in jail for a year, and the price is expected to rise rapidly as the size and average age of the prison population increases. Fewer people would object to the abolition of capital punishment if our society were rich enough to maintain as many maximum security prisons as are needed to accommodate the criminal population. The moral dilemma arises when we ask ourselves whether the money required for this purpose would not be better spent taking care of the poor in our midst or supporting other worthwhile causes. There are no ready answers to the question, especially since our system of appeals allows cases to drag on for years, with the result that it can cost twice as much to put a convict to death as it does to keep him alive for the rest of his life.

Because of the fallibility of human justice and the possible effect of punishment on the punisher himself, the biblical tradition has always stressed the need to season justice with mercy, the virtue whereby, as Portia puts it in *The Merchant of Venice,* earthly power shows itself most like God's. In Matthew's gospel, the apostle Peter, who wants to know whether he should forgive "as many as seven times," is told by Jesus that he should do so not seven times but "seventy times seven" (Matt. 18:21-22). The implicit allusion to the story of Lamech, to which reference has already been made, tells us a great deal about the spirit that informs the New Testament teaching.

For all that, the greatest Christian thinkers have always taught that unbounded leniency was not an adequate response to the unbounded appetite for personal revenge; for, practiced without discrimination, it will inevitably be construed by some elements of society as an invitation to further crime. Saint Augustine thought that nothing was more injurious to the common life than to allow the wicked to prosper and use their prosperity to oppress the good. Pascal was of the same opinion. Right without might, he said, was "helpless," and might without right "tyrannical." Both authors could appeal to the authority of the Bible. In the Hebrew scriptures, capital punishment is prescribed for a fairly wide variety of sins ranging from murder and adultery to witchcraft, blasphemy, and idolatry. Only in the latter Rabbinic tradition do we find a tendency to apply it as little as possible, if not to do away with it altogether. The New Testament itself, while much less specific when it comes to legal matters, nevertheless reminds us that rulers do not "bear the sword in vain," having been appointed by God "to execute his wrath on the wrongdoer" (Romans 13:4). It thereby implies that, properly administered, punishment is neither morally wrong nor necessarily un-Christian.

IS THE UNIVERSE MORALLY CONSISTENT?

These few remarks about the dark side of criminal justice bring us back to the question raised earlier concerning the existence of any cosmic or suprahuman support for justice. Do we live in a morally consistent universe, that is, a universe in which everyone sooner or later receives his due, if not in this life, at least in the next; or should we resign ourselves to the fact that the world is not entirely fair in that it leaves all kinds of moral ends untied?

The problem is as familiar to the biblical tradition as it is to the ancient philosophic tradition. The author of the Book of Ecclesiastes had already observed that "under the sun the race is not always to the swift, nor the battle to the strong, nor bread to the wise, nor riches to the intelligent, nor favor to the men of skill; but time and chance happen to them all" (9:11). Plato, who alludes to the same problem in the *Laws*, leaves it at saying that the wicked who go unpunished on earth will have their just deserts meted out to them in Hades. The notion that justice is good in the sense that it always redounds to benefit the one who practices it would thus seem to rest on the belief in a solicitous and providential God, a belief that Plato himself may not have personally shared.

When it came to such matters, the Christian tradition was more definite. It spoke of the natural law, which, as a law of the cosmos presumably enforced by an all-knowing and omnipotent God, guarantees that justice will prevail and that all wrongs will eventually be righted. The difficulty is that the strict natural law theory has never played a significant role outside the Christian world or been universally accepted within Christianity itself.

There is a simpler version of that theory which holds that the punishment in question need not be administered by an external agent, that it is intrinsic to the evil deed and takes the form of an inner torment from which the culprit cannot escape until such time as he has atoned for his crime. But is this always or necessarily the case? Experience tells us that a life wholly dedicated to the pursuit of crime is not an enviable one—nobody in his right mind thinks of Al Capone as having been a happy man—and it is equally evident that some forms of behavior, such as overeating and alcohol or drug abuse, carry their punishment with them. All well and good, except that we are left with the possibility that some lucky person might commit a single unjust act by means of which he obtains the fortune or the honor on which his heart is set and then, without repenting or giving up any of his ill-gotten gains, lives honestly and happily thereafter.

We occasionally hear of criminals who turn themselves in because they can no longer live with themselves, haunted as they are by the feeling that the original injustice is replicated, like a computer virus, with every new benefit that it brings. There may be others who would never dream of doing as much and show no signs of being similarly afflicted by the pangs of a guilty conscience. The happy crook, if such there be, is not likely to brag about his success, since it is in the nature of the case that he cannot do so without jeopardizing his position. Nietzsche was right when he remarked in *Beyond Good and Evil* that this is a species which moralists are only too eager to "bury in silence."[5] We shall never know, save through an act of faith, whether the universe is structured in such a way as to satisfy fully the deepest longings of all moral decent human beings.

This appears to be the reason for which, without explicitly rejecting the notion of retribution, the classical philosophers focused above all on rehabilitation. They knew that some and perhaps most criminals are incorrigible, but, unlike their modern counterparts, they resisted the temptation to limit the function of punishment to deterrence. There was at least a chance that some of their hearers would prove sensitive to the love of the good and the beautiful, learn to shun evil of their own accord, and become better human beings as a result of it. Anything that might be accomplished along those lines, they thought, was well worth the effort.

For all anyone knows, they may have been right. It is a sign of a profound dissatisfaction with our state of affairs that thoughtful people have again begun to ask whether, in addition to containing vice, society should not do more to encourage virtue and whether its laws can ever command general respect unless they are seen as part of a moral order that deserves to be upheld for its own sake. A fuller treatment of this matter, although indispensable to an adequate understanding of criminal justice, would take us well beyond the scope of the present inquiry.

NOTES

1. Pascal, *Pensées*, translated with an Introduction by A. J. Krailsheimer (Middlesex, England: Penguin Books, 1966), 46 (no. 60); (in Brunschwieg edition, no. 294).

2. Henry Fielding, *The Journal of a Voyage to Lisbon* in *The Works of Henry Fielding*, vol. 11 (New York: George D. Sproul, 1903), 202.

3. Jean-Jacques Rousseau, *Emile or On Education*, introduction, translation and notes by Allan Bloom (New York: Basic Books, 1979).

4. Nietzsche, *On the Genealogy of Morals*, Second Essay.

5. Nietzsche, *Beyond Good and Evil*, aphorism 39.

II

CHRISTIANITY, SCIENCE,

AND THE ARTS

8

AUGUSTINE, THE ARTS, AND
HUMAN PROGRESS

Anyone who is impressed with the innumerable benefits conferred upon us by the achievements of modern science and technology will find little support for his enthusiasm in the literature of classical or Christian antiquity. It is hardly necessary to recall that the word "technology" is itself a technical term, formed on the basis of the Greek *technē,* to be sure, but for the purpose of expressing a reality that is largely alien to the spirit of Greek philosophic thought. Thus, to speak of the attitude of the classical authors and their Christian counterparts toward technological progress is to raise a question that did not pose itself in the same terms or with the same acuteness to any of them. Our task, then, is not to look for any specific answer that they might have offered to this question; rather it is to uncover the principles that undergird their understanding of the connection between science and society and the reasons for which they would have eyed with suspicion any project geared essentially to the improvement of human life through the use of scientific inventions. What classical and Christian authors have to say on this subject is bound to come as a source of disappointment to those among us who instinctively look to science for the solution to all or most of our problems, including the ones created by science itself. This alone, however, is not a sufficient reason for dismissing these authors out of hand. Only by reflecting on that disappointment and the causes which provoke it do we stand any chance of recovering the alternatives which the external triumph of modern science has successfully repressed, and of ridding ourselves of some of the deepest prejudices of our age.

The issue is further complicated by the variety of competing theories abroad regarding the nature of modern science and its problematic relationship to ancient science. For present purposes only two of the more important ones need to be mentioned. The first of these, which may be traced back to the Enlightenment, holds that modern science finds its origins in fifth century Greece, where the spirit of scientific inquiry initially manifested itself in embryonic form. This development marks a new departure in the history of human thought in that it exhibits a willingness on the part of human beings to be guided by nature rather than by ancestral custom and thus breaks sharply with the religious or, as some would prefer to say, superstitious mentality that once characterized all of the ancient Near-Eastern civilization. Its hero is Socrates, whom everyone remembers as the misunderstood and ill-fated victim of the narrow-mindedness of his unenlightened fellow citizens. Unfortunately, its growth was stunted by the subsequent rise of revealed religion and its antagonism toward all forms of independent inquiry. The centuries that followed were for the most part centuries of fanaticism and virtually unrelieved obscurantism. No doubt, one finds during that long period a few towering geniuses who managed to keep Greek rationalism alive against the opposition of a hostile society. One of them is Averroes, who, according to Ernest Renan's still widely held interpretation, championed the cause of "critical science" and "civilized humanity" in an age that did everything to discourage them.[1] But these men were the exception rather than the rule. One has to await the Renaissance for the definitive flowering of science and the fulfillment of the promises held out by early Greek science. Modern science is to ancient science what the adult is to the child. It is science "come of age."[2] Its completion, to which one could already look forward, was nothing less than the perfection of our natural understanding of the universe.

This once popular argument has since given way to an altogether different view which, far from denouncing the biblical tradition as antiscientific, sees in it the very foundation of modern science. Any attempt to apply the criteria of reason to the study of nature rests on the prior assumption that nature is governed by rational principles and hence amenable to scientific treatment. Were it not for the belief in the orderliness, the inner consistency, and the predictability of nature, the "incredible labours" of scientists would be without hope and probably would never have been undertaken in the first place. What gives to scientific research its motive power is the "instinctive conviction" that there is in nature "a secret that can be unveiled."[3] As the precondition rather than the product of science, this conviction cannot be thought to have originated in science itself. It has one and only one source, namely, "the medieval insistence on the rationality of God, conceived as with the personal energy of Jehovah and with the rationality of the Greek philosopher."[4] This, according to Whitehead, is the greatest contribution of medievalism to the formation of the scientific movement. Faith and science, the two great antagonists of the modern world, have more in common than the age-old rivalry between them suggests, to such an extent that the latter is inconceivable without the former. In short, "faith in the possibility of science, generated antecedently to the development of scientific theory, is an unconscious derivative from medieval theology."[5]

There is reason to doubt, however, whether either of these two views, whatever element of truth they may contain, can stand as an adequate explanation of the phenomenon at hand. If we go back to the originators of the modern project, who must be our first and may still be our best witnesses in the case, we gain the impression that for all its indebtedness to ancient and medieval thought, the new science which comes to the fore in the seventeenth century is radically different from it in both its intention and method of procedure. One does not do full justice to it simply by saying that it proceeds inductively rather than deductively or that it places greater emphasis than ever before on experience, for it could then be construed as little more than a perfected version of premodern science. What distinguishes it from all previous scientific endeavors is its abandonment of the ideal of contemplation and the transformation of the whole of science into an enterprise that is self-consciously directed toward the conquest of nature. Unlike premodern science, which Francis Bacon compares to a sterile virgin, it is deliberately meant to issue in works. Its sole aim is to promote human well-being by increasing the store of earthly goods at our disposal beyond anything that was formerly thought possible and by so doing, to render us "like masters and possessors of nature."[6] Since the main object of its study is not nature in the free state in which it normally discloses itself to us but nature under the constraints and "vexations of art,"[7] it calls for the collaboration of large numbers of experts and technicians, whose function is to carry out the sophisticated experiments required for the implementation of its goal. As such, it is inseparable from and, indeed, unthinkable without the arts that have come to be associated with our notion of technology.

Its emergence coincides with, or, better still, presupposes a new understanding of human nature as well, which is no longer defined in terms of its highest aspirations but rather of such lower goals as self-preservation, peace, and the tranquil if somewhat pedestrian enjoyment of the amenities of life. Accordingly, it ceases to look upon man as a rational and hence a naturally social being and simply regards him as a being which, in Hobbes's phrase, is capable of "reckoning consequences" or calculating his advantages.[8] Since the ends pursued by the new science are identical with those to which the whole of humanity would one day be dedicated, the long-standing conflict between its demands and those of civil society could be considered resolved once for all. All scientific truths were salutary and conducive to the good of society as a whole. At long last, one could look forward not only to the harmonious collaboration of scientists among themselves but also of the scientific and political communities. Society owed it to itself to promote and subsidize scientific research in return for the advantages accruing to it from the discoveries to which the research gives rise.

Such, in barest outline, are the foundations of the modern theory of progress, which was explicitly formulated for the first time in the eighteenth century and which takes it as axiomatic (a) that the development of human thought is on the whole a continuous development which, once begun, leads from lower to ever higher stages; (b) that there is a fundamental and necessary parallel between intellectual and moral progress and, hence, a built-in guarantee that humanity will never sink below the level of attainment that has thus far been reached; and (c) that the nature of this development is such that there are no assignable limits to the degree of intellectual, moral, and social progress of which human beings are capable.[9] "Progress," Herbert Spencer was to write with astonishing boldness a century or so later, "is not an accident but a necessity." What we call "evil and immorality must disappear." It is certain that "man must become perfect."[10]

On all three counts the hopes engendered by the new trend appear to have suffered a series of severe setbacks. Less than fifty years ago, J. B. Bury could still speak of the "commanding position" in which Progress "with apparent security" was enthroned.[11] Chastened by a second world war and the sudden revelation of the unspeakable atrocities perpetrated on millions of human beings by the leaders of some of the most advanced nations of the world, our own generation has learned to view the prospects for the future of civilization with less equanimity. It has not lost faith in science altogether, far from it; but it is less likely to rule out the possibility of a return to barbarism and is not averse to raising doubts even about its ability to escape total annihilation. As the Einstein-Russell *Manifesto* of the early fifties put it:

> There lies before us, if we choose, continual progress and happiness, knowledge and wisdom. Shall we, instead, choose death because we cannot forget our quarrels? We appeal as human beings to human beings: remember your humanity and forget the rest. If you can do so, the way lies open to a new

paradise; if you cannot, there lies before you the risk of universal death.[12]

Since, according to all estimates, the "new paradise" to which our authors allude remains as elusive as ever, we may be inclined to take more seriously the reservations expressed by all or most premodern writers in regard to the naturally beneficent character of scientific progress. Given the limitations of space and the complexity of the subject matter, I shall confine myself to a few exploratory remarks about St. Augustine, the author to whom we owe what is perhaps the most complete treatment of the arts transmitted to us from Christian antiquity.

Although in Augustine the word "art" is sometimes used to designate any intellectual capacity, its application is generally restricted to those intellectual activities which involve some sort of production, whether it be that of a material object, such as a house, a bench, or a cart, or a purely mental production of the kind that we find in mathematics, grammar, logic, and the other liberal arts.[13] As a rational capacity, art is the prerogative of human beings and is not to be encountered as such in brute animals, remarkable as their own constructions may be in other respects.[14] It differs from nature in that its principle is in the agent in a rational and fully conscious manner.[15] The artist is thus responsible not only for the coming into being of the product in question but for its very structure. His domination over it is as complete as it can possibly be insofar as he himself has given it the form that it takes, either in his mind or in external matter.

In all such cases, it does not suffice that one know what should be done and how to do it; he must also be able to state the reasons for which it should be done, and done in this or that fashion. A mason may have learned from repeated experience that mortar binds stones better than clay, but the competence that he possesses can at best be described as a "vulgar art."[16] The true artist is the one who has mastered the principles of his craft, and is thus able to apply them consistently and to good effect.

All art, whatever form it may take, derives from and shares in the supreme art or wisdom of God himself. But whereas God created all things from nothing and is responsible not only for their structure but for the totality of their being, human artists are limited to working with preexisting materials and conforming to preexisting models. Their art is not, strictly speaking, creative but imitative.[17]

Thus understood, art serves a twofold purpose. In the first instance, it prepares the mind for an understanding of the truth and constitutes a propaedeutic to the acquisition of wisdom. This is particularly true of the liberal arts, through whose study the rational soul is led to the discovery of those principles from which all science proceeds but which are not themselves the object of any science.[18] These principles are the objects of what Augustine habitually calls "memory," as distinguished from intellect and will. Memory in this higher sense is not to be confused with the faculty that functions as the storehouse of past events whose images may be recalled at will. A close analysis of the learning process reveals that it, too, is a component of the rational soul. All newly acquired knowledge is based on previous knowledge and ultimately on such knowledge as is not itself acquired or handed down through education but inherent in human nature. If one did not know anything at all and hence were unaware of his own ignorance, one would never embark upon the quest for new knowledge; and if one already knew everything, one would be dispensed from the effort altogether. Neither alternative is acceptable, for it is an undeniable fact of experience that we do learn new things. Lest we should fall into an infinite regression, however, we have no choice but to postulate the existence of certain non-acquired or naturally known principles from which the mind moves by degrees toward a better, though perhaps never complete, apprehension of the truth. To use Augustine's own example, the spectator who beholds the facade of a building can usually tell at a glance whether it is symmetrical or not without ever having studied geometry or architecture.[19] This he would never be able to do if the notions of magnitude and proportion implied in his judgment were not in some mysterious way already operative in him.

It is nevertheless true that the notions and the principles that derive from them remain for the most part buried in the recesses of the mind, from which they need to be actively summoned or recalled before any further advance in knowledge can be made. To learn is, therefore, to remember, according to the view that Augustine inherits from the Platonic tradition, but which he is careful to dissociate from the literally interpreted doctrine of the preexistence of the soul.[20] Among the disciplines that may be pressed into service for the accomplishment of this task, none are better suited to the needs of the beginner than the liberal arts, by reason of their greater closeness to the human mind and their ability to awaken it to its latent possibilities. It is in the nature of these arts that they point beyond themselves to something higher than themselves, whose grasp eludes most people but on which the well-trained mind is eventually able to fix its gaze.[21] Through their exercise, the student discovers his ignorance and begins his laborious ascent toward wisdom. He recognizes his opinions as mere opinions and gradually acquires the strength to lay hold of certain truths whose dazzling brilliance the feeble eye of his mind was previously unable to withstand.[22] At the same time, he is subjected to a series of tests calibrated in such a way as to measure his intellectual capacity and encourage him to intensify his efforts in the pursuit of the final goal of knowledge, to wit, the contemplation of the truths of the Christian faith, the end to which in the last analysis all arts and disciplines are directed.

Whereas the liberal arts are first and foremost in the service of knowledge, other arts are essentially ordered to the procurement of the necessities of life and the satisfaction of one's bodily needs.[23] Such are the so-called mechanical arts, whose role is to furnish the implements with which much of the ordinary business of life is conducted, and those arts which, like medicine and agriculture, come to the assistance of nature either to restore health or to increase the productivity of the soil. One finds in nature alone or in nature aided by art all that is required for one's survival and nurture.[24] If nature were stingy and failed to supply material goods in sufficient abundance, human beings would necessarily be drawn into a life and death struggle for their possession; war and crime would become inevitable, and the goal of virtue would prove illusory. But then one could no longer be blamed for one's wicked deeds and God's wisdom and goodness would be shown to be at fault. Under the circumstances, it is perhaps not surprising that Augustine should go out of his way to extol the merits of agriculture, which he does not hesitate to call "the most innocent of the arts," in response to the Manichaeans who impugned it on the ground that it inflicts suffering on plants and commits murder.[25] Since human wants are not autonomous, however, the arts devised to meet them must themselves be subject to regulation. Necessary utility rather than artificial need is the touchstone by which all of them are to be judged.

It should be added that the goal served by these arts is not entirely foreign to the liberal arts themselves, which have their own practical uses insofar as they promote human sociability and contribute to the development of a sound political life. Indeed, there is "nothing so social by nature, so unsocial by its corruption, than this [human] race."[26] For this reason alone, human beings must be restrained, not only by law, but by instruction and persuasion, both of which call for the use of rhetoric and the arts related to it.[27]

De civitate Dei sums up all of these points in what is by far the most positive statement to occur anywhere in Augustine regarding intellectual and artistic progress, that of Book XXII, chapter 24, which expatiates in rhapsodic terms on the resourcefulness of the human mind and the splendor of its accomplishments:

> For, over and above those arts which are called virtues . . . has not the genius of man discovered and applied countless astonishing arts, partly the result of necessity, partly the result of exuberant invention, so that this vigor of mind . . . betokens an inexhaustible wealth in the nature that can invent, learn, or employ such arts? What wonderful—one might say stupefying—advances has human industry made in the arts of weaving and building, of agriculture and navigation! With what endless variety are designs in pottery, painting, and sculpture produced, and with what skill exe-

cuted! What wonderful spectacles are exhibited in the theatres, which those who have not seen them cannot credit! How skillful the contrivances for catching, killing, or taming wild beasts! . . . To provoke appetite and please the palate, what a variety of seasonings have been concocted! To express and gain entrance for thoughts, what a multitude and variety of signs there are, among which speaking and writing hold the first place! What ornaments has eloquence at its command to delight the mind! What wealth of song is there to captivate the ear! How many musical instruments and strains of harmony have been devised! What skill has been attained in measures and numbers! With what sagacity have the movements and connections of the stars been discovered! Who could tell the thought that has been spent upon nature, even though, despairing of recounting it in detail, he endeavored to give only a general view of it![28]

Lest one be carried away by Augustine's lyricism, one should bear in mind that these observations belong to the context of a general discussion of God's solicitous care for the whole of creation and the "countless blessings" that he has seen fit to bestow upon it. They do nothing to alter the perspective established earlier, which refuses to consider art as an independent activity, subordinating it instead to the lofty ends of wisdom and virtue whose attainment it is destined to facilitate. If art can be said to occupy a privileged position in the hierarchy of human goods, it is chiefly by reason of its unique capacity to mediate between the world of the senses and that of the mind.

The same conclusion is borne out by Augustine's analysis of the various operations of the soul as it appears, for example, in *De quantitate animae.* On the seven operations listed there in ascending order, three are directly related to the body: animation, sense perception, and art. The first has the body as its subject (*de corpore*); the second is exercised by means of it (*per corpus*); the third revolves around it (*circa corpus*), producing the goods required for its subsistence and one's general well-being. These are followed by virtue and repose, the two operations by which the soul returns to itself (*ad seipsam*) and is gathered together in itself (*in seipsa*). To these must be added the operations through which it ascends to God (*ad Deum*) and finally comes to rest in him (*apud Deum*).[29] Easily discernible in this elaborate classification, with its tripartite subdivision into activities pertaining to the body, to the soul, and to God, is the familiar Augustinian pattern according to which the road to human perfection is one that leads, horizontally, so to speak, from the outer area of the senses to the inner area of the mind, and thence, vertically, to the God in whom alone the restless heart discovers its true home.

This much being said, one cannot fail to notice in the scheme that has been laid out certain tensions to which Augustine himself was forced to devote greater attention as time went on. A more cautious note is sounded in the *Retractationes*, where he reproaches himself for having laid too much stress on the liberal arts in his early dialogues.[30] Although there is a sense in which it is true to say that "without the good arts" the soul is "famished" or that without scientific training nobody reasons accurately,[31] one does well to remember that some souls are thoroughly ignorant of the arts, though none the less pious for it, and others thoroughly grounded in them, though not pious at all. There is no denying that the knowledge of the arts and sciences has increased enormously over the centuries, but this would betoken unqualified progress only if it could also be shown that nothing has been lost in the process. That human wisdom is not cumulative is obvious among other ways from the spread of heresy and the not infrequent use of philosophical doctrines to obfuscate or pervert the teachings of the Christian faith, rather than to deepen our understanding of them.

What is more, nothing assures us that progress in the intellectual sphere, even when it does occur, is always and necessarily accompanied by a corresponding increase in moral goodness. The plain fact of the matter is that there are few artists who do not display their talents for the sake of money or glory rather than for wisdom and virtue, and few orators who are not more eager to lie to their hearers than to lead them to the truth.[32] As for the list of achievements at which Augustine marvels in the long passage just quoted from *De civitate Dei,* it would be more impressive yet if it did not include a mandatory but less than reassuring reference to the "poisons," "deadly weapons," and "engines of war" that have likewise been invented, not for the enhancement of life, but this time for its destruction. If all the arts, however good in themselves, can just as easily be diverted to evil purposes, one is at a loss to see how they could serve as a reliable criterion of the progress accomplished by the human race in the course of the ages.[33]

The problem would be less acute were it not for the inherent conflict between the requirements of the intellectual life and those of the political life. The arts thrive on novelty and are always, so to speak, on the move. With every discovery that is made comes the possibility of new discoveries, and so on *ad infinitum.*[34] Society, on the other hand, relies for its well-being on laws; and laws, in contradistinction to the arts, draw their power of persuasion less from reason than from ingrained habit or custom.[35] They are effective to the extent to which the way of life they prescribe remains relatively stable and is not forever being disrupted by a constant desire to innovate. This means that what is good for the sciences and the arts is not ipso facto good for society and vice versa. The only ancient political philosopher ever to come out in favor of unrestricted technological progress was Hippodamas, who, interestingly enough, happens to be the most ridiculous of them all.[36]

Some social changes are, of course, desirable or otherwise unavoidable, but in such cases it is all the more important that they be introduced only gradually and with an eye to their potentially disruptive effects. For one thing, the very survival of society is contingent on its ability to defend itself against invaders and hence on its willingness to maintain a military capacity that is at least equal to theirs, even if its own way of life should be adversely affected by the development of such a capacity. Furthermore, since civil society, no matter how perfect it may be, is never in a position to satisfy all the needs and aspirations of its members, it has to give some encouragement to the so-called fine arts.[37] Yet full participation in these arts requires talents that are extremely rare. Most human beings will not find in them the gratification of their strongest desires and will not be induced by them to curb their self-indulgence or their appetite for private gain; and the few who do will not necessarily develop into better citizens, inasmuch as the attraction that they experience for the fine arts tends to breed contempt for the concerns of civic virtue.

Differently stated, moral progress requires that human beings disregard whatever belongs to them as private individuals in order to devote themselves entirely to the pursuit of justice and the common good. But to demand that people gladly renounce the attachments springing from their bodily nature and, if need be, sacrifice what they most cherish in a spirit of unstinting generosity or self-forgetfulness may be asking for more than human nature can usually bear. There is only so much that can be expected of most human beings, and the little that they can be persuaded to give up is likely to decrease rather than increase with the availability of ever larger amounts of material goods. Whether one traces the problem back to the classical distinction between body and soul or—in Augustine's Christianized version of that doctrine—to original sin, one is forced to conclude that the prospects for the betterment of the social life are strictly limited. One can never be sure that the gains registered on one front will not be offset by serious losses on another or that the advances of today will not be wiped out by the setbacks of tomorrow.

It is interesting to note that the sober and less than optimistic view of human progress that emerges from *De civitate Dei* was precisely meant to counteract the theory to which, since Theodor Mommsen, scholars often refer as the Christian idea of progress.[38] The studies published in the last thirty years or so have shown that, although ostensibly written for the purpose of vindicating the Christian faith against the attacks mounted by its pagan adversaries in the wake of the disaster of 410, Augustine's *magnum opus* is concerned as much if not more with refuting the melioristic conception of history propounded by Eusebius and his Latin followers, including Augustine's own would-be disciple, Orosius, all of whom forecasted an end to the evils that had hitherto plagued human existence and the long-awaited instauration of the kingdom of God on earth.[39]

Against this background, one is in a better position to appreciate the novelty as well as the ambiguity of the modern ideology of progress and its resolutely this-worldly solution to the problem of human society. Without going into any of the details of this vast question, one has only to point to Bacon's *New Atlantis* as an early model of what the new society might become. Not the least distinctive feature of this revolutionary work is its implicit claim to being the first specifically modern utopia. Like its predecessors, the *New Atlantis* presents us with an ideal that is not to be found anywhere, but, unlike them, it views the ideal society as a realizable though still distant goal. What in the past could not have been more than the object of an idle dream suddenly looms before us as a definite possibility. Bacon's ideal society is now possible thanks largely to an unprecedented emphasis on scientific "works" and the epoch-making substitution of the distinction between theoretical and applied science for the old distinction between theoretical and practical science, with the express understanding that henceforward all theory would be inherently practical.[40] The question that Bacon could not or did not raise is what the conquest of nature might do to the conqueror or how the quality of human life might be affected by the dynamic hedonism that it fosters.

It suffices to read one newspaper today to become aware of the doubts that have since arisen concerning the ultimate viability of the Baconian project, and to read two newspapers to realize the extent to which opinions differ concerning the nature of the crisis that afflicts it. In the words of one perceptive critic, what is called for is nothing short of a "total critique" of the dominant and predominantly scientific consciousness of our time, one that would lay bare the long-forgotten presuppositions on which it rests.[41] Such a critique would not of necessity compel us to turn our backs on the stupendous advances that modern science has made thus far and is likely to make as time goes on. It demands only that we be willing to consider with an open mind the possibility of restoring science to a less constricting and more properly human context.

Equally symptomatic of the modern predicament is the rise of aestheticism as an independent branch of philosophy and the elaboration of the "modern system of the arts," with its sharp distinction between the fine arts on the one hand and the practical or technological arts on the other. One has no trouble understanding that, faced with a growing crisis not only *in* the sciences but *of* the sciences (as Husserl once called it), so many people should be attracted to the fine arts as a form of relief from the impoverishment and desolation of human life. The result, however, has not been a new integration of the personality but the emergence of the aesthetic state as a general disposition capable of determining one's existential attitude, alongside other possible states, such as the scientific state or the political state. One is merely concerned with bringing about a harmony of the feelings on the basis of which one's sentimental life and one's view of the world is formed. Where this leads, nobody knows or even bothers to ask. Culture is taken for granted as a requirement of human progress, but the crucial question of the goal of that progress receives no answer. Aesthetic man is content to feel himself justified within the whole of culture and to be allowed to appreciate and contemplate the world from that particular vantage point. All of this, as Heidegger has pointed out, is still only a facade, behind which lurks the nihilism that Nietzsche was the first to perceive and unmask.[42]

It is fashionable to criticize Augustine, as some recent scholars have done, for his less than sanguine assessment of scientific progress, his alleged disparagement of aesthetic values, and his desensualized approach to the realm of art. Still, in the light of recent and not so recent experience, one may well ask whether the emancipation of art from the tutelage of wisdom and the service of virtue has brought us any closer to the goal of wholeness for which we continue to long. One surely does not find in Augustine a simple answer to the problems of the hour. For, the relative success of the modern scientific-technological worldview has produced a new kind of society in which the old principles, valid as they may be, are no longer immediately applicable. Only those living today can find the solutions to the problems of the day. However, this is a far cry from saying that Augustine's struggles are irrelevant to ours or that there is nothing to be gleaned from the insights that his incomparable genius was able to achieve through these struggles.

NOTES

1. E. Renan, *Averroès et l'averroïsme. Oeuvres complètes d'Ernest Renan*, vol. 3, edited by H. Psichari (Paris: Calman-Levy, 1903), 134ff.

2. Francis Bacon, Preface to *The Great Instauration*, near the beginning. Cf. Bacon, *New Organon* I.78 and II.52.

3. A. N. Whitehead, *Science and the Modern World* (New York: Macmillan, 1931), 18.

4. Ibid.

5. Ibid., 19. As it has come to be formulated, the idea that modern science is ultimately rooted in and supported by the biblical tradition is at least as old as Nietzsche; e.g., *Genealogy of Morals* III, 24-25. See, among its recent defenders, Stanley Jaki, *The Road of Science and the Ways to God* (Chicago: University of Chicago Press, 1978), 13ff., et passim and *The Origin of Science and the Science of*

Its Origins (South Bend, IN: Regnery/Gateway, 1979). Also, E. McMullin, "Medieval and Modern Science: Continuity or Discontinuity?" *International Philosophical Quarterly* 5 (1965): 103-129. For a survey of some of the recent literature on this topic cf. C. Mitcham, "Questions of Christianity and Technology: A Bibliographic Introduction," *Science, Technology and Society* (Curriculum Newsletter of the Lehigh University STS Program), 14 (November 1979): 1-7.

6. Descartes, *Discourse on Method*, part VI; cf. Bacon, *New Organon* II.52.

7. Bacon, Plan of *The Great Instauration*, near the middle.

8. Hobbes, *Leviathan* I.4; *De cive* I.1.2.

9. Cf. J. B. Bury, *The Idea of Progress: An Inquiry into Its Growth and Origin* (New York: Macmillan, 1932), 127ff. Bury credits the Abbé de Saint-Pierre with being the first to have proclaimed "the new creed of man's destinies, indefinite social progress" (p. 143). For an analysis of Greek and Roman views of progress, with numerous references to the contemporary literature on the subject, see E. R. Dodd's *The Ancient Concept of Progress and Other Essays on Greek Literature and Belief* (Oxford: Oxford University Press, 1973), 1-26.

10. H. Spencer, *Social Statistics,* Abridged and Revised (New York: Appleton, 1897), 32.

11. J. B. Bury, *The Idea of Progress*, 352.

12. For a commentary on this text, cf. Gerard Piel, *Science in the Cause of Man*, 2nd edition (New York: Knopf, 1964), 144ff.

13. St. Augustine, *De doctrina christiana* (*On Christian Doctrine*) II.30.47.

14. St. Augustine, *De musica* (*On Music*) I.3.4, and 4.6.

15. St. Augustine, *De immortalitate animae* (*On the Immortality of the Soul*) IV.5; *De Genesi contra Manichaeos* (*On Genesis, Against the Manichaeans*) I.8.13.

16. St. Augustine, *De vera religione* (*On the True Religion*) XXX.54.

17. St. Augustine, *De diversis quaestionibus LXXXIII* (*On Eighty-three Different Questions*) q. 78; *De musica* (*On Music*) I.3.4, and 5-6.

18. St. Augustine, *De quantitate animae* (*On the Greatness of the Soul*) XXVI.51.

19. St. Augustine, *De vera religione* XXX.54; XXXII.59; *De ordine* (*On Order*) II.11.34.

20. St. Augustine, *Retractationes* (*Retractations*) I.4.4 and VIII.2. Cf. *De trinitate* (*On the Trinity*) XV.24; *De quantitate animae* (*On the Greatness of the Soul*) XX.34; *Soliloquia* (*Soliloques*) I.20.35; *De libero arbitrio* (*On Free Will*) III.20.56; *Epistula* (*Letter*) 7.

21. St. Augustine, *De quantitate animae* VIII.12; *De ordine* II.13.38 and 16.44. Also, on the dangers of philosophy for the untrained or less gifted student, *De ordine* II.5.17.

22. St. Augustine, *De quantitate animae* XV.25.

23. St. Augustine, *De ordine* II.16.44.

24. St. Augustine, *De civitate Dei* XXII.24. If nature sometimes appears less than generous, it is only because human beings need to be reminded that the happiness to which they are called is not to be sought in this life; cf. ibid., XXII.22.

25. St. Augustine, *De haeresibus* (*On Heresies*) XLVI. See also *Confessiones* (*Confessions*) II.10; *Contra Faustum Manichaeum* (*Against Faustus, the Manichean*) VI.4; *Enarrationes in Psalmos* (*Expositions on the Psalms*) 140.12.

26. St. Augustine, *De civitate Dei* XII.27.

27. *De civitate Dei* XXII.22.

28. *De civitate Dei* XXII.24.

29. St. Augustine, *De quantitate animae* XXXV.

30. St. Augustine, *Retractationes* (*Retractations*) I.3.2.

31. St. Augustine, *De beata vita* (*On the Happy Life*) I.8; *De immortalitate animae* I.4.

32. E.g., St. Augustine, *De civitate Dei* VII.iii; *De doctrina christiana* IV.3.

33. St. Augustine, *De civitate Dei* XXII.xxiv.3. Robert A. Nisbet, *History of the Idea of Progress* (New York: Basic Books, 1980), conveniently overlooks these reservations and thinks that Augustine's encomium "could have been written by Condorcet" in the eighteenth century (p. 54). For a clear statement of Augustine's ambivalent attitude toward progress, see H.-I. Marrou, *L'ambivalence du temps de l'histoire chez saint Augustin* (Montreal and Paris: Institut d'etudes mediévales, 1950).

34. See, for example, Aristotle, *Sophistic Refutations* 34.183b16f.

35. Augustine's views on the role of custom in human life deserve a separate study. It is obvious that some customs are good and others bad, some tolerable and others intolerable; cf. *De civitate Dei* XV.xvi; *Epistula* (*Letter*) 40. While one cannot always be guided by them, a degree of caution must be exercised lest, by liquidating them, people should be left with nothing to fall back on. Thus, "in avoiding those actions which are offenses against the customs of men we must take due consideration of the diversity of customs. What has been laid down as a general rule, either by custom or by law, in any city or nation must not be violated simply for the lawless pleasure of anyone, whether citizen or foreigner" (*Confessiones* II.8). The importance of custom is stressed through Augustine's discussion of Varro's reforming activities in *De civitate Dei*; cf. E. L. Fortin, "Augustine and Roman Civil Religion: Some Critical Reflections," in *Classical Christianity and the Political Order: Reflections on the Theologico-Political Problem*, edited by J. Brian Benestad (Lanham, MD: Rowman and Littlefield, 1996), chapter 4. Part of Augustine's problem stems, of course, from the fact that Christianity was itself a tremendous novelty, whose impact on a society in which paganism was still a recognizable force could not be disregarded.

36. Cf. Aristotle, *Politics* II.8.1267b22-29. This appears to be the only passage in all of Aristotle to describe (with an unmistakable touch of humor) the character of the person whose opinions are being criticized. Max Scheler notes perceptively but without any further argument that "if the Greeks had no technical civilization, it is not because they were incapable (or, as yet, incapable) of creating one, but because they

did not *wish* to create it. It would have run counter to their morality." In *Ressentiment*, translated by Wm. Holdheim (Glencoe: Free Press, 1961), 80.

37. See, in regard to this matter, Augustine's penetrating assessment of Roman mythical or poetic theology in books II and VI of *De civitate Dei*.

38. T. E. Mommsen, *Medieval and Renaissance Studies*, edited by Eugene F. Rice, Jr. (Ithaca, NY: Cornell University Press, 1969), 266-298: "St. Augustine and the Christian Idea of Progress" did much to explode the myth of early Christian progressivism when it was first published in 1951. The myth survives in R. A. Nisbet, *History of the Idea of Progress*, 47ff., 69, 71, 74.

39. See, *inter alia*, R. A. Markus, *Saeculum: History and Society in the Theology of St. Augustine* (Cambridge: Cambridge University Press, 1970), and, for a summary presentation, E. L. Fortin, "Augustine's *City of God* and the Modern Historical Consciousness" in *Classical Christianity and the Political Order: Reflections on the Theologico-Political Problem*, edited by J. Brian Benestad (Lanham, MD: Rowman and Littlefield, 1996), chapter 7.

40. R. M. Unger, *Knowledge and Politics* (New York: Free Press, 1975), 5.

41. Martin Heidegger, *Nietzsche*, vol. 1: *The Will to Power as Art*, translated by D. F. Krell (New York: Harper & Row, 1979), 88-90.

42. See, most recently, R. J. O'Connell, *Art and the Christian Intelligence in St. Augustine* (Cambridge, MA: Harvard University Press, 1978), esp. 143-172.

9

SCIENCE AS A POLITICAL PROBLEM

There are few more obvious symptoms of the malaise that has come to afflict our society than the ambivalence of its present attitude toward science and scientific research. No one, not even the most dedicated ecologist or anti-nuke protester, seriously believes that the complex world in which we live can dispense with the benefits of modern science, and no one doubts that this same science, which has already given us so much, could one day annihilate us. The situation, needless to say, has no exact parallel in previous history. Earlier theorists may have indulged in scientific or pseudo-scientific speculations about devices powerful enough to blow up the planet, but neither they nor anyone else had seen them with their own eyes. Ours is a different world. It is a post-Baconian world of uncertain expectations and badly shaken rational convictions, a world that continues to be informed by Baconian principles but no longer shares Bacon's optimism and is daily becoming more skeptical about the inherently beneficent character of the science on which, for better or for worse, it depends for its survival and well-being. This, as much as anything else, is what distinguishes us from our predecessors. The disillusioned romantics of the first half of the nineteenth century longed nostalgically for the "glory" and "grandeur" of former ages, which they had lost all hope of recovering. If our own disillusioned contemporaries can be said to long for anything, it is less for a heroic but irretrievable past than for some more perfect future that the science in which they had learned to place their trust is not about to give them. The problem lies to some extent with modern science itself, whose "metaphysical foundations," as E. A. Burtt pointed out thirty years ago, have yet to be fully understood;[1] but it also has much to do with the delicate relationship between science and society, which the Enlightenment was supposed to have settled once and for all and which the potentially destructive character of the latest breakthroughs in such fields as nuclear physics and molecular biology has again brought massively to the fore. It is to the latter aspect of the problem that my sketchy and somewhat tentative remarks are addressed.

The people of my generation grew up with the image of a lonely Frankenstein who conducts his weird experiments in secret and without the slightest concern for the manner in which they might affect the lives of others. But the monstrous Frankenstein appears to be the exception rather than the rule. Recent studies on the social psychology of natural scientists reveal that most of them tend to be loyal and devoted servants of the society to which they belong. There is reason to think that the overwhelming majority of the 1,250,000-odd scientists who worked in the Soviet Union were not dissidents but staunch supporters of the goals of their regime, and the same is undoubtedly true of their counterparts in the West. Along the same lines, one cannot fail to be impressed by the enthusiasm with which most scientists rallied to the cause of their respective governments during World War II. We know from a secretly recorded conversation that, when the news of Hiroshima was leaked to a group of interned German scientists in August, 1945, Werner Heisenberg was conscience-stricken at the thought, not that the bomb had been dropped, but that it was vastly more powerful than the one on which he and his associates had been working. His only regret was that he had lacked the moral courage to press the Nazi bureaucracy for a larger payroll. He soon recovered from the shock, but, interestingly enough, only upon learning that the victorious scientists were having second thoughts about the morality of the means by which their triumph had been secured. What the example proves, if it proves anything, is not that scientists have no sense of responsibility toward the government that subsidizes their efforts but that they are not always particularly reflective about the nature of their commitment to it, or, more bluntly, that the most sophisticated theoretical knowledge sometimes goes hand in hand with an all but total lack of self-knowledge.

It suffices to pose the problem in these terms to realize that one cannot leave it at speaking about the immediate political context in which the scientist has his being and his movement. The peculiarity of science is that, by its very nature, it transcends all national or political boundaries. Although originally a product of the Western countries, it has long since been exported to the rest of the world, to such an extent that there is scarcely a country anywhere today that has not been touched, and in most cases profoundly transformed, by it. One is thus prompted to inquire about the responsibility of the scientist not only to his country and his fellow citizens but to the global human community.

To this question present-day natural science has by its own admission no clear answer to offer. In its original form, modern science was inseparable from philosophy and guided by the natural needs that it sought to satisfy. Having since emancipated itself from the tutelage of its partner, it is finding it more and more difficult to give an account of its own doings. Quarrels among scientists have been known to occur in the past but, in retrospect, they were barely more than lovers' quarrels, reinforcing the faith of the antagonists in the redeeming virtues of their profession and confirming them in their resolve to get on with their business. To be sure, the methods employed to solve our old problems had managed to create a few new ones in the process, but this meant only that we were still not scientific enough. The simple answer to the dilemma generated by science was "more science." That once commonly accepted view has since come under severe criticism from scientific as well as nonscientific quarters. The famous Einstein-Russell *Manifesto* of the early fifties is one, but only one, illustration of the critique of science. There have been others since then. The gist of practically all of them is that science, the giant of the modern era, is a blind giant. Severed from philosophy, it cannot teach wisdom. By reason of its self-professed metaphysical and ethical neutrality, it has in fact nothing to say about the rightness or wrongness of the uses to which it may be put.

The impasse into which this state of affairs has led us is well illustrated by two contemporary examples, among others, that of the late Jacques Monod, whose reflections on this topic issue in a call for a wholly unsupported existential decision on the part of the researcher regarding matters of right and wrong, and that of Louis Althusser, whose reformed Marxism sacrifices human freedom to the blind determinisms of scientific law. It was inevitable, I suppose, that science and freedom, the two great forces of the modern world, should sooner or later collide and thus bring into full view a fundamental contradiction that had been submerged rather than solved by centuries of heated debate. No wonder that, faced with a choice between moral freedom and scientific necessity, so many of our leading scientists should have begun to experience a failure of nerve. What we really need, as Roberto Unger suggests in the introduction to his book, *Knowledge and Politics*, is not a further radicalization but a "total criticism" of the modern liberal and scientific enterprise, a critique grounded in a thorough reexamination of the basic and long-forgotten assumptions on which, from its inception, that enterprise was made to rest.[2]

Science, after all, is a human activity, whose goals cannot reasonably be divorced from those of human life as a whole. As such, it presupposes a measure of clarity about the end or ends that, like every other intellectual pursuit, it was intended to serve. If it is to achieve any degree of self-understanding, it has no alternative but to allow itself to be guided by the pre-scientific or common sense knowledge out of which it emerged and from which, wittingly or unwittingly, it continues to live. It makes all the difference in the world, even to the scientist, whether one defines the human being as a rational animal who finds his highest perfection in the contemplation of the truth, or simply as an animal capable of calculating his own interest. But the choice of one definition rather than the other is not a matter to be decided on modern scientific grounds. It is precisely at this juncture that the concerns of science and those of theology or philosophy intersect without necessarily encroaching upon one another.

As has often been noted, most recently by Mary Hesse, theologians and philosophers would be ill-advised to engage science on the level of empirical fact. The last thing anybody wants is a replay of the noisy and sometimes acrimonious disputes that have marred the relations between the practitioners of these various disciplines in the past. This is not to imply that facts are of no interest to theologians and philosophers but only that what we call a "fact" is usually more than a mere fact, inasmuch as it has little significance apart from the purpose for which it is gathered or the specific frame of reference within which it comes to light. If I say that Boston has a population of slightly more than half a million, I am stating a fact, but a fact that is utterly trivial unless it can be shown to have some bearing on, say, the city's traffic jams, its housing shortage, the logistics of its water supply, or any number of more or less vital issues that the census taker probably did not have in mind when he assembled his statistics. Nobody nowadays thinks that theology or philosophy ought to dictate the answers to strictly scientific problems, but they would fall considerably short of their own prescribed goals if they failed to engage in a comprehensive reflection on the ends from which all of our activities, scientific or otherwise, take their ultimate meaning. Viewing the problem in this perspective obviously does not require that we turn our backs on the stupendous advances that science had made thus far or is likely to make as time goes on. It does, however, demand—as mentioned in the previous chapter—that we be open to the possibility of inserting the whole of modern science into a more suitable framework.

The tragedy is that theology itself has been overtaken by the crisis that engulfs all of the human disciplines at the present moment. Whether or not it is actually in a position to discharge its proper responsibilities is, to say the least, a matter of serious doubt. The bewildering variety of conflicting theologies to which we have been treated of late bears sufficient testimony to the disarray into which the former "queen of the sciences" has fallen. Crudely stated, we have a plethora of laborers in the theological vineyard—"functional specialists" as they are now politely called—but few theologians. The "total type" is largely absent from the scene and not too many calls are being issued for his return. The result is that, instead of leading the way, theology has all too readily acquiesced in its new role as the humble servant of its externally more successful rivals. The Middle Ages used to speak of philosophy or science (the two terms were then synonymous) as the handmaiden of theology. As Kant was later to observe, however, and as the medievals already knew, the crucial question is whether the said handmaiden is destined to precede her mistress with a torch or follow bearing her train.[3] My suspicion, based on recent experience, is that the answer is even less clear now than it was in Kant's day.

To its own long-range detriment, contemporary theology has either vainly tried to ape the methods of modern science or else retreated into myth, the general category under which matters pertaining to religion now tend to be subsumed. By so doing, it has hallowed the disastrous rift between what C. P. Snow has labeled, somewhat hyperbolically, the "two cultures."[4] It, too, has come to exhibit that Laputian blend of abstract science and unrefined taste which so typifies our modern outlook. If there is something radically wrong with a science that produces Nobel prize winners who are Nazis, one wonders whether there might not also be something wrong with a theology that is no longer able to distinguish between a religious truth and a pious myth. Everyone else has suffered from the consequences. It is hard to fault our scientists for paying little heed to a theology that has done such a poor job of addressing their human needs and providing the guidance that many of them honestly seek. Few spectacles are more distressing than that of the scientist who attempts to satisfy his longing for wholeness by turning to a religion that reason cannot touch or language describe, and whose self, as Unger puts it, is split in two, "each side finding the other incomprehensible, then mad." Assuming that what I have been saying is true, and I am certain that at least some of it is, theology will have to put its own house in order before very much can come about.

Since there seems to be no relief in sight from either end of the academic spectrum, what is the solution? If I knew, I probably would not be sitting at this desk. Surely, a committed Christian will have no trouble discovering in the resources of his faith the elements of a consistent and personally satisfying view of human existence. Still, as a thoughtful Christian, he cannot remain blind or indifferent to the problem of articulating that faith in terms that come to grips with what others perceive as an insuperable obstacle to it. As was hinted at earlier, one way to proceed is to go back to the prescientific and as yet undifferentiated world of common sense in which all of our abstract notions about the universe and human life are finally rooted. The difficulty is that we are ourselves total products of the world in which we live. What we take to be our common sense world is already a second-story world, shot through with ideas derived from the prevailing scientific world view. In this respect, our predicament is not unlike that of the fabled Baron Munchausen, who had stumbled into a pit of quicksand into which he would have soon vanished had he not had the bright idea of pulling himself out by his wig. Hopeless as the situation may appear to be, it nevertheless has a relative advantage over other, more stable situations. In so far as it is characterized by the shaking of all traditions and cultural horizons, it allows for a reconsideration of the fundamental human alternatives in ways that might have been unthinkable at other moments in our history. The sense of disintegration of which Unger speaks so eloquently is itself an invitation to undertake a fresh or nontraditional assessment of the tradition to which the ruling consciousness of our day is the mostly unconscious heir.

I do not mean to suggest that the recovery of these basic alternatives, if it were to be achieved, would automatically yield a series of recipes with which to deal with our problems. But there is at least a chance that such a study would give us a better insight into the true character of modern science by allowing us to situate it within the context of its conflict with ancient science, a conflict which, perhaps under the influence of Nietzsche and Max Weber, most contemporary writers on this subject have had a tendency to blur. It might also encourage us to view that conflict within a still larger context, this one defined by the rapport between reason and revelation, those two distinct and irreducibly different guides to life that human consciousness at its highest level has disclosed.

One positive indication that something like this is already happening is the frequency with which, after a century and a half of virtual oblivion (thanks largely to the nineteenth-century historian, H. T. Buckle), the name of Francis Bacon is once again being sounded as the author to whom we are indebted for what is still perhaps the most lucid and penetrating analysis ever given of the essential thrust of the modern scientific project and the spirit that informs it. There is no telling what direction our intellectual world and the myriad factors that impinge upon it might yet take. Cultural cycles do not conform to the immutable patterns of nature; and besides, our culture critics, their critics, and the critics of their critics do not all read the signs of the times in the same way. Which one will eventually prove to be correct is something that can only be demonstrated in the event. Despite my own reservations about the viability of the solution proposed by Unger in the book to which I have referred more than once, I cannot help thinking he is on the right track when he says that the present chaos is a parody of the transcendence for which our trendy theologians are groping and for which our worldly scientists have likewise begun to thirst. However much one may deplore the parody, one can still glimpse in its distorted features the grandeur of the lost original to which it points. The extremity into which we have fallen could indeed be our greatest opportunity.

NOTES

1. E. A. Burtt, *The Metaphysical Foundations of Modern Physical Science: A Historical and Critical Essay* (New York: Harcourt and Brace, 1932), esp. 11-22.

2. Roberto Unger, *Knowledge and Politics* (New York: The Free Press, 1975), 1-28.

3. Immanuel Kant, *Perpetual Peace*, Second Supplement.

4. C. P. Snow, *The Two Cultures and a Second Look* (New York and Toronto: New American Library, 1963).

10

THE BIBLE MADE ME DO IT:
CHRISTIANITY, SCIENCE, AND
THE ENVIRONMENT

The task of assessing the Christian tradition's understanding of our relationship to the natural environment, never an easy one, has not been made easier by the proliferation of ecological studies in our time. For one thing, environmentalism as a special discipline is barely more than thirty years old. Hence the problems on which it focuses were not dealt with thematically or with the same sense of urgency by any of our predecessors, Christian or otherwise. I refer to such highly publicized and politicized issues as air, water, and ground pollution, the depletion of the planet's natural resources, the accumulation of toxic waste, the deterioration of the biosphere or the world's great ecosystems through deforestation and urbanization, the extinction of species, the hole in the sky, global warming, and various other threats to our physical well-being about which, until recently, little was known or being said. For another thing, the new discipline began to take shape just as Christian theology was entering a period of seemingly endless fragmentation.

The trouble is not that theologians need to be awakened to the gravity of the situation—many of them have been only too eager to jump into the fray—but that they have yet to agree on the appropriate response to it.

What we find, instead, is a wide range of positions that claim to be Christian but mostly reflect the confusion that prevails in the larger academic community. This much is evident from the titles of two well known books on the subject: *Cry of the Environment: Rebuilding the Christian Creation Tradition*, edited by P. N. Joranson and K. Butigan,[1] which suggests that the tradition in question has suffered severe damage, for otherwise it would not have to be "rebuilt," and *The Travail of Nature: The Ambiguous Ecological Promise of Christian Theology* by P. Santmire,[2] which suggests that the notion of nature, the other pivot of the once standard Christian approach to these matters, is likewise in trouble. My aim is not to propose a new solution to the problem—I have none to offer—but to lay out as briefly and as clearly as possible the general principles by which classical Christianity would have been guided had it been confronted with issues of the kind we face today. These principles were derived from two distinct and originally independent sources: the Bible, which supplies us with the notion of creation, and Greek philosophy, to which we are indebted for the notion of nature. It so happens that on the question at hand, these sources, different as they may be in other respects, are more convergent than divergent.

THE BIBLE AND CREATION

The key biblical text relating to our subject is chapter 1 of Genesis, which tells us that heaven and earth and all they contain come from God, who, having completed his work, declared the whole of it to be not only good but "very good" (1:31). Each part taken separately had already been pronounced good, with two notable exceptions, the sky and the man Adam. What Genesis means by "good" is not spelled out, but the reference appears to be to the fact that each of the elements so described is perfect in the etymological sense of the term, that is to say, complete and fit to do its work. To express the same thought in philosophic rather than biblical terms, a thing is perfect to the extent that it possesses all that belongs to it by reason of its nature. Goodness is wholeness: *bonum ex integra causa*, as the schoolmen used to say. "Good" in that sense adds nothing to the noun that it qualifies. A chair is good if it has what it requires to fulfill its function. To say that it has even the slightest defect is to imply that it is already less than a chair. There is ultimately no difference between a chair and a good chair. If the sky, conceived as a solid vault stretched like a tent over the earth, is not individually called good at the moment of its creation, it is because, with the earth still uncreated, it has not yet demonstrated its capacity to carry out its assigned task, that of separating the waters above the vault, which occasionally come down to us in the form of rain, from the waters here below, such as those found in rivers, lakes, and oceans.[3]

We further learn that Adam, the only being made in the image and likeness of God, is the peak of creation and the being for whose sake all other material beings exist. Adam is to have dominion over them and be served by them. His superiority to them is corroborated by the fact that he is the one to give to each its name. The habitat prepared for him by God is not a desert but a garden capable of satisfying all his needs without his having to compete with anybody else for his livelihood, engage in violence, or do any hard work. The original plan thus calls for perfect harmony among human beings and between human beings and God. It also calls for perfect harmony between man and irrational animals, a condition highlighted by the vegetarianism of the early humans.

Needless to say, the Bible, which frowns on disinterested knowledge and discourages its pursuit—as the deuterocanonical Book of Sirach puts it, "Seek not what is above you or hidden from you" (3:22)—does not offer anything that comes close to a scientific account of nature. Its concerns are strictly religious. In accordance with these concerns, the chapters that follow the creation narrative in Genesis concentrate on the origin and spread of evil in the world. We soon realize that the insistence on the perfection of the prelapsarian state has as its purpose to absolve God of any responsibility for this worsening state of affairs. Had material goods been in short supply from the start, human beings would have been caught up in a life and death struggle for survival, and injustice could have been viewed as a natural necessity, calling God's goodness into question. As matters stand, Adam and his descendants have themselves to blame for the evils they commit or from which they suffer. Adam is not called "good" apart from the rest of creation precisely because he was capable of going astray and, as the rest of the narrative shows, did in fact go astray.

Once introduced into the world, human wickedness will increase as time goes on, until God, noticing that "every imagination of the thoughts of [man's] heart was evil continually," decides to "blot him out" (6:5-7), sparing only Noah and his family, along with a male and female of every species of animals. Shortly thereafter, God establishes an explicit covenant with all living creatures, vowing never again to destroy them by means of a flood and reconfirming the human being as master of the earth, but this time, under less than ideal conditions.

An attempt is nevertheless made to approximate the original conditions by promoting a strong sense of reverence for life. For the first time, murder is expressly forbidden and provisions are made for the punishment of the murderer. No doubt, man's relations with the animal kingdom have become less peaceful inasmuch as the rule of vegetarianism is allowed to lapse (9:3); but, unlike the myths of the Golden Age with which it shares this rule, the Bible specifies that with the arrival of the messianic era vegetarianism will be restored not only for human beings but for carnivorous beasts as well. In Isaiah's words, "the lion shall eat straw like the ox" (11:7).

A similar concern for dumb animals and the rest of nature reappears in various other biblical books. Exodus 23:12 sets the seventh day of the week aside as a day of rest for one's asses and oxen. Leviticus declares that each seventh year is to be a sabbath or year of solemn rest during which one's field is not to be sown (25:2-5). Deuteronomy 20:19-20 stipulates that, when a siege is laid to a city, care should be taken not to destroy fruit-bearing trees; only trees that bear no fruit are to be used to build siegeworks. Such ordinances are motivated neither by a belief in the sacredness of nature nor by a sentimental attachment to it. The Bible's outlook is rather more down to earth. Animals and plants are to be kept in good health lest their utility to human beings should be lost or diminished. Less evident but by no means absent is a desire on the part of the sacred writers to inculcate habits of gentleness among human beings by discouraging cruelty to animals.

Also to be noted is the Bible's profound admiration for the splendor and orderliness of nature. Nowhere is this more evident than in the so-called "creation psalms." Psalm 8 speaks of the heavens not simply as the work of God's hands but as the work of his "fingers," thereby stressing its surpassing delicacy. Psalm 104 dwells on the solicitude of a loving God who looks after all his creatures, both human and subhuman, providing for each its shelter and its food in due season: trees for birds to nest in, high mountains for wild goats, rocks as a refuge for badgers, bread to strengthen our hearts, and wine to gladden them.

None of this adds up to a full-blown ecotheology, nor, since sacred Scripture is not long on speculations of this kind, can any such theology be expected of it. Still, if these few texts and others like them signify anything, it is that there is ample biblical support for the reasonable demands of the present-day ecological movement. Human beings are not to worship nature, but neither have they been given a blank check to do whatever they please with it. As the stewards or custodians rather than the owners of creation, they are to care for it and "guard" it (Gen. 2:15). Besides, if human beings are urged to respect life in general and forbidden to take any human life unjustly, their own or someone else's, it follows that they are not to destroy or inflict unnecessary damage on the physical environment needed to sustain that life.

THE GREEKS AND NATURE

By and large, the same perspective was shared by the tradition of philosophy that came to dominate in the West and that traces its origin to Plato and Aristotle. I limit myself to a few remarks about Aristotle, the first philosopher to come up with a bona fide science of nature and the author of a book on the subject, the *Physics*, to which, in an essay that did much at the time to rekindle interest in Aristotelian philosophy, Heidegger refers as "the hidden, and therefore never adequately studied, foundational book of Western philosophy."[4]

Heidegger's statement is all the more pertinent as Aristotelian physics was long thought to have been dealt a fatal blow by the stupendous advances of modern physical science. Certainly, much of what Aristotle says about nature needs to be reformulated in the light of more recent scientific discoveries. The question is whether the same is true of his highest principles as laid out in the *Physics*, a work less dependent on a detailed observation of natural phenomena than the ones that follow it in methodical order in the Aristotelian corpus; or, more generally, whether the whole of Aristotle's natural science has been superseded and rendered obsolete by modern science.

The heart of the Aristotelian enterprise is the well-known and now almost universally contested thesis that nature acts for an end. This principle applies to inanimate as well as animate beings, all of which are said to be ordered to a determinate end or ends, and endowed with a constitution suited to the attainment of these ends, along with an inclination to pursue them and rest in them once they have been attained. As used in the *Physics*, nature is defined as the primary and intrinsic principle of motion and rest in beings that are subject to change,[5] that is, beings composed of matter and form and thus capable of motion of one kind or another: local, quantitative, or qualitative. It is remarkable, for instance, that plants, animals, and human beings grow until they reach a certain size and then just as mysteriously stop growing. Mobility, the most universal characteristic of such beings, is the formality under which all of them can and must be studied. If everything around us were perfectly stationary, there would be no natural science. Only in the light of the end to which this mobility is ostensibly directed can we make sense of the regularities that we observe in the operations of nature.

As distinguished from, say, mathematics, which involves neither common matter nor motion and limits itself to formal and efficient causes, physics carries out its demonstrations by means of arguments based on all four causes, of which the final cause is not only the highest but the one that commands the other three. To use Aristotle's example, taken from the more familiar realm of art, a saw is made of metal and given a serrated form because its purpose is to cut wood, something it would not be able to do if it lacked teeth or were itself made of wood. The end is what dictates the form and the matter, and not vice versa.[6]

These commonsensical remarks should not blind us to the fact that, of Aristotle's four determinate causes, the final cause is the one that is hardest to understand and most vulnerable to attack, as is clear from the ridicule that has been heaped upon it in modern times. Although the principle that all of nature acts for an end is taken by Aristotle to be self-evident and hence uninferable from prior premises, it still demands the most careful examination. We know that nature acts for an end not because it always succeeds but because it sometimes fails. If the result of its operations were identical in every single instance, it could be attributed to the necessity of matter. There would be no reason to think, for example, that rain falls from the clouds because it is needed to make corn grow and not simply as a consequence of the interplay of the mechanical laws of condensation, gravity, and the like.[7] The appeal to any other kind of agency to explain its occurrence would be superfluous. Likewise, if nature exhibited no regularity whatever, if anything could come about at any time as in fairy tales, if acorns gave birth to oaks one day and to elephants the next, if all events were chance events (in which case we could not identify them as chance events), in short, if there were no "natures," the question of finality would not arise. The clue to the existence of this finality is that nature sometimes falls short of the mark. But it also hits it, accomplishing for the most part what is manifestly good for the individual or the species. It is "better" for human beings to be born with eyes that see and feet that enable them to walk, as a majority of them are, than to be born blind or with a club foot. The habitual though not infallible recurrence of something that redounds to the good of a particular being or kind of being is what compels us to speak of it as intended by nature.

From these sketchy remarks one begins to glimpse the chasm that separates classical science from modern science. For classical science, all natural beings *tend* toward a specified end, regardless of whether they attain it or not. Their behavior may differ on occasion, but the end remains the same. For modern science, there is typically no preordained end, but the behavior is universally the same. All that counts is the regularity and unfailing predictability with which it occurs. The difference between these opposing views is well brought out by two texts from the early modern period, one by Hooker, which reflects the old view, and the other by Spinoza, which reflects the new. The first occurs in Hooker's *Laws of Ecclesiastical Polity*:

All things that are have some operation not violent or casual. Neither does anything ever begin to exercise the same without some fore-conceived end for which it works. And the end which it works for is not obtained unless the work be also fit to obtain it by. For unto every end every operation will not serve. That which assigns to each thing the kind, that which moderates the force and power, that which appoints the form and measure of working, the same we term a *Law*.[8]

It matters little in this instance that the end is not always achieved. What does matter is that all things, whether they be human beings or stones, have such an end, and it is with a view to this end that we are able to grasp their nature.

The second text makes no reference whatever to a possible end and stresses only the universal sameness of the operations. Such is the view that underlies the whole of modern science. What it ultimately entails is the absence of any essential difference between animate and inanimate or between rational and nonrational beings. If there is no end, there is no nature that corresponds to it. In Spinoza's words:

> The term law, taken in the abstract, means that by which an individual, or all things, or as many things as belong to a particular species, act in one and the same fixed and definite manner, which manner depends either on natural necessity or on human decree.[9]

Let us assume that nature is ordered to an end. What end? Upon analysis, that end can be shown to be twofold. There is, first of all, the end that is intrinsic to the being itself and which is none other than its perfection or full development, both physical and, in the case of human beings, moral and intellectual. This much is implied in the definition of nature as a principle of motion and rest inherent in all natural beings. The end to which in the first instance the process is ordered is the form of the whole, a being that is complete insofar as it is endowed with all the powers, properties, or characteristics of its species. It is in this sense that Aristotle is able to say in the *Politics* that "what each thing is when fully developed we call its *nature*,"[10] and, in the *Poetics*, that "having gone through many changes, tragedy stopped when it attained its *nature*."[11]

Beyond this intrinsic end, there is the extrinsic end to which all non-human species are ordered, namely, the human or rational being, who is the measure of all things in the material universe. From an Aristotelian perspective, the human being cannot be a mere accident on the face of the earth, an aberrant primate species, or the product of a blind evolutionary process; for without the presence of such a being the world in which we live becomes unintelligible. Since mind is clearly at work in nature, understood as an extrinsic participation in reason on the part of nonrational beings, the material universe requires for its perfection that it contain within itself that by means of which it is able to attain its end or return to its principle. At least one of its parts must be gifted with reason and hence essentially different from all the other parts in that it is the only one capable of knowing the whole. Thus, from the principle that nature does nothing in vain, Aristotle infers that plants exist for the sake of animals and nonrational animals for the sake of rational animals.[12] What human beings can appropriately do in their dealings with subrational nature is determined either by nature itself or by reason guided by nature. An example of this is hunting, a species of the natural art of acquisition, which, as Aristotle is careful to point out, is not unlimited but limited by human need.[13]

One final remark on this subject. If nature is indeed permeated with mind or reason, we can study it with the expectation that, even though its material component rules out the possibility of our ever attaining an exhaustive knowledge of it, nothing a priori prevents our being able to learn more and more about it. Given the manifest presence of finality in some of its parts, it is hard to see why it would not be present in other parts as well.[14] I know of no text that gives us a better feel for the spirit of Aristotle's natural science than the passage in the treatise *On the Parts of Animals*, where one reads:

> Of the things that exist by nature some are ungenerated, imperishable, and eternal, while others are subject to generation and decay. The former are excellent beyond compare and divine, but less accessible to knowledge. The evidence that might throw light on them, and on the problem that we long to solve respecting them, is furnished but scantily by sensation; whereas respecting perishable plants and animals we have abundant information, living as we do in their midst, and ample data may be collected concerning all their various kinds, if only we are willing to take sufficient pains.
>
> Both departments, however, have their special charm. The scanty conceptions to which we can attain of celestial things give us, from their excellence, more pleasure than all our knowledge of the world in which we live; just as a half glimpse of persons whom we love is more delightful than a leisurely view of other things, whatever their number and dimensions. On the other hand, in certitude and in completeness our knowledge of terrestrial things has the advantage. Moreover, their greater nearness and affinity to us balances somewhat the loftier interest of the heavenly things that constitute the object of the higher philosophy.
>
> Having already treated of the celestial world, as far as our conjectures could reach, we proceed to treat of animals, without omitting, to the best of our ability, any member of the kingdom, however ignoble. For if some have no graces to charm the senses, yet even these, by disclosing to intellectual perception the artistic spirit that designed them, give immense pleasure to all who can trace links of causation and are inclined to philosophy.
>
> Indeed, it would be strange if artistic representations of them were attractive, because they disclose the mimetic skill of the painter or sculptor,

and the original realities themselves were not more interesting, to all at any rate who have eyes to discern the reasons that determined their formation.

We must therefore not recoil with childish aversion from the examination of the humbler animals. Every realm of nature is marvelous; and, as Heraclitus, when the strangers who came for a visit found him warming himself at the furnace in the kitchen and hesitated to go in, is reported to have bidden them not to be afraid to enter, as even in that kitchen divinities were present, so we should venture on the study of every kind of animal without distaste; for each and all will reveal to us something natural and something beautiful. Absence of randomness and conduciveness of everything to an end are to be found in nature's works in the highest degree, and the resultant end of her generations and combinations is a form of the beautiful.[15]

Medieval theologians may have been uneasy with some of Aristotle's assumptions about nature, but they could not fail to be struck by the affinity between his views on this matter and the teachings of their faith. What Aristotle saw as the work of an eternal nature they could and did attribute to the wisdom of the divine artisan of chapter 1 of Genesis.

WESTERN CHRISTIANITY AND THE ECOLOGICAL CRISIS: LYNN WHITE'S BOMBSHELL

In view of the enormous respect shown for nature by both the theological and the philosophic traditions, it comes as something of a surprise to learn that the blame for the ecological catastrophe for which we appear to be headed is now being laid at the door of Western Christianity. Yet this is exactly what assorted philosophers and historians of science have been doing for the past twenty-five years or so. The landmark document in this regard is Lynn White, Jr.'s celebrated essay, "The Historical Roots of Our Ecological Crisis," originally delivered as an address to the American Association for the Advancement of Science in 1966, published as an article in the March 10, 1967, issue of *Science*, and reprinted over and over again in the years that followed.[16]

The success of White's essay was due in large part to the bluntness with which it makes its point, namely, that Christianity "bears a huge burden of guilt" for the devastation of nature in which the West has been engaged for centuries. By not only allowing but commanding human beings to "subdue the earth" and "exercise dominion over every living thing" (Gen. 1:26-28), the Bible bred an "exploitive mentality" that made the devastation inevitable. As the chief purveyor of that mentality, Christianity declared open season on nature, to which it denies any intrinsic worth and which it subordinates entirely to the fulfillment of human purposes. No religion, we are told, was ever more "anthropocentric." Moreover, by teaching that time takes its course along a straight line rather than in cycles, Christianity laid the groundwork for the idea of progress, thus accrediting one of the main presuppositions of modern science, the tool par excellence of man's subjugation of nature.

The disenchantment of nature effected on a more popular level by the triumph of biblical monotheism led to the same dire consequences. For centuries, pagan "animism," which looked upon trees, rivers, fountains, and various other natural sites as the dwelling places of divinities, had functioned as a curb on the human propensity to ravage nature. With the removal of these taboos, the old inhibitions crumbled. The "sacred grove" vanished and, for the first time, human beings felt free to "exploit nature in a mood of indifference to the feelings of natural objects."[17] This prompts White to describe Christianity's victory over paganism as the "greatest psychic revolution in the history of our culture."[18]

Nor does it make any difference to White that we now live in a post-Christian age, for, even though many in the West have rejected Christianity, their daily habits continue to be tacitly informed by it. The faith in perpetual progress that supports them is "rooted in, and indefensible apart from, Judeo-Christian teleology."[19] Marxism, for all its open hostility to religion, is a Christian heresy. Galileo was more of a theologian than a scientist, condemned not because of his "errors" but because the theologians of his time objected to his poaching on their grounds—further proof that the whole of "modern Western science was cast in a matrix of Christian theology."[20]

White's stinging diatribe imparted an ironic twist to a story that had been unfolding since the beginning of the nineteenth century. In an effort to counter the Enlightenment slogan that religious belief was an obstacle to scientific progress, a number of apologists for the Christian faith had embarked on a project the aim of which was to demonstrate that Christianity was not only open to modern science but had actually given birth to it. Joining their ranks toward the end of the century was the eminent French physicist and philosopher of science, Pierre Duhem, whose pioneering research on John Buridan and Nicholas Oresme, two hitherto little known fourteenth-century Aristotelians, seemed to have uncovered the source of Newton's first law of motion.

A devout Christian, Duhem ran into stiff resistance on the part of the anticlerical establishment of his day. His work attracted little attention in his own lifetime and was virtually ignored for many years after his death in 1916 at the age of fifty-six. Official opposition to him was such that the last five volumes of his major work, *Le Système du monde*, were not published until the 1950s, some forty years later.[21] The overall accomplishment was nonetheless impressive, especially in view of the fact that the information contained in this work as well as in the three hefty volumes of his *Etudes sur Léonard de Vinci* and several other important studies, had to be painstakingly dug out of unedited manuscripts to which Duhem, stranded as it were in a provincial university (Bordeaux) where he taught for the last twenty-one years of his life, did not have ready access.

Duhem's thesis, which is also part of White's argument, is that the old cyclical notion of time had to be jettisoned before modern science could come into its own and that Christianity is the religion that had finally "laid it to rest."[22] Duhem went so far as to identify the moment of its birth with the famous condemnation of what is now called Latin Averroism by the bishop of Paris, Etienne Tempier, in 1277. Modern science, he wrote, "was born, one may say, on March 7, 1277, from a decree issued by Monsignor Tempier, bishop of Paris. A principal objective of the present work will be the justification of this assertion."[23] The meaning of that obscure statement is clarified by what Duhem had said earlier about the impact of revealed religion on the cosmological theories of the Ancients. In volume 2 of the same work, he states, with considerable eloquence:

> We hear it said that the very slow changes of the earth are linked to the almost imperceptible movement of the fixed stars, whose revolution is supposed to measure the Great Year. To the building of this system, all the adepts of Greek philosophy—Aristotelians, Stoics, Neoplatonists—have contributed one after another. To it Abou Masar offered the tribute of the Arabs. Such is the system that, from Philo of Alexandria to Maimonides, the most illustrious rabbis have adopted. To condemn it as a monstrous super-stition and lay it to rest, Christianity was needed.[24]

The problem to which Duhem adverts was the central theological problem of the Middle Ages, namely, the conflict that pitted Aristotelian necessitarianism against divine freedom, or, more correctly stated, Aristotle's eternal world against the created world of biblical revelation. In point of fact, the notion of the Great Year had been discarded by important thinkers long before the advent of Christianity. Aristotle disposed of it with the observation that people believe in the cyclical nature of time because of the circular motion of the heavenly bodies by which we measure it.[25] Yet Duhem's argument was not without force. By ascribing the origin of the universe to the will of an omnipotent God, medieval theology demanded that it be studied, not as it reveals itself to us in the light of nature's own immutable principles, but as it actually is by divine *fiat*. For this, a much greater reliance on sense experience was needed. It is in this sense that modern experimental science can be said to rest squarely on a foundation of theological voluntarism and nominalism.

The thesis to which Duhem was committed—let us call it the "continuity thesis"[26]—later received indirect support from no less a figure than Alfred North Whitehead, who defended it on slightly different philosophical grounds. The new argument, which is set forth in chapter 1 of *Science and the Modern World*, entitled "The Origins of Modern Science," is discussed in chapter 8.[27] Whitehead's main point is that "faith in the possibility of science" depended on the medieval belief in the rationality of God as well as in the orderliness and predictability of nature.[28]

Although Whitehead was probably unaware of it—he was not an avid reader of nineteenth-century German philosophy—the same point had already been made by Nietzsche who argued that science, long looked upon as Christianity's mortal enemy, is not a genuine alternative to Christianity but a parasite that lives off of it. "Strictly speaking," he wrote, "there is no such thing as science 'without any presuppositions' . . . a philosophy, a 'faith,' must always be there first, so that science can acquire from it a direction, a meaning, a limit, a method, a right to exist . . . It is still a *metaphysical faith* that underlies our faith in science [Nietzsche's emphasis]."[29] By attacking Christianity, science was not destroying an adversary, but was merely cutting itself off from its roots.

Nietzsche's analysis is particularly valuable insofar as it brings to light the hidden premise of Whitehead's position: in a world that considers reason to be superficial, denies that there is any part of the human being that is not intrinsically dependent on matter, and claims to have replaced the "will to truth" with the "will to power," the foundations of everything become by definition nonrational or "religious" in the Nietzschean sense. All "values," science included, originate in the mysterious or godlike recesses of the creative self.

Both approaches, Duhem's and Whitehead's, were later rolled into one by the British scholar and Oxford fellow, Michael B. Foster, in a series of widely read articles, the most important of which appeared in *Mind* between 1934 and 1936.[30] Like many of his predecessors, Foster endorses the continuity thesis but limits it to the impact of a "certain form of Christian theology" on the "methods of the modern a priori sciences," such as those of Bacon and Descartes. His overarching principle is that any change in one's natural theology entails a corresponding change in one's understanding of nature. More sensitive than either Duhem or Whitehead to the difference between the premodern and the modern conceptions of nature, Foster finds in the advent of Christianity, with its radical monotheism and its notion of divine omnipotence, the only factor capable of accounting for this difference.

If I refer to Foster's provocative but not unproblematic articles,[31] it is only to indicate that the continuity thesis had gradually become a common staple of twentieth-century scholarly literature.[32] Duhem was vindicated at last. Posthumously, anonymously, and in circuitous ways, his zeal had paid off. I say "anonymously" because he is not mentioned by either Whitehead or Foster and appears to have been kept at a safe distance by most historians of science.[33] The conspiracy of silence around him was not officially broken until 1974 when, in its fifteenth edition, the *Encyclopedia Britannica* devoted a brief twenty-five line biographical notice to him.

By that time, the ecological wars touched off by White's bomb had been raging for several years and both Christianity and modern science were under fire for the part they allegedly played in provoking the crisis. The attack left Duhem's major premise untouched—Christianity is the mother of modern science—but not the rest of the syllogism, which now ran as follows: But modern science is bad; therefore, Christianity, too, must be bad. It is providential that Duhem was not alive to witness the paradoxical triumph of the proposition to the defense of which he had dedicated his life.[34]

In fairness to White, himself an avowed Christian, it should be added that the target of his broadside was not Christianity as such, but the form it began to take when Aristotle was introduced into the fold in the course of the thirteenth century. Hence his plea for the return to an earlier form of Christianity, the one exemplified by St. Francis and the Franciscan movement. For some reason, few people were ready to second White's nomination of St. Francis as the patron saint of ecologists. Other factors, such as Francis's alleged anti-feminism, may have intervened to block it.

THE CRISIS OF MODERN SCIENCE AND THE REINTEGRATION OF MAN INTO NATURE

This is not the place to engage in a point by point discussion of the skewed argument by which White sought to nail down his thesis, especially since it has already been criticized by competent scholars from the standpoint of both its logic and its historical accuracy.[35] One is surely entitled to ask whether Galileo was taken to task more for his unauthorized forays into the domain of theology than for his scientific theories.[36] I, for one, find more fruitful Jacques Monod's astute remark that the Catholic Church was right to condemn Galileo but that it condemned him for the wrong reason, namely, for his heliocentrism rather than for his rejection of Aristotelian teleology.[37] The same goes for the theory that, by ridding us of animism, biblical monotheism unleashed a campaign of destruction against nature. Refurbishing G. E. Stahl's early eighteenth-century animistic speculations at this late date and attributing feelings to inanimate nature, as White does, is not likely to advance the cause of either science or the ecological movement. Unless one reads the myth of the *Timaeus* as a scientific treatise, one cannot possibly impute to Plato the notion that the whole of nature is a huge animal.[38] Timaeus's fanciful account of the origins of the cosmos appears to be nothing but a protracted metaphor for the unity, beauty, and intelligibility of the universe, problematic as these may ultimately be. Finally, one is at a loss to explain how the biblical and medieval conception of the visible world as an elaborate system of symbols all pointing to the existence of an invisible and infinitely more beautiful reality could have been construed by Christians as a signal to embark upon a systematic desecration of nature. The image of creation as God's handiwork calls for the exact opposite, to wit, a much higher regard for nature than one might otherwise have for it.[39]

White's argument not only gravely distorts premodern thought, but also obscures the true nature of modern science. The latter, properly understood, is not a product of the Christian tradition, however hard its promoters may have tried for political reasons to give that impression. It arose by way of a reaction against it and was militantly opposed to it from the start as regards both its goals and its method of procedure. Unlike Aristotelian and Scholastic science, modern science was ordered to the production of results, i.e., inventions that would contribute to the relief of man's estate.[40] Its ambition was not to contemplate nature but to conquer it as one conquers an enemy; to enlarge our power over it;[41] to establish human beings, not indeed as its "guardians" or stewards, as the Bible had done, but as its "masters and possessors."[42] Therein lies the ambiguity of White's essay, which confuses the two meanings of "mastery" or "dominion." When the Bible uses the word, it is always in its primary sense, the *dominus* or "master" being the one who rules his subordinates for the good of the whole rather than for his own private good. Baconian mastery is different. It is achieved by observing nature under conditions other than those under which it habitually finds itself, doing violence to it, and "putting it to the question," an expression meant to evoke the horrors of the Inquisition. The real difference between premodern and modern science is not, as is often said, that one is deductive and the other inductive, that is to say, based on experience, but that modern science relies on a new kind of experience, one that takes nature out of its normal state and subjects it to the "constraints" and "vexations" of an endless series of artificially devised experiments.[43] The Moderns, not Ancients, are the ones who put nature on the rack and thereby set up an adversarial relationship between the investigator and the object of his investigation.

My point is well illustrated by an incident that occurred when, for the first time (on May 28, 1953), Sir Edmund Hilary and his associate, Tenzing Norkay, reached the summit of Mount Everest, a feat that made headlines throughout the world. A prominent London daily ran an account of it under the title, "EVEREST CONQUERED," printed in bold letters across the top of its front page; to which an Indian student living in England at the time responded spontaneously: "Why 'Conquered'? Why not say instead, Everest 'Befriended'?"

Underlying the new outlook is the view that nature is inimical or at least indifferent to human purposes and hence something to be beaten into submission and forced to yield to the will of a shrewder opponent. Human beings cannot take their bearings from nature and are not directed by it to any good in the attainment of which they might find their perfection. As Hobbes puts it with characteristic bluntness in *Leviathan*, chapter 11, "There is no such *finis ultimus*, utmost aim, nor *summum bonum*, greatest good, as is spoken of in the books of the old moral philosophers." There is only a supreme evil, which we have no choice but to leave as far behind us as possible.

The same negative attitude toward nature shows up again, among other places, in Hegel, about whom Emile Meyerson has written:

> [H]is contempt for nature is absolute. We do not believe that in the entire immense Hegelian corpus, although it pretends to embrace the whole of man's spiritual activity, can be found one sentence, one expression testifying that the spectacle of nature moved him or provoked his admiration in the slightest. That was, surely, an innate predisposition, and his correspondence reveals that already, at the age of twenty-five, when he visited the Bernese Alps, the spectacle left him indifferent. The sight of the glaciers "has nothing of interest about it," it offers "nothing great or pleasing." The traveler finds in it "no satisfaction except that of having approached such a glacier" and judges that the bottom of the glacier resembles a very muddy street. In general, he observes that "neither the eye nor the imagination discovers in these shapeless masses any point whatever where the former could alight with pleasure and the latter find a subject of occupation or play." Reason perceives in them "nothing awe inspiring, nothing that imposes astonishment or admiration." The sight of those eternally dead masses, he adds, gave me "nothing more than the monotonous and, in the long run, boring impression: it is so [*es ist so*]."
>
> But perhaps even more characteristic of his attitude is the way he speaks of the starry sky. He finds the admiration Kant professed for this sublime spectacle to be foolish; it is a subject of constant derision for him, one to which he returns on several occasions. The immensity of the celestial spaces is an example of "bad infinity"; one must beware of any admiration or even surprise on its account, and as for the stars, they are quite simply comparable to an eruption on the skin.[44]

So much for the wonderful harmony that was once thought to exist between human and nonhuman nature. Little wonder that Hegel, a major contributor to the development of modern aesthetics, should have seen fit to deny the imitative character of art and proclaim its independence from nature.

If the preceding analysis is correct, the problem with which the ecological movement concerns itself is symptomatic of a deeper issue affecting the overall status of science in our society. Edmund Husserl argued fifty years ago that what we are witnessing today is not just a crisis *in* the sciences but a crisis *of* the sciences.[45] Few thoughtful people still believe down deep that the answer to the problems generated by science is more science. At the same time, not too many of us are prepared to renounce the benefits of modern science, not only because we have been spoiled by them but because we are no longer able to survive without them. The only feasible alternative, as mentioned, is to look for a way of restoring modern science to a properly human framework. This would require, among other things, a reconsideration of its metaphysical foundations on the one hand and of the fundamental principles of Aristotelian science on the other. Specifically, it would force us to reopen, on the basis of our vastly increased knowledge of the physical world, the key issue of teleology to which reference was made in an earlier part of this essay.

It is reasonable to suppose that, were he to come back, Aristotle himself would be the first to revise some of his hypotheses regarding the structure of universe, the laws of physics, the nature of matter, and any number of similar issues. Even the axiom that nature acts for an end has to be reformulated in a manner that takes into account the astronomical and other discoveries of more recent centuries. This is not to say, however, that it has been disproved by these discoveries. Spinoza was content to dismiss it as a pure and simple anthropomorphism, without making the slightest effort to come to grips with Aristotle's argument in favor of it, so certain was he that the achievements of modern science had consigned it to the dustbin of history. True, one would never begin to look for final causes in nature if one were not acquainted with them from one's own experience, but this alone says nothing about whether or not they exist independently of one's experience.

To this must be added the fact that Aristotelian teleology is often grossly oversimplified by its modern interpreters. Nature, as Aristotle sees it, operates on many levels and displays characteristics that vary greatly when we pass from one to another. The clearest example of its directedness to an end is the fall of a heavy body, which always takes place in the same way unless it is impeded by some obstacle. If I let go of the stone that I am holding in my hand, it invariably and uniformly falls to the ground. The same is not true of plants, which send off shoots or limbs in various directions, and it is even less true of animals and human beings, whose patterns of behavior are subject to even greater variations. From this standpoint, plants are less natural than inanimate objects, animals less natural than plants, and human beings less natural than animals.[46]

Further complications arise when one considers Aristotle's teaching concerning the fixity of the species, another point on which his thought is frequently misconstrued. Because of its relatively high degree of indeterminacy, nature is capable of all sorts of false starts, mistakes, failures, and monstrosities.[47] The evolutionary process, if that is the right word for it, is not reversible after the manner of a mathematical equation. We have no way of deducing the various species one from the other, but neither is it necessary that we be able to do so. All that is required to guarantee the intelligibility of the natural order is that it have the capacity to sustain rational life. For all I know, we could get along without cockroaches and hippopotamuses, but we do need other things—food from plants and animals, water, air, heat, wood, and so forth—without which human life would be impossible or greatly impoverished. The amazing thing is not that there are useless or contingent species around[48] but that the human being is more highly determined in his physical make-up than all other natural beings. We have multiple subspecies of dogs, snakes, spiders, or whatever it may be; to my knowledge no biologist has yet spoken convincingly of a subspecies of man. The reason for the greater determination in this instance is that, although the human being is less "natural" than all other natural beings, the exercise of his intellectual powers requires a certain bodily disposition that nature, which "never fails in what is necessary," could not neglect to provide for him.[49]

The further implication of this whole doctrine is that even in their most spiritual activities human beings remain dependent on the operations of their lower nature. Not that the higher is reducible to the lower—nothing could be more contrary to Aristotle's thought—but only that it is so rooted in the lower as never to be able to emancipate itself totally from it. Approaching the issue in this manner would help to reintegrate the human being into nature and overcome the radical mind-matter dualism to which modern thought since Descartes, for whom animals were nothing but machines, has accustomed us.

It is significant that the liveliest intellectual debates of our time have not been between rival schools of modern thought—rationalists and empiricists, Kantians and Hegelians, Realists and Idealists, and so on—but between Ancients and Moderns. The impetus came from Martin Heidegger, who, in an attempt to "uproot" a philosophic tradition that had supposedly gone astray from the beginning and was ultimately responsible for the crisis we now face, was led to study Aristotle with a thoroughness that had not been seen since the thirteenth century. The project yielded unexpected results. To a good number of his students, the exhumed Aristotle proved more persuasive than Heidegger's critique of him. A plausible alternative had been uncovered in the light of which all of the fundamental problems could be reexamined.

Enter the ecological movement, which, inconsistent or misguided as it may be at times, lent further credence to the notion that nature itself has something to teach us about our relationship to it. Our most urgent need is what I am tempted to call a new "Socratic turn" that, like the original Socratic turn, would both preserve science and restore it to its native human context. Modern science has explained many things and will no doubt explain many more as time goes on, although there is reason to think that it will never be able to explain everything.[50] Natural beings are unfathomable, if only because one of their principles is matter, which is refractory to mind. There will always be more to them than we are capable of knowing. If one were to eat carrots as a biologist defines them, one would probably end up with a terrible indigestion. Above all, modern science has yet to tell us how a mindless universe gave rise to mind. It can neither explain the scientist nor make room for him in its own universe. I hope scientists will forgive me if, in support of my modest proposal, I quote from a novel by the late Walker Percy, who writes in *The Moviegoer*, the book that made his reputation:

> Until recent years, I read only "fundamental" books, that is key books on key subjects. . . . Schroedinger's *What is Life?*, Einstein's *The Universe as I See It*, and such. During those years, I stood outside the universe and tried to understand it. I lived in my room as an Anyone living Anywhere and read fundamental books and only for diversion took walks around the neighborhood and saw an occasional movie. Certainly it did not matter to me where I was when I read such a book as *The Expanding Universe*. The greatest success of this enterprise, which I call my vertical search, came one night when I sat in a hotel room in Birmingham and read a book called *The Chemistry of Life*. When I finished it, it seemed to me that the main goals of my search were reached or were in principle reachable. . . . The only difficulty was that though the universe had been disposed of, I myself was left over. There I lay in my hotel room with my search over yet still obliged to draw one breath and then the next. But now I have undertaken a different kind of search, a horizontal search. As a consequence, what takes place in my room is less important. What is important is what I shall find when I leave my room and wander in the neighborhood. Before, I wandered as a diversion. Now I wander seriously and sit and read as a diversion.[51]

This is as good a recapitulation of the Socratic turn as one is likely to come across anywhere today. Socrates's turn toward human things did not necessitate the abandonment of his quest for the knowledge of nature. The two must somehow go hand in hand. Our task is to find a way of bringing them together again. The issue may be formulated more pointedly by means of another quotation from Percy's book:

> A regular young Rupert Brooke was I, "—full of expectancy." Oh the crap that lies lurking in the English soul. Somewhere it, the English soul, received an injection of romanticism which nearly killed it. That's what killed my father, English romanticism, that and 1930 science. A line from my notebook: 'Explore connection between romanticism and scientific objectivity. Does a scientifically minded person become a romantic because he is a left-over from his own science?'"[52]

It took a novelist trained as a physician and a natural scientist to bring home to us in such vivid fashion the full import of the divorce between our dehumanized natural science and our concerns as human beings. Percy was not thinking of the ecological movement, which was still in its infancy when *The Moviegoer* was written, but the question that his book raises about the typical modern scientist's lack of self-knowledge sheds far more light on the roots of our ecological crisis than either the Bible or medieval theology.

NOTES

1. *Cry of the Environment: Rebuilding the Christian Creation Tradition*, edited by P. N. Joranson and K. Butigan (Santa Fe: Bear & Co., 1984).

2. P. Santmire, *The Travail of Nature: The Ambiguous Ecological Promise of Christian Theology* (Philadelphia: Fortress Press, 1985).

3. Cf. U. Cassuto, *A Commentary on the Book of Genesis*, Part I: *From Adam to Noah*, translated by I. Abrahams (Jerusalem: Magnes Press, 1978), 34.

4. Martin Heidegger, "On the Being and Conception of *Phusis* in Aristotle's *Physics* B, 1," *Man and World* 9 (August 1976): 224.

5. Aristotle, *Physics* II.1.192b22.

6. Ibid., II.8.199b34ff.

7. Ibid., II.7.198b17f.

8. Hooker, *Laws of Ecclesiastical Polity*, I.2.

9. Spinoza, *Theologico-Political Treatise*, ch. 4, *init.* Cf. Descartes, *Discourse on Method*, part 5, *init.*: ". . . I have also discovered certain laws, which God has so established in nature, and the notion of which he has so fixed in our minds, that after sufficient reflection we cannot doubt that they are exactly observed in all which exists or which happens in the world," translated by L. J. Lafleur (Indianapolis, 1960), 31.

10. Aristotle, *Politics* I.2.1252b32.

11. Aristotle, *Poetics* IV.1449a5.

12. Cf. *Politics* I.8.1256b15f. Aristotle's notion of extraspecific finality is proper to the *Politics*. It is not mentioned in the *Physics*, which deals only with things that are common to all of nature, both animate and inanimate.

13. *Politics* I.8.1256b28f.

14. Cf. *Physics* II.4.196a25f.

15. *The Basic Works of Aristotle*, translated by R. P. Hardie and R. K. Gaye, edited by R. McKeon (New York: Random House, 1941), 656-57.

16. The article appears in White's book, *Machina ex Deo: Essays in the Dynamism of Western Culture* (Cambridge: MIT Press, 1968), 75-94. See also *Creation: The Impact of an Idea*, edited by D. O'Connor and F. Oakley (New York: Charles Scribner's Sons, 1969); *The Environmental Handbook*, edited by Garrett De Bell (New York: Ballentine Books, 1970).

17. *Machina ex Deo*, 86.

18. Ibid.

19. Ibid., 85.

20. Ibid., 89.

21. For a lively account of this strange story, cf. S. Jaki, "Science and Censorship: Hélène Duhem and the Publication of the *Système du monde*," *The Intercollegiate Review* 21 (Winter, 1985-86): 41-49. See also the recent biographies of Duhem by S. Jaki, *Uneasy Genius: The Life and Work of Pierre Duhem* (The Hague: Martinus Nijhoff, 1984), and R. N. D. Martin, *Pierre Duhem: Philosophy and History in the Work of a Believing Physicist* (LaSalle, Illinois: Open Court, 1991).

22. Pierre Duhem, *Le système du monde: Histoire des doctrines cosmologiques de Platon à Copernicus*, vol. 6 (Paris, 1954), 66.

23. See on this point the remarks by S. Jaki, *Science and Creation: From Eternal Cycles to an Oscillating Universe* (Edinburgh/London: Scottish Academy Press, 1974), 229-30. The sections of Tempier's decree that are relevant to the history of medieval science are analyzed in *A Source Book in Medieval Science*, edited by E. Grant (Cambridge, MA: Harvard University Press, 1974), 45-50. Vol. 3 of Duhem's *Etudes sur Léonard de Vinci* is dedicated "To the glory of the Faculty of Arts of Paris, the true mother of our experimental science." Buridan had taught at the University of Paris and Oresme spent some time there as a younger man.

24. Ibid., vol. 2 (reprinted, 1954), 390. For a brief analysis of various classical texts pertaining to the Great Year, cf. P. R. Coleman-Norton, "Cicero's Doctrine of the Great Year," *Laval Théologique et Philosophique* 3 (1947): 293-302.

25. Cf. *Physics* IV.223b24-34. As for Plato, it is hard to draw any definite conclusion from the discussion of the "perfect year" in the *Timaeus* (38b-39e) and the parallel discussions in *Statesman* (269c-273e) and *Epinomis* (991c). Nor do we find any agreement among ancient writers concerning the length of the Great Year.

26. On the highest philosophic level, that thesis can be traced back to Hegel, for whom "modernity originated in an historical development out of Christianity," as is rightly noted by R. Faulkner, *Francis Bacon and the Project of Progress* (Lanham: Rowman and Littlefield, 1993), 12.

27. Cf. *Science and the Modern World* (New York: Macmillan, 1931), and Fortin's discussion of Whitehead, in chapter 8, pp. 88-89.

28. *Science and the Modern World*, 19.

29. Nietzsche, *Genealogy of Morals* III, 23-24. Also, along the same lines, *The Gay Science*, aph. 344.

30. M. B. Foster, "The Christian Doctrine of Creation and the Rise of Modern Natural Science," *Mind* 43 (1934): 446-68; "Christian Theology and Modern Science of Nature," Part I, *Mind* 44 (1935): 439-66; Part II, *Mind* 45 (1936): 1-27. The three articles are reprinted in *Creation, Nature, and Political Order in the Philosophy of Michael Foster (1903-1959): The Classic Mind Articles and Others, with Modern Critical Essays*, edited by C. Wybrow (Lewiston, N.Y., 1992).

31. There is some question as to whether on its own terms Foster's argument is entirely consistent. It holds that the modern scientific view of nature is "more truly Christian" than the Aristotelian view, and, almost in the same breath, that modern natural science was designed as an instrument for the domination of nature and, hence, destructive of the respect owed to nature as the work of God. See Foster's later article, "Some Remarks on the Relations of Science and Religion," *The Christian Newsletter,* Supplement to no. 299 (26 November 1947): 5-16; reprinted in Wybrow, 149-60.

32. By 1934, Lynn Thorndike was ready to concede Duhem's point about some of the medieval antecedents of modern science, although he was careful to dissociate himself from any of the larger conclusions that one might be tempted to draw from that fact; cf. *A History of Magic and Experimental Science*, vol. 4: *Fourteenth and Fifteenth Centuries* (New York: Columbia University Press), 615: "Then presently seeds that had lain dormant since the fourteenth century were, as Duhem has shown, to sprout and fructify in modern philosophy and science. Frankly, it is not for this contribution towards modernity that we most prize these writings of two remote centuries which we have been at some pains to decipher and to set forth. We have taken them as we have found them and we esteem them for what they are in their totality, their fourteenth and fifteenth century *complexio*—a chapter in the history of human thought. Read it and smile or read it and weep, as you please. We would not credit it with the least particle of modern science that does not belong to it, nor would we deprive it of any of that magic which constitutes in no small measure its peculiar charm."

33. On Duhem's treatment at the hands of George Sarton and other historians of science, cf. S. Jaki, *The Origin of Science and the Science of Its Origins* (South Bend: Regnery, 1978), 74-79; *The Road of Science and the Ways to God* (Chicago/London: University of Chicago Press, 1978), 12-14. According to Sarton, Duhem was "carried away by his enthusiasm and by his desire to magnify the Schoolmen and belittle Galileo," *Introduction to the History of Science*, III, Part 1, 146n.

34. Duhem was not the only one to fall into the trap. To mention only one other name, the same thing seems to have happened to the well-known Protestant theologian Langdon Gilkey. See, for example, *Maker of Heaven and Earth: A Study of the Christian Doctrine of Creation* (Garden City: Doubleday, 1959).

35. For a survey of the abundant literature spawned by White's thesis, cf. R. A.

Gray, "Theological Responses to Environmental Decline: An Annotated Bibliography," *Reference Services Review* 22 (November, 1994): 69-94. Also, on this general subject, J. Passmore, *Man's Responsibility for Nature: Ecological Problems and Western Traditions*, 2d edition (London: Duckworth, 1980; New York: Scribner, 1974); and, for a balanced, penetrating, and up-to-date assessment of the ecological movement as a whole, Ch. T. Rubin, *The Green Crusade: Rethinking the Roots of Environmentalism* (New York: The Free Press, 1994).

36. On Galileo's relations with the theologians of the Roman College, cf. Wm. Wallace, *Galileo and His Sources: The Heritage of the Collegio Romano in Galileo's Science* (Princeton: Princeton University Press, 1984). Wallace's book, like so many others, stresses the kinship rather than the break between Galileo and premodern science.

37. "On Values in the Age of Science," *Aspen Institute Readings*, edited by M. Krasney (New York: Aspen Institute for Humanistic Studies, 1978), 86.

38. The charge of animism leveled at Plato by Foster in the above-mentioned articles is based on just such a fundamentalist interpretation of the Platonic myth.

39. On this score, we do well to remember that the Bible attributes the discovery of the arts and sciences, according to White the instruments of man's callous treatment of nature, to the progeny of the reprobate Cain; cf. Gen. 4:17-22. For Timaeus's account of the origin of the universe, cf. Plato, *Timaeus* 29d-33c.

40. See, for example, Bacon, *New Organon*, I.129, where inventors are ranked above statesmen as humanity's greatest benefactors.

41. Cf. *New Organon*, II.52, end.

42. Cf. Descartes, *Discourse on Method*, Part VI, *init.* To glimpse the antibiblical thrust of Cartesian philosophy, one has only to look at Descartes's reworking of Genesis 1 in Part 5 of the *Discourse on Method*. The main elements of the original story are clearly present: the chaos with which it begins, followed by the creation of the sky, light, the sun and the stars, and the earth. At that moment, the story is interrupted by a lyrical account of the miraculous properties of fire and how it can be used to turn ashes into glass, for Descartes a phenomenon as admirable as any that occurs in nature and which he confesses to have taken untold pleasure in describing. Inserting a purple patch such as this one just prior to the creation of animals and the human being is Descartes's way of insinuating that science can do a better job of explaining the origin of the human soul than does the Bible, which has God breathe the breath of life into the nostrils of the man he had formed from the dust of the earth (Gen. 2:7).

43. Cf. F. Bacon, Plan of *The Great Instauration*, near the middle. The difference between the two forms of induction is stressed by Bacon himself, who writes: "But the greatest change I introduce is in the form itself of induction and the judgment made thereby. For the induction of which the logicians speak, which proceeds by simple enumeration, is a puerile thing, concludes at hazard, is always liable to be upset by a contradictory instance, takes into account only what is known and ordinary, and leads to no result" (ibid., beginning). My intepretation of Bacon's

relationship to Christianity accords in the main with that of R. Faulkner, op. cit., 61-65 et passim. For a more "pious" interpretation within the context of the contemporary ecological debate, see Wm. Leiss, *The Domination of Nature* (Boston: Beacon Press, 1972).

44. E. Meyerson, *Explanation in the Sciences*, translated by M.-A. and D. A. Siple (Dordrecht: Kluwer Academic Publishers, 1991), 357-58.

45. See esp. the introduction to Husserl's *The Crisis of European Sciences and Transcendental Phenomenology* (Evanston: Northwestern University Press, 1970).

46. A similar distinction can be made among man's spiritual activities. In the *Ethics*, Aristotle refers to moral habit as something that does not change easily "because it is like nature" (VII.1152b31), a "second nature," as was later said, thereby implying that moral virtue is more natural than intellectual virtue.

47. Cf. *Physics* II.8.199c34f.

48. Paleontologists calculate that, of the billions of species that have populated the earth at one time or another over the past three or four billion years, fully 99.9 percent have disappeared, sometimes as a result of the mass extinctions that have occurred at periodic intervals. See on this subject the recent books by N. Eldredge, *The Miner's Canary: Unraveling the Mysteries of Extinction* (New York: Prentice Hall Press, 1992), and D. M. Raup, *Extinction: Bad Genes or Bad Luck?* (New York: W. W. Norton, 1992).

49. This much is suggested by Aristotle, *De anima* II.421b16-25, which stresses the greater perfection of the sense of touch in human beings and its peculiar relation to the life of the mind. Also Thomas Aquinas, *Summa theologiae* I, qu. 76, a. 5, entitled: "Whether the Intellectual Soul is Conveniently United to this Kind of Body."

50. Unlike Kant, Aristotle and his followers did not think it possible to draw in advance a firm line between what the human intellect can or cannot do. All virtue, including intellectual virtue, is demonstrated in its employment. One discovers what reason can do when it does it.

51. Walker Percy, *The Moviegoer* (New York: The Noonday Press, 1960), 69-70.

52. Ibid., 88.

III

ANCIENTS AND MODERNS

11

OTHERWORLDLINESS AND SECULARIZATION IN EARLY CHRISTIAN THOUGHT: A NOTE ON BLUMENBERG

Our generation may well go down in history as the one that reopened the long-forgotten quarrel between the Ancients and the Moderns. It may even be remembered as the first generation to glimpse the full import of that famous quarrel, the larger dimensions of which could come to light only once "modernity" had had a chance to play itself out and unfold its hidden consequences. Barely more than fifty years ago, no one in academic circles would have predicted that the liveliest intellectual debates of our time would be tied to, if not actually triggered by, a passionate interest in the writers of classical antiquity. Nor did it occur to any but the most perspicacious minds to suggest that virtually all of our basic problems need to be reassessed in the light of the break with ancient and medieval thought that was consummated in the course of the early modern period.

Unlike its fifteenth- and sixteenth-century predecessor, the new "renaissance," if the term is not too pretentious, owes nothing to the West's more or less accidental recovery of a large number of hitherto unknown classical philosophic and literary texts. It was mandated by modern philosophy itself, the most advanced form of which insists that all human thought be understood "historically," that is, as part of an ongoing process, the seeds of which were sown in a remote premodern past.

This said, there is no universal agreement as to how the relationship between Ancients and Moderns can best be articulated, and even less agreement as to which of the two groups is likely to emerge as the winner in the contest that continues to pit them against each other. The still most widely held position among contemporary scholars is the one that sees modernity as a simple product of premodernity, and hence, denies that there is any profound breach of continuity between them. Such is the position that was developed for the first time by Hegel in the early-nineteenth century and later propagated under his direct or indirect influence by a host of prominent twentieth-century historians of political thought—the brothers Carlyle, Charles McIlwain, George Sabine, Edward Corwin, and others—through whose works it became part of the conventional wisdom of our age.

Not the least interesting feature of Hans Blumenberg's book, *The Legitimacy of the Modern Age,*[1] is that it boldly takes issue with that position and, reverting to a pre-Hegelian understanding of the problem, stresses the radical newness of modern thought. In this respect, if not in all others, Blumenberg's assessment of the situation comes much closer to that of the founders of the modern movement, all of whom were firmly convinced that, like Columbus, they had discovered a new continent and stood on fundamentally different ground.[2] Modernity, they thought, was not a secondary or derivative phenomenon. It has its own indigenous roots and is not to be viewed as the natural outgrowth of what came before it. At the level of its highest principles, it represents a "distinct event"[3] and owes nothing of importance to either the classical or the Christian tradition, from which it is separated by an abyss.

As Blumenberg is quick to admit, however, no one would think of making a case for the "legitimacy" of the modern age if that legitimacy was not being called into question by modern critics of modernity. Accordingly, the main target of his attack is not the continuity thesis as such but a modified version of it, usually referred to as the "secularization thesis," of which Karl Löwith, with whom Blumenberg was engaged in a feud for a number of years, stands as a typical representative.

The defenders of this thesis look upon modernity as nothing more than a desacralized version of the biblical and Christian world from which it descended. At their hands, the term "secularization," which had formerly been used to designate either the forcible expropriation of church goods by the civil authority, or else the canonical process whereby a cleric who failed to live up to his religious obligations could be laicized, was suddenly and unexpectedly applied to the realm of ideas. The foundations of the modern world, it was now argued, are to be sought in the biblical tradition, which our contemporaries have repudiated or otherwise lost sight of, but from which they fraudulently or duplicitously continue to live. Ours is a civilization whose misguided opinions are not simple errors but "heresies"—Christian doctrines gone mad, as some have called them. Modernity as a whole is at bottom a "pseudomorph," a form that has been emptied of its substance and that could regain its erstwhile vitality only by means of a self-conscious return to its biblical roots.

Thus understood, the secularization theory stands or falls by a combination of three distinct criteria, namely, original ownership, ascertainable derivation, and unilateral removal, by which is meant the appropriation of a given notion by a second party without the owner's consent or approval. For, unless it can be shown that the "secularized" notion, whatever it may be (there are numerous candidates for the honor), originated with the biblical tradition from which it was subsequently pilfered and transferred to a nonreligious context by someone who acted without the authorization of the proprietor, any talk of secularization is meaningless. Blumenberg's contention is that on all three counts the secularization theorem fails to meet the test.

This much is evident from a rapid look at the notion of history or historical consciousness, on which Blumenberg's debate with Löwith focuses. Contrary to what Löwith maintains, the modern notion of history is not ultimately traceable to Christianity and has nothing to do with it if only because early Christianity, like the Jewish apocalypticism out of which it grew, was strictly nonhistorical and otherworldly. The first generation of Christians lived in the expectation of an imminent end of the world and had no incentive to work for the improvement of the conditions of life on earth. Only when the anticipated end failed to materialize was the Christian Church forced to rethink its relationship to the world and develop a more positive stance toward it.

Secondly, the theory of history that Christianity eventually came up with as a result of the delay of the Parousia—a theory known to theologians as the "history of salvation"—bears only a tenuous resemblance and no relation of paternity whatever to what nowadays would pass for a bona fide philosophy of history. The events of the end-time, now relegated to an indeterminate future, will not be brought about by forces intrinsic to the historical process and do not in any known way depend on developments occurring within that process. They depend on God's unfathomable will and will come about solely as the result of his direct intervention in this world. The theological doctrine of divine providence asserts only that God, as the supreme ruler of the world, has his own plan for the redemption of mankind and that, as a rational agent, he is capable of ordering even contingent events to the end that he has prescribed for them. Since that plan has not been revealed to us, human beings cannot be guided by it in their actions and are required to take their bearings from the moral law. Concretely stated, the end does not justify the means. Contrary to one of the key premises of the modern philosophy of history, evil means are never sanctioned by the grandeur of the overall goal to the attainment of which they might contribute. No, the modern notion of history was not spawned by Christianity. It arose in opposition to it and, like so many other modern doctrines, is predicated on a radical departure from it.

As for the third criterion, the consent of the owner, there is a serious question as to whether early Christianity did not in fact acquiesce in its own secularization, setting a precedent that modern-day secularizers would follow and carry to its logical conclusion. Faced with the necessity of coping with the problem created for it by the indefinite postponement of the end-time, the Church had no choice but to make its peace with the world or at least relax its opposition to it, at the risk of compromising its own identity. For the otherworldly outlook that had characterized its early existence a new worldliness was substituted that, as time went on, would induce Christians to become more deeply involved in the affairs of this life. The danger was all the greater as the tools employed for this purpose were mostly borrowed from classical philosophy. Inevitably, the question arose as to whether, by plundering the works of the pagan philosophers, the Church was not diluting the wine of the gospel with the water of secular thought. That the problem was perceived as a real one by Christians themselves is clear from the fact that a number of them tried to defuse it by contending, implausibly and with little success, that Plato had learned his philosophy from Moses. If one adds to all of this that the notion of proprietary rights to, or the legal "ownership" of, ideas, as distinct from their origin, is a dubious modern conceit, one is bound to have serious misgivings about the viability of the secularization theory.

Yet Blumenberg's position, for all its hostility to that theory, is anything but a simple return to the pre-Hegelian or Enlightenment account of the rise of modernity. Its novelty lies not so much in its reassertion of the autochthonous character of the modern development as in its attempt to show how this development is incidentally related to what preceded it. Part of the difficulty with the Enlightenment version is that its proponents were misled by Descartes into thinking that modernity stood as a kind of absolute beginning, unconditioned as it supposedly was by anything that had gone before. Historicism has since reminded us of the difficulty involved in any attempt to locate absolute beginnings anywhere, least of all in the realm of philosophic thought. Descartes himself is unintelligible apart from the tradition that produced him and on which he remains extrinsically dependent.

To say this, however, is not to take anything away from the originality of modern thought. The link between it and the older view would be supplied by the notion of "reoccupation," Blumenberg's personal contribution to the elucidation of this long-standing problem. "Reoccupation" in this context refers to a variety of attempts aimed at filling the vacuum created by the abandonment of positions whose untenability had become, so to speak, visible to the naked eye, but which in the meantime had given rise to intellectual expectations that would henceforth have to be taken into account and, if possible, met by any new theory.

The key issue in this regard is the problem of evil, which Blumenberg, together with Löwith, singles out as the weakness of the Christian tradition. If the history of early Christian and medieval theology proves anything, it is the futility of any and all attempts to reconcile the doctrine of creation, which emphasizes the goodness of the universe as a whole, with the doctrine of redemption, which is predicated on an implicit denial of that goodness; or, what amounts to the same thing, to reconcile belief in divine omnipotence with the apparent ineradicability of evil in the world. Therein would lie the contradiction at the heart of the Christian religion. Gnosticism, in the struggle against which the Church redefined its orthodoxy, was more consistent: it traced all evil to an independent and coeval demiurgic principle, thereby exonerating God, but not without denying his infinite power.

Neither Augustine, the champion of orthodoxy against Manichaeanism (the form taken by Gnosticism in his day, although at the time the two movements were not perceived as related to each other), nor the Nominalists of the late Middle Ages, whom Augustine had saddled with the indefensible notion of divine absolutism or omnipotence, fared any better in their efforts to deal with this nagging problem. The failure of both ancient and medieval theology to come up with an adequate solution to it precipitated the demise of Christianity as a respectable intellectual force in the West and set the stage for the new beginning represented by what we now call "modernity." In short, the principles of modern thought were not precontained in premodern thought and hence could not have been educed from it by a process of logical inference. Modern thought has its own substance; it did not have to steal it from its predecessors. The only real points of contact between the two traditions are the ones related to the expectations built up by the old and now vacated positions, expectations related to the notion of infinity, for example, for which more promising applications were to be found in the typically modern project geared to the mastery of an infinitely malleable nature and its scientific presuppositions, the concept of infinite space among them.

Doing justice to the richness and complexity of this provocative book, the first, as far as I know, to subject the notion of secularization to what purports to be a rigorous philosophic analysis, is not an easy task. Being in no position to undertake such a task, I shall limit myself to a few desultory remarks about Parts I and II, the ones devoted to a study of the inner tensions, irresolvable difficulties, and outright contradictions that in Blumenberg's view plagued early Christianity and eventually proved fatal to it.

To begin with some fairly obvious considerations, it is entirely possible that Blumenberg's overriding concern with "function" as opposed to "content," coupled with an excessive reliance on the constructions of modern biblical and historical scholarship, has led him to look for contradictions where none exists. A simple example will illustrate the point. As a strictly otherworldly religion, primitive Christianity, we are told, was interested only in personal salvation and imbued with a sense of urgency stemming from the conviction that the Parousia was imminent. Its followers were simple people for whom the end of the world, the return of Christ, and the final consummation were events to be prayed for and ardently desired. Yet, before it knew it, the Church had made an about-face. Its outlook had changed from one of outbounded hope to one of extreme fear. It was now praying for the *mora finis*—not for the coming of the end but for its delay. As Blumenberg puts it, citing a marginal note by Karl Holl, "No apologist hopes for an early return of the Lord!"[4]

Did the early Christians contradict themselves in such blatant fashion? Maybe, but the story is more complicated. Those among them who begged God to stay his hand and postpone the day of judgment seem to have had something quite different in mind. In a spirit that is very much that of Scripture, they were concerned that human beings be given the time they needed to repent. They knew from the Book of Genesis that Abraham's prayer for the preservation of Sodom and Gomorrha would have been heeded had there been any hope for the reform of the inhabitants of those cities. The New Testament is even more explicit. It teaches that God is patient, that he waits for the conversion of sinners, just as in the days of Noah, and that he is also waiting for the entry of the Gentiles into the Church (2 Peter 3:4-9; 1 Peter 3:20; Rom. 11:25-27). Blumenberg would probably reply that such considerations are foreign to the first generation of Christians and reflect a mentality that crept into the New Testament only at a later date under the influence of Paul and John. Yet Paul's letters are generally thought to be the earliest of the New Testament texts. No one denies, of course, that apocalyptic ideas played an important role in the mind of the first community. Whether they exhaust the thinking of that community is another matter. Ideally, one would likewise have to explain why the shift in perspective occurred in this particular apocalyptic community and none of the others whose general orientation it seems to have shared.

A similar problem arises in connection with the treatment of Gnosticism, this time with graver consequences, insofar as we are brought closer to the central thesis of the book, encapsulated in a sweeping statement to the effect that "The modern age is the second overcoming of Gnosticism."[5] Blumenberg can hardly be faulted for attaching great significance to Gnosticism, the most formidable adversary that the budding Christian Church had to face as it headed into the second century. Unfortunately, historians of Western thought labor under a special handicap when it comes to this matter: none of the works of the major Gnostic thinkers, Marcion and Valentinus, has survived (their contents are known to us only through the refutations of their orthodox adversaries), and the newly discovered Nag-Hammadi texts, the analysis of which was held up for years by scholarly intrigues of the most disgraceful kind, have yet to reveal all their secrets, assuming that they contain any. As a result, the specific origins of the movement remain obscure. To the best of my knowledge, even specialists are divided over the issue of whether Gnosticism should be regarded as a specifically Christian heresy or whether it simply came into contact with Christianity after it had developed in the Roman empire under the influence of dualistic doctrines emanating from the East.

Be that as it may, one wonders whether the one-sided concentration on Gnosticism has not caused Blumenberg to overlook other equally important facets of the problem. As more or less contemporary religious movements whose appearance on the scene coincides roughly with the disintegration of Roman political life and the decline of civic religion, Christianity and Gnosticism exhibit a number of common traits and present themselves, among other things, as transcendent or nonpolitical religions. Where they differ conspicuously is in their response to the political life. What characterizes Gnosticism is its radical otherworldliness, that is, its aversion to and repudiation of the material world. The same is not true of Christianity, which does not demand that its followers turn their backs on civil society but rather binds them in love to the service of their fellow human beings. It is this ability to relate to the political order while claiming to transcend it that constitutes its genius and determined both the course of its history and the development of its doctrines. Blumenberg's "otherworldliness" and "secularization," as applied to the early Church, are just misleading labels for what is better described as the "transpolitical" nature of the Christian faith. Blumenberg is right when he says that "Christian theology always returns to its classical heresies,"[6] but he is wrong about the heresies. It could be shown, although this is not the place to do it, that all of the major heresies of the fourth and fifth centuries—Arianism, Nestorianism, Monophysitism, to mention only three—were in one way or another challenges to that transpolitical character. Gnosticism may or may not have invented a more coherent solution to the problem of evil, but one thing is certain: its failure to face squarely the question of its relationship to civil society, on which like every other group it depended for its survival, sealed its doom. The Gnostics, not the orthodox Christians, were the fanatics or the "crazies" of that troubled but endlessly fascinating age.

Further questions arise when one subjects Blumenberg's intriguing notion of "reoccupation" to closer scrutiny. Is it true, as Blumenberg contends, that the Moderns merely succumbed to a dangerous and perhaps tragic temptation when they mistook the problems inherited from their predecessors for real problems, calling for a response on their part? The argument would be more compelling if Blumenberg had considered alternative explanations and shown them to be inadequate. It is conceivable that at least some of the Moderns were not at all interested in reoccupying territory once staked out and later abandoned by their predecessors and used theological language for the sole purpose of accommodating themselves to the prejudices of their supposedly less enlightened contemporaries. Blumenberg, although not unaware of that possibility, has chosen to ignore it. As a result, he ends up by conceding more than is necessary to the Christian tradition, which he dimisses as incoherent, and at the same time diminishes the stature of the Moderns, whom he has set out to defend. Interestingly enough, the present-day opponents of the Moderns are the ones who give them credit for being more intelligent and more prudent than the rest of us.

I leave it to others to explore at length and at greater depth the philosophic underpinnings of Blumenberg's argument. I note only by way of conclusion that almost nothing—certainly nothing of substance—is said about Machiavelli in Blumenberg's book. This is all the more surprising as there is mounting support among scholars for the view that Machiavelli is the true founder of modernity. By neglecting the larger political and human context within which modernity made its appearance and gradually took shape, Blumenberg has deprived us of the means of forming an enlightened opinion regarding its nature and assessing its relative merits. If, as may be argued, Machiavellian "realism," that is, the decision to study human beings as they are rather than as they ought to be, is the soul of the modern project, more attention should have been paid to the type of human being that is likely to become dominant if the project should succeed. But these are obviously not the kinds of questions that haunt Blumenberg's mind. Had he given serious thought to them, he might have had more to say, not only about the relationship of modern thought to its Christian antecedents, but about its relationship to the pre-Christian classical tradition, on which by the seventeenth century the Moderns had resolutely begun to train their sights.

NOTES

1. Hans Blumenberg, *The Legitimacy of the Modern Age*, translated by R. M. Wallace (Cambridge, MA and London: MIT Press, 1983).

2. See, for example, Machiavelli, *Discourses*, book I, Introduction: "Although the envious nature of men, so prompt to blame and so slow to praise, makes the discovery and introduction of new modes and orders as dangerous almost as the exploration of unknown seas and continents, yet, animated by the desire that impels me to do what may prove for the common benefit of all, I have resolved to open a new route, one that has not yet been followed by anybody and may prove difficult and troublesome, but may also bring me some reward in the approbation of those who will kindly appreciate my efforts."

3. Blumenberg, 49.
4. Ibid., 45.
5. Ibid., 126.
6. Ibid., 40.

12

THOUGHTS ON MODERNITY

Any insight into the recurrent debate about the vitality of modern Western civilization or, as others prefer to think, its sickness and possible demise presupposes among other things a reasonably clear understanding of what is meant by "modern." Since the word occurs in a wide range of contexts and is interpreted differently by different people, an attempt to elucidate its content and uncover its hidden premises may be in order. Obviously not everything that is said or done today is regarded as typically modern. The fact that some contemporary movements are labeled "modern," as opposed to, say, "traditional," "reactionary," or "old-fashioned," already suggests as much. The distinction implied by the very use of the term "modern" is thus more than a purely chronological one. The following remarks purport within broad limits to define modernity in this pregnant or emphatic sense, sketch briefly its historical development, and explore the nature of the problem with which it confronts us. To avoid unnecessary complications, I shall restrict myself to the moral and political dimensions of the question, which are both more urgent and more readily accessible to most of us.

THE NOTION OF MODERNITY

Modern thought in the sense in which I have chosen to speak of it first comes to sight toward the beginning of the sixteenth century as a critique of premodern thought, both classical and Christian, in opposition to which it defines itself and which it eventually replaced as the dominant intellectual force in the Western world. This is not to say that all early or later representatives of the new movement stand together on every major issue, but only that, great as the differences between them may be, they remain united in a common hostility to, and final repudiation of, premodern thought. To that extent at least, one is justified in referring to them as a distinct and more or less homogeneous group.

It should be noted that the foregoing interpretation is not accepted by all or even most contemporary scholars. The still most frequently encountered opinion is that modern thought was already latent in ancient thought, that it was educed from it by a process of logical inference, and hence that there is no real dichotomy or breach of continuity in the Western tradition from roughly the time of the Stoics to the nineteenth century. Far from signaling a break with the past, the modern development would be nothing but an effort to refurbish the time-hallowed principles of classical antiquity and bring them into line with the needs and circumstances of our own time. Such was the view propounded in the early decades of this century by R. W. and A. J. Carlyle in their monumental and enormously influential *A History of Mediaeval Political Theory in the West,* where one reads:

> There are no doubt profound differences between the ancient mode of thought and the modern—the civilization of the ancient world is very different from that of the modern; but, just as it is now recognized that modern civilization has grown out of the ancient, even so we think it will be found that modern political theory has arisen by a slow process of development out of the political theory of the ancient world—that, at least from the lawyers of the second century to theorists of the French Revolution, the history of political thought is continuous, changing in form, modified in content, but still the same in its fundamental conceptions.[1]

More recent research has shown that the harmony between Ancients and Moderns for which the Carlyles argue is more apparent than real. It is undoubtedly true that the modern writers frequently adopt the language of their predecessors and even profess to side with them, and it is also true that many of their teachings are to be found in one form or another in the older scheme. But neither consideration warrants the conclusion that the two groups share the same basic assumptions. The very newness of the modern teaching enjoined on its proponents the utmost caution in the public expression of that teaching. In the words of the foremost early representative of the modern project, "there is nothing more difficult to carry out, nor more doubtful of success, nor more dangerous to handle, than to initiate a new order of things."[2]

More precisely, the writings of these authors exhibit a strange blend of restraint and boldness, which experience had shown to be the only effective way to accredit new ideas before the bar of commonly received opinion. The conventional surface of most of these writings should not blind us to the novelty of their core, whose revolutionary character it partially masks. Any chance of success that their authors may have had hinged on their ability to clothe their thoughts in traditional garb, something they managed to do, for example, by deliberately expressing contradictory views in different works or different parts of the same work, by maintaining an eloquent silence on issues that are expected to figure prominently in any thematic discussion of a given problem, or by resorting to such other literary techniques as are apt to hide their real thought from the uninformed or unwary reader. Francis Bacon put the matter nicely when he said that one should always begin by telling people what they most want to hear and that what they most want to hear is what they are accustomed to hearing. Only once their confidence has been won is it safe to insinuate unfamiliar notions into their minds.[3]

Even the presence, in the works of the modern writers, of innumerable carry-overs from premodern thought does not in any way derogate from their originality or constitute an objection to it. To say nothing of other considerations, one could effect a definitive break with classical thought merely by emphasizing, emancipating, or suppressing certain elements that were explicitly or implicitly present in it. The modern writers themselves were obviously aware of their dependence on their predecessors—indeed, each successive wave of modernity may be seen as an attempt to purge modern thought of any lingering vestiges of the past—but none of this kept them from casting themselves in the role of daring adventurers, eager to explore "new seas" and discover "new continents." One thing is certain: in the eyes of its originators who are still its most reliable witnesses, the modern project involved not just the necessary adaptation of the principles of classical thought to an unprecedented situation, but a dismissal of those principles in favor of an entirely new set of principles and the "new modes and orders" based on them.

ORIGIN AND DEVELOPMENT OF MODERNITY

At first glance one is tempted to look upon the new philosophy of the Renaissance as a direct result of the emergence of modern natural science, in which it eventually sought its theoretical foundation. Closer examination reveals, however, that the achievements of modern experimental science, stunning as they may be in comparison with those of ancient and medieval science, do not of themselves contradict the highest principles of ancient and medieval science. Nor can they be said to have been wholly unforeseen by our ancient forebears, who were duly impressed by "stupefying advances" registered by the sciences of their time and were well aware of their virtually unlimited possibilities. The specifically new phenomenon concerns rather the change that the notion of science undergoes at this time or, to speak with greater accuracy, the rise of an independent or metaphysically neutral science, now ordered exclusively to the relief of man's estate and destined to supersede philosophy as the perfection of our natural understanding of the universe. Be that as it may, it is a matter of historical record that the birth of modern philosophy antedates that of modern experimental science by some seventy-five years and that its first manifestations occurred on the level of moral and political thought rather than that of properly speculative thought. If such is the case, one must seek to explain modern science in terms of modern philosophy and not vice versa.

The man chiefly responsible for the intellectual revolution of the sixteenth century is Machiavelli, whose sweeping critique of the classical tradition may be summarized as follows. Classical philosophy studies human behavior in the light of man's loftiest goals, that is, in the light of virtue, and it claims to be able to show the way to those goals. It culminates in a discussion of the best life and, on the political level of the "best regime" or the type of rule that is most conducive to the best life. In so doing, it is compelled to speak about an ideal whose actualization is contingent on the unlikely confluence of an untold number of auspicious circumstances and which for that reason is rarely if ever seen among human beings. Such an approach is inherently defective insofar as it fails to take into account the "human, all too human" character of people's lives. By prescribing goals that are unattainable for all people most of the time and for most people all of the time, it makes impossible and hence unreasonable demands on human nature. In the words of an early disciple of Machiavelli, its discourses are "like the stars, which give little light because they are so high."[4] Little wonder that its efforts to improve human behavior should have ended in failure. Any philosophy that prescribes unfulfillable duties is untenable. It thus becomes necessary to redefine the horizon within which political discourse can appropriately take place.

The new political philosophy was in its most laudable aspirations an attempt to provide adequate guidance in human affairs by self-consciously lowering its sights. It ceased to take its bearings by man's highest excellences and focused instead on the way people actually live. In this manner it broke in the sharpest way with the ancient tradition.[5]

Machiavelli's statements on this matter need to be interpreted cautiously. By referring to what people actually do rather than to what they ought to do, Machiavelli is not suggesting, after the manner of our present-day "value-free" social science, that there is no way of distinguishing one from the other. People do many different things, some of them objectively better or worse than others. Like his ancient and medieval predecessors, Machiavelli thinks that certain things ought to be done and others avoided. The difference concerns the "ought." Machiavelli's "ought" is different from that of the classical philosophers. It is lower. Because Machiavelli's ideal makes fewer demands on people, its chances of success are vastly enhanced. Fortune has been all but conquered. Its role in human affairs has been drastically reduced.[6] The good society is now thought to be possible always and everywhere.

Machiavelli's aim was to provide effective political guidance by lowering the standards of human action. This epoch-making transformation of our moral horizon forced him to take a decisive step affecting the status of wisdom. He questioned the supremacy of wisdom. With Machiavelli, wisdom ceases for the first time to be a major theme of political philosophy. The "virtue" that would henceforth be demanded of people has its sole source in the needs of society rather than those of the mind.

As profound as Machiavelli's impact may have been on the evolution of modern political thought, his views do not by themselves account for every aspect of that thought. The shocking character of some of these views led to their eventual modification by later theorists. For the sake of brevity, we can leave it at saying that the form in which the trend inaugurated by Machiavelli finally crystallized and gained widespread acceptance is the typically modern natural right doctrine of John Locke.

The easiest way to understand Locke's political philosophy is again to compare it to that of his premodern predecessors. For the bulk of the classical tradition, there is no preestablished harmony between the individual and the civil society to which he belongs or between self-interest and the common good. If that harmony is to exist at all, something will have to bring it about. That something is moral virtue, which requires a painful conversion from the premoral concern with worldly goods to a concern with the good of the soul. Only through such a conversion does the individual come to look upon the good of society as his own and dedicate himself entirely to its pursuit. Locke for his part claims to have found a safer way to achieve a similar result. The trick consists in using private vice or selfishness to procure public benefits. Since self-interest means, politically at any rate, the accumulation and preservation of property, we are led to the view that acquisitiveness and, ultimately, unlimited acquisitiveness is more conducive to the good of society than such virtues as moderation and liberality. In their desire to make money for themselves, entrepreneurs create jobs for others, who likewise become richer. Everyone is better off. Two things suffice to produce the desired result: enlightenment and positive law. People need only to understand that it is in their own interest to abide by the law. They will shun injustice if they see it as unprofitable. Honesty is the best policy and nothing more. Moral virtue as something choiceworthy for its own sake is replaced by the mere transition from unenlightened to enlightened self-interest. There is no need to be concerned with the common good, which is procured automatically by the fact that each individual is now free to pursue his selfish goals and encouraged to do so. Instrumental or bourgeois virtue is all that is required of anyone. Mandeville, one of Locke's early disciples, stated the issue in a nutshell when, in *The Fable of the Bees; or, Private Vices, Publick Benefits*, he argued that the day bees started worrying about moral virtue the hive would be ruined and would recover its prosperity only when each one returned to its vices.[7]

Locke managed to provide what appeared to be the most convenient solution yet offered to the problem of civil society. It is in its Lockean form that the revolution initiated by Machiavelli finally triumphed. The very elegance of his scheme was a decisive factor in its success. Whatever its deficiencies, and these are no doubt more evident to us than they were to Locke's contemporaries, it became the most influential political theory of the modern period.

Its triumph, it should be added, was never complete. Throughout the centuries that followed, we find, side by side with the partisans of the modern outlook, a handful of staunch and embattled critics who were determined to buck the trend. One of the most vocal, not to say vociferous, among them is Jonathan Swift, all of whose works may be read as a relentless satire on every conceivable aspect of modernity. Swift's portrayal of the Ancients as giants and the Moderns as pygmies in *Gulliver's Travels* leaves little doubt as to where his own sympathies in the matter lay. The problem as Swift saw it is admirably summed up in the following passage from *The Battle of the Books*:

> This quarrel first began . . . about a small spot of ground, lying and being upon one of the two tops of the hill Parnassus; the highest and largest of which had, it seems, been time out of mind in quiet possession of certain tenants called the Ancients; and the other was held by Moderns. But these, disliking their present station, sent certain ambassadors to the Ancients, complaining of a great nuisance, how the height of that part of Parnassus quite spoiled the prospect of theirs, especially towards the east; and therefore, to avoid war, offered them the choice of this alternative: either that the Ancients would please to remove themselves and their effects down to the lower summit, which the Moderns would graciously surrender to them, and advance in their place; or else that the said Ancients will give leave to the Moderns to come with shovels and mattocks and level the hill

as low as they shall think it convenient. To which the Ancients made answer: how little they expected such a message as this from a colony whom they had admitted out of their own free grace to so near a neighbourhood. That as to their own seat, they were aborigines of it and, therefore, to talk with them of a removal or surrender was a language they did not understand. That if the height of the hill on their side shortened the prospect of the Moderns, it was a disadvantage they could not help, but desired them to consider whether that injury (if it be any) were not largely recompensed by the shade and shelter it afforded them. That as to levelling or digging down, it was either folly or ignorance to propose it if they did or did not know how that side of the hill was an entire rock which would break their tools and hearts without any damage to itself. That they would therefore advise the Moderns rather to raise their own side of the hill than dream of pulling down that of the Ancients, to the former of which they would not only give license but also largely contribute.[8]

A similar denunciation of modern life and habits is found a century or so later in Flaubert, with an added touch of bitter disillusionment and a romantic longing for a glorious and irretrievable past, but without the airy sense of humor that pervades everything that Swift wrote.

Swift and Flaubert had little positive effect on their contemporaries, probably because neither one was able to offer them an alternative to which they might be attracted. The same cannot be said of Rousseau, the first modern critic of modernity and one of the truly seminal writers of the entire modern period. With Rousseau, modernity reached its first great crisis. Of particular significance in this regard is the *Discourse on the Arts and Sciences*, which is nothing less than a declaration of war on the Enlightenment, the then most advanced form of the modern project.

The immediate question to which the *Discourse* addresses itself is whether the restoration of the arts and sciences after centuries of religious obscurantism had contributed to the purification of morals, or, simply put, whether moral progress necessarily goes hand in hand with intellectual progress. Rejecting a key premise of the Enlightenment, Rousseau proceeds on the assumption that intellectual progress poses a threat to moral progress, which is the only true progress. His specific thesis is that, far from improving society and its citizens, the rebirth of the arts and sciences had corrupted their manners, destroyed civic virtue, and paved the way for the greatest of all social evils, despotism. Between democracy, which Rousseau champions, and popular enlightenment, so the argument ran, no reconciliation whatever is possible. Contrary to what would later be said by John Dewey and others, the rule or method of science is emphatically not the rule or method of a free society.

Rousseau not only single-handedly crushed the Enlightenment, his dissection of it laid the foundation for virtually every significant development in modern thought from the end of the eighteenth century down to our time. For, although he assailed the *philosophes* on what were largely classical grounds, he himself did not push for a return to premodern thought or take seriously the possibility of such a return. In the final analysis, his love of freedom proved stronger than his concern for virtue. The position upon which he settled was, in effect, even more modern than those of his predecessors and did more than any of them to enshrine what had become, by that time, a well-established and increasingly powerful tradition.

THE CONTEMPORARY CRISIS of MODERNITY

In the course of his all-out attack on the Enlightenment, Rousseau had called for a return to nonmercenary morality over against the utilitarianism or instrumentalism of the modern natural right theorists. According to his account, this morality had its remote source in a prehuman rather than a human "state of nature."[9] To that extent it was highly problematic, for it is not clear how the "stupid animal" that inhabits Rousseau's state of nature could serve as a guide to properly human behavior. It thus became apparent that any thought of establishing human morality on the bedrock of nature would have to be abandoned altogether. The plunge was taken shortly after Rousseau by a man whose intellectual debt to Rousseau cannot be overestimated, Immanuel Kant. Kant's moral philosophy is best described in his own words as an effort to emancipate human beings once for all from the tutelage of God and nature. In Kantian philosophy, universalized reason alone, as opposed to nature, becomes the standard by which all moral activity is to be judged.

The new moral requirements or "imperatives," as Kant called them, were in many ways more stringent than the most stringent requirements of classical philosophy. Yet, and this is not the least of its paradoxes, Kant's political scheme remains anchored in the modern doubt concerning man's ability to behave morally. His perfect city is no longer a city that exists in speech alone and whose pattern is laid in the sky, as was Plato's; nor does its actualization require that its citizens be gods or children of the gods, that is, that they be virtuous.[10] The one thing needful is not moral virtue, which it is not the business of government to promote, but the right kind of laws, by which Kant means laws that will grant to everyone as much freedom as is compatible with everyone else's freedom. Differently stated, the good city does not need good citizens. In theory, it can be made up entirely of "devils," as long as they are intelligent![11] Never before had the realms of ethics and politics been so sharply set apart.

Kant's philosophy stands as the last great attempt to restore something like classical morality, albeit on the basis of the modern premise, that is, without the benefit of the teleological understanding of nature that supported this morality. The futility of such an endeavor was pointed out with unsurpassed vigor and clarity by Nietzsche toward the end of the nineteenth century. Nietzsche's argument in simple terms is that one cannot get rid of the architect and keep the building or do away with the lawgiver and claim the protection of the law. If classical speculative philosophy goes, then classical morality will have to go too, to be replaced not only by a new morality but by a new kind of morality, one that has little more than the name in common with the old morality.

Nietzsche's self-proclaimed superiority over all other philosophers, living or dead, is founded on his alleged insight into the radical "historicity" of human thought. Although the leading philosophers of the past had never been able to reach a consensus on the fundamental issues, they nevertheless shared a crucial premise: they all assumed the existence of an objective truth or a preexisting order, which each one claimed to have discovered and set forth more perfectly than any of his predecessors. This, says Nietzsche, is an illusion. Philosophy is not the theoretical enterprise for which it has always been taken. Its practitioners do not "discover" the truth, they "create" it. The same goes for Nietzsche's own philosophy, as he himself admits. The only difference is that he is aware of this fact, whereas his predecessors were not. They were "dogmatic" and "naive." Their philosophies are nothing but a variety of perspectives that pretend to be inspired by the "will to truth" but are just a subtle manifestation of a more basic impulse that Nietzsche calls the "will to power." There is about all of them an invincible stupidity that prompts Nietzsche to compare them to beautiful and strong asses.

What Nietzsche says about speculative philosophy applies to all other branches of knowledge as well, including modern natural science, now viewed as only one of any number of possible interpretations of the universe, none of which is demonstrably superior to any other. It is equally applicable to all moral principles, which are neither supplied by nature nor established by reason but imposed upon us by a mysterious historical dispensation over which we have no control. Like all other would-be eternal or objective truths, they rest on a primordial option in the light of which one's subsequent choices make sense but which is not itself in the light. Modern man looms before us as a blind cyclops, deprived of any horizon that could lend permanent meaning to his life and left with no criterion that would enable him to distinguish between the right and wrong use of the immense power that the emancipation of technology has placed at his disposal. To borrow a phrase from Tillich, the Western world had reached the brink of the abyss.

Nietzsche himself was fully apprized of the portentous character of his doctrine and indicated as much when he predicted that his name would one day be associated with "a crisis the like of which had never before existed on earth."[12] The same apocalyptic mood would soon pervade the works of a whole generation of writers for whom the future of Western civilization *tout court* was suddenly fraught with the greatest uncertainty. The clarion call was sounded by Paul Valéry in a celebrated essay entitled "The Crisis of the Mind" (1919), which began with the ominous words, "We civilizations, we know now that we are mortal."[13] In the years that followed, Nietzsche's views gained widespread currency on the philosophical level through the work of his greatest "disciple," Martin Heidegger, and on a more popular plane through Spengler's *Decline of the West*, written almost entirely under Nietzsche's aegis and shot through with the same foreboding of an impending global catastrophe.

The rest of the story is still too fresh in our minds to bear retelling at this juncture. Suffice it to say that recent developments, while less dramatic than those of the first half of the century, do not appear to have altered the situation appreciably or done much to alleviate its pathos. This does not mean that everyone is equally sensitive to the mortal danger so clearly perceived for the first time by Nietzsche. It could mean, however, that those who are not have simply chosen to fiddle while Rome burns without any real awareness of the fact that it is burning.

CHRISTIAN FAITH AND MODERNITY

Thus far I have spoken exclusively of the rapport between modern thought and classical thought. It is obvious that one cannot deal adequately with the problem of modernity without raising the issue of its relation to the other great fountainhead of the premodern tradition, biblical religion. Just as modernity, as I have defined it, implies a rejection of classical philosophy, so it has come to be identified with a rejection of revealed religion and, at the limit, of any form of transcendence. By the end of the eighteenth century, this fact, long acknowledged by a small group of perspicacious thinkers, had become a matter of common concern among theologians. From this moment, unsurprisingly, date the first significant attempts to bridge the widening gap between Christianity and secular thought.

The distinctive feature of the theology that makes its appearance at this time is precisely its openness to modern thought and its willingness to be guided by it in its investigations. Its first great exponent was Schleiermacher, the title of whose book, *On Religion: Speeches to Its Cultured Despisers*, conveys as well as anything else the author's main preoccupation.[14] Schleiermacher's work inaugurated a trend that has persisted in one form or another down to our time and that has yielded in turn a romantic, an idealist, a positivist, and an existentialist theology, to mention only a few of its better known avatars. Its most illustrious and by far its most numerous representatives were Protestants, but they had their opposite numbers on the Catholic side in such scholars as Moehler, Sailer, Hirscher, Scheeben, and Passaglia, whose pioneering work has lately begun to attract a good deal of attention.

The new theology enjoyed its euphoric moments, particularly during its heyday in the nineteenth century, but it has also experienced its stations of the cross and silent agonies. For one thing, its efforts to come to terms with the modern world appear to have left the intellectual giants of the period (Goethe, Hegel, Nietzsche) singularly unimpressed. For another, it did not always escape the danger to which it exposed itself by engaging in debate with secular thought. In its desire to be timely, it often ceased to be timeless; in its quest for relevancy, it tended to lose sight of its traditional substance. Nineteenth-century evangelical theology is thought to have reached its peak in 1900 with the publication of Harnack's *What is Christianity?* It died fifteen years later when its new leader, Ernst Troeltsch, gave up his chair of theology at the University of Heidelberg for a chair in philosophy at Berlin. What had begun as a generous attempt to win back a prodigal and long-lost child had seemingly ended in an all but complete sell-out of Christianity to the modern world. In the opinion of many, a similar fate befell certain segments of the Roman Catholic community at the time of the Modernist crisis toward the beginning of this century.

The confrontation between Christianity and the modern world undertaken by Schleiermacher and pursued by his followers represents only one response to the challenge of modernity in the nineteenth century. Throughout the same period one finds, within Roman Catholicism particularly, a second group of theologians who tried to deal with the problem of modernity, not by settling their accounts with it, but by ignoring it in favor of a return to medieval thought in its pristine or premodern integrity. From its obscure beginnings circa 1810, the movement grew to sizable proportions and gradually took the form of a massive counteroffensive against the encroachments of modern thought. It received a new and powerful impetus when it was officially approved by the Roman authorities with the publication of Pope Leo XIII's encyclical, *Aeterni Patris* (1879), which established Thomas Aquinas as the preeminent authority in matters pertaining to Catholic philosophic and theological thought.

Unlike their liberal counterparts, the new conservatives were more concerned with preserving a threatened heritage than with accommodating themselves to the spirit of an estranged and, in their judgment, hopelessly devious modern mind. On the surface, they seemed for a time to fare better than almost everyone else. Their singleness of purpose, coupled with the fact that the movement was largely confined to the Roman Catholic world, gave it an inner strength that liberalism by its very nature was incapable of generating. But it also had its blind spots, which became more apparent as time went on. It never aroused more than a passing interest on the part of outsiders and failed to muster universal support within the Roman Catholic fold itself. Nor, with a few notable exceptions, did it manage to produce scholars of the very first rank. Worse still, the movement had inherited from its romantic origins a touch of archaism that it never completely shook off. By extolling tradition as a prescriptive right, it tended to breed in the minds of its followers a prejudice in favor of the past that is absent from and alien to the thought of the past itself. Its longing for the "reason" of the Middle Ages over against the "unreason" of the modern period appears to have been motivated by a tacit traditionalism that precluded any fresh insight into the nature of premodern thought. The result was an impoverished scholasticism that is typical of much of the nostalgic Christianity of the nineteenth and twentieth centuries and that further served to discredit medieval thought in the eyes of its detractors.

Despite momentary and sometimes impressive triumphs, neither the conservatives nor the liberals managed to provide a generally acceptable solution to the problems posed by the rise of modern secularism. The results were not all negative, however. The very impasse at which both modern conservatism and modern liberalism had arrived pointed to the need for a fresh start as well as to the direction in which such a start could be made. It thus paved the way for a new approach that might eventually lead to a better understanding of certain basic problems, if not to greater certainty about the solutions to them.

Modernity had sprung up as a revolt against the past and retained its revolutionary character throughout the centuries that followed. Its outstanding feature had been and remained its antitraditionalism, an antitraditionalism revealed by a constant readiness to question everything, old and new, ancient and modern. The only thing it took for granted is modernity itself toward which it adopted an uncritical stance that proved to be its Achilles' heel. As late as the eighteenth century, modernity still called for a decision on the part of its devotees. The latter-day descendants of the great modern thinkers no longer saw the need for such a decision. By a strange irony, the fight against tradition had degenerated into a tradition that was easily as tyrannical as the traditionalism against which it had begun by pitting itself. Like the restored scholasticism of the nineteenth century, it stood convicted by it own standards.

The paradoxical situation in which modernity suddenly found itself called for a reassessment of the modern enterprise in all its breadth and scope. It pointed to the necessity of a dialectical return to the moment at which modern thought had deviated from premodern thought. Only by reopening, in the light of the experience that has since accrued to us, the long forgotten but unresolved quarrel between the Ancients and the Moderns, and by seriously reexamining the alternatives with which it confronts us, was it possible to achieve any clarity about our own aims and purposes, along with the conditions under which they might suitably be carried out.

NOTES

1. R. W. and A. J. Carlyle, *A History of Mediaeval Political Theory in the West* vol. I (New York: Barnes and Noble, and London: Blackwood, 1950), 2. This is a six volume history.

2. Machiavelli, *The Prince*, chapter 6.

3. Francis Bacon, *Plan of The Great Instauration*, ca. init.

4. Francis Bacon, *The Advancement of Learning* II.23.49.

5. Cf. Machiavelli, *The Prince*, chapter 15; *Discourses*, book I, introd.

6. Cf. *The Prince*, chapter 25.

7. Cf. Bernard Mandeville, *The Fable of the Bees; or, Private Vices, Publick Benefits* (New York: Capricorn Books, 1962).

8. Jonathan Swift, *The Battle of the Books*, in *Gulliver's Travels and Other Writings* (New York: Modern Library, n.d.), 370-71.

9. J.-J. Rousseau, *The First and Second Discourses*, edited by Roger D. Masters (New York: St. Martin's Press, 1964).

10. Cf. Plato *Laws* V.739a.

11. Immanuel Kant, *Perpetual Peace*, First Supplement.

12. Friedrich Nietzsche, *Ecce Homo* IV.1.

13. Paul Valéry, "La crise de l'esprit" in *Oeuvres*, vol. I (Paris: Gallimard, 1957), 988-1014.

14. Friedrich Schleiermacher, *On Religion: Speeches to Its Cultured Despisers*, introduction, translation, and notes by Richard Crouter (Cambridge, England and New York: Cambridge University Press, 1988).

NATURAL LAW

The expression "natural law" is often loosely applied to any theory that holds to objective standards of morality. In its proper sense, it designates a moral law that both exists by nature and is known by nature to be binding on everyone. Thus understood, the natural law is distinguished from the civil law, whose statutes are enacted and enforced by some human legislator, and from the divinely revealed law, the knowledge of which is communicated to us through the sacred scriptures. It is also distinguished from the Kantian moral law, which presents itself as a law of "reason" rather than a law of "nature."

The natural law was for centuries the cornerstone of Western ethical and political thought. Among other things, it played a crucial role in the emergence of modern liberalism and inspired some of its most famous documents, including the Declaration of Independence (1776) and the different versions of the Declaration of the Rights of Man and the Citizen (1789, 1793, 1795). Today it is for the most part an object of historical study rather than a source of authoritative moral judgment, a recollected rather than a living doctrine. Although efforts to revive it have not been lacking, nothing indicates that it is about to regain the position of prominence it formerly occupied in our tradition. Even Roman Catholicism, where until the 1960s it continued to hold sway, appeals to it less frequently in its official documents. Pope John Paul II's recent use of natural law in his encyclical *Veritatis Splendor* might very well indicate the beginning of a revival within the Catholic Church.

ANCIENT AND MEDIEVAL NATURAL LAW

It should be noted that the doctrine informing our modern constitutions is the dynamic and reformist natural law doctrine of the Enlightenment period. The original natural law theory was considerably more conservative. It comes to sight as a modification of the natural "right" teaching of Plato and Aristotle, from which it differs most conspicuously by the fact that it not only points to what is intrinsically right or wrong but commands the performance of the one and the avoidance of the other under pain of sanction.

Modern scholars usually attribute its invention to Stoicism. However, except for a single unconfirmed report by Cicero (*De natura deorum* I.36), there is no evidence that the Old Stoics ever used the expression "natural law." Stoicism nevertheless furnished one of the key premises of that theory, namely, the notion of a providential God who guarantees the moral consistency of the universe by seeing to it that the just are always rewarded and the wicked punished, if not in this life at least in the next. The mode and severity of the punishments in question are not specified by the natural law itself. They are left to the determination of the civil law and may vary according to circumstances. Capital punishment, for example, is neither imposed nor forbidden by the natural law.

The earliest extant accounts of the natural law date back to the first century B.C.E. and are to be found in Cicero's *Republic* III.32-41, and *Laws* I.18-32. In the *Republic*, the jurist Laelius, one of the dialogue's main characters, uses it to defend the justice of Rome's conquest of the civilized world. It is doubtful whether Cicero, an academic skeptic and a critic of the Stoic notion of divine providence, personally subscribed to Laelius's theory, which he seems to have endorsed less as a true doctrine than as a rhetorical tool with which to curb the excesses of Roman imperialism.

From Cicero, the natural law was absorbed into the Roman legal tradition and, soon thereafter, into the theological tradition of the Christian West, of which it became a standard feature. St. Augustine (354-430) invoked it not only to defend his own just war theory[2] but also to absolve God of the blame that he would have incurred for the sins of his creatures had he left them invincibly ignorant of the basic principles of moral behavior.[3] For Augustine, compliance with the natural law, which requires the subordination of the lower to the higher both within the individual and in society at large, is synonymous with the whole of human perfection.

The premodern natural law doctrine achieved its classic expression in the works of St. Thomas Aquinas (1225-1274), who provided its most precise formulation with the aid of the newly rediscovered Aristotelian philosophy, albeit not always in complete agreement with it.[4] Whereas Aristotle teaches that all of natural right is variable[5] Thomas distinguishes between the primary principles of the natural law, which are invariable, and its secondary principles, which are subject to change. The primary principles are regarded as self-evident and form the object of an intellectual virtue called *synderesis*, a term that Thomas inherits from his Christian predecessors. They are intimated to human beings through the natural inclinations by which they are directed to the most general ends of human existence, the highest of which are the knowledge of the truth and life in society. Nowhere are we given an exhaustive list of these principles, although Thomas does say that the moral law of the Old Testament, which is summed up in the Ten Commandments, belongs to the natural law.

Thomas's natural law teaching met with a good deal of resistance on the part of other medieval philosophers and theologians. Marsilius of Padua rejected it on Aristotelian grounds. Duns Scotus and William of Ockham objected to it on the ground that, by binding God himself to the precepts of the natural law, it conflicted with the biblical notions of divine freedom and omnipotence. Accordingly, Scotus reduced the natural law to a single negative precept, the one that prohibits the hatred of God. Ockham went even further, asserting that God could command us to hate him if he so desired. The sixteenth and seventeenth centuries witnessed a return to the Thomistic teaching, spearheaded by a number of influential commentators among whom the names of Francisco de Vitoria (1483-1546) and Francisco Suarez (1548-1617) stand out, along with that of Hugo Grotius (1583-1645). The motivation was furnished by the wars of religion that were ravaging Europe at the time and by the cruelties inflicted on the American Indians by the Spanish conquistadors. The rules of just warfare were spelled out in ever greater detail and the old Roman notion of *ius gentium*, or "law of nations," was transformed into what became known as international law.

MODERN NATURAL LAW

Of the utmost significance for the development of modern democratic theory is the appearance of the natural law teaching propounded by the seventeenth-century founders of modern liberalism, especially Hobbes and Locke, who broke decisively with the previous tradition and sought to establish the whole of political thought on a new and supposedly more solid foundation. Human beings were no longer said to be naturally political and social. They are solitary individuals who once existed in a prepolitical "state of nature" and are actuated, not by a desire for some preexisting end or ends in the attainment of which they achieve their perfection, but by a premoral passion, the desire for self-preservation, from which arises the "right" of self-preservation. Individual rights, conferred by nature, replace duties as the primordial moral phenomenon. Civil society is not something natural and desirable for its own sake. It comes into being by way of a contract made by human beings who freely "enter" into it for no other reason than to escape from the dangers that threaten them in the state of nature. As for the natural law, it has nothing to do with the self-evident principles of which the medievals spoke, but is identified with the sum of the conclusions that human reason arrives at by a process of deduction from the right of self-preservation. In *Leviathan*, chapters fourteen and fifteen, Hobbes lists nineteen such principles, all of them calculated to insure the individual's security and physical well-being.[6]

Unlike the old natural law doctrine, which is not essentially egalitarian and whose compatibility with any decent regime, whether democratic or aristocratic, was never doubted, the new doctrine holds that there is but one just and legitimate regime, namely, liberal democracy or popular sovereignty, in the name of which authoritarian or nondemocratic regimes may rightfully be overthrown, as in fact they were in America and France during the last quarter of the eighteenth century.

The last and the most radical of the modern natural law theorists is Jean-Jacques Rousseau, who accused the Hobbesian and Lockean scheme of fostering a bourgeois mentality that undermines civic virtue and therewith the freedom and equality for the protection of which civil society is established. The solution to the problem, Rousseau thought, lay in the formation of small societies in whose life all citizens would participate on an equal basis in accordance with the principle of "one man, one vote."

THE DEMISE OF THE NATURAL LAW

Rousseau's search for the true state of nature, which, he argued, neither Hobbes nor Locke had reached for the simple reason that they took as their standard a human nature that was already corrupted by society, led him backward in time to a stage that antedates not only the formation of civil society but the emergence of rationality and, indeed, of humanity itself. But if the state of nature is not a properly human state, one fails to see how it could serve as a reliable guide to human behavior. The critical next step was taken by Kant, who abandoned nature as the touchstone of the moral rightness of one's actions and replaced it with the categorical imperative, or the laws of universal reason.

Further challenges to the natural law theory came from two of the most powerful intellectual forces of modern times, historicism or historical relativism and social science positivism. The former originated as a reaction against the atrocities spawned by the French Revolution and tried to forestall the possibility of such revolutions in the future by rejecting the notion of a permanently valid "natural" or "higher" law to which a direct appeal can be made from existing civil laws. Social science positivism for its part denies scientific status to any proposition that is not empirically verifiable and relegates all natural law principles, none of which is subject to this kind of empirical verification, to the realm of subjective "value judgments."

This is not to say that the natural law has simply vanished from our midst. The Neo-Thomistic movement of the nineteenth and twentieth centuries did much to keep its memory alive, even though the literature that it produced is more notable for its abundance than for its originality. Other scholars, such as R. W. and A. J. Carlyle and Edward S. Corwin, have argued for the essential and unbroken continuity of the natural law tradition from its inception in pre-Christian antiquity to the end of the eighteenth century. Jacques Maritain and John Finnis have since offered revised versions of both the medieval natural law doctrine and the initially antithetical modern rights doctrine in a valiant attempt to bring the two of them into line with each other. The impact of these efforts has thus far been somewhat limited. Rightly or wrongly, the majority of our contemporaries remain suspicious of the natural law, finding it either too vague to be of any real use, or potentially subversive, or incompatible with the flexibility required for the proper conduct of the affairs of state.

Yet experience has shown that dispensing with it altogether is not an easy matter, for without it one is deprived of a valuable moral argument against the blatantly unjust laws under which people are often made to live. Some Germans resorted to it in the 1940s to justify their resistance to the Nazi regime and the attempt on Hitler's life. The same argument was pressed into service on the occasion of the Nuremberg International War-Crimes Trials (1945-46), the legality of which could not be established on the basis of existing national and international laws. Martin Luther King referred to it in his famous "Letter from the Birmingham City Jail," citing Thomas Aquinas in support of the view that a human law not rooted in eternal and natural law is an unjust law. Similar problems arise when a country, either alone or with the help of others, intervenes in the domestic affairs of another country in order to put a stop to what are now called crimes against humanity. Finally, there is a serious question as to whether, in upholding or striking down civil laws, judges can consistently avoid becoming involved in some kind of implicit natural law reasoning.

NOTES

1. A. J. Carlyle and R. W. Carlyle, *A History of Mediaeval Political Theory in the West*, 6 vols., 2d edition (New York: Barnes and Noble, and London: Blackwood, 1950). Edward S. Corwin, *The "Higher Law" Background of American Constitutional Law* (Ithaca: Cornell University Press, 1955). Michael B. Crowe, *The Changing Profile of the Natural Law* (The Hague: Martinus Nijhoff, 1977). John Finnis, *Natural Law and Natural Rights* (Oxford: Clarendon Press, 1980). Harry V. Jaffa, *Thomism and Aristotelianism: A Study of the Commentary by Thomas Aquinas on the Nicomachean Ethics* (Chicago: University of Chicago Press, 1952). Jacques Maritain, *The Rights of Man and Natural Law* (New York: Charles Scribner's Sons, 1943). Arthur Nussbaum, *A Concise History of the Law of Nations* (New York: Macmillan [1947], 1954). Heinrich A. Rommen, *The Natural Law: A Study in Legal and Social History and Philosophy* (St. Louis and London: Herder Book Co., 1947). Paul E. Sigmund, *Natural Law in Political Thought* (Cambridge, MA: Winthrop Publishers, 1971). L. Strauss, *Natural Right and History*, 2d edition (Chicago: University of Chicago Press, 1965). L. Strauss, "Natural Law," *International Encyclopedia of the Social Sciences* (New York: Macmillan, 1968), vol. XI, 80-85.

2. Augustine, *Contra Faustum Manichaeum* (*Against Faustus, the Manichean*) 22.72-79.

3. Augustine, *De libero arbitrio* (*On Free Choice of the Will*) book I.

4. Cf. esp. Thomas Aquinas, *Summa theologiae* I-II, qu. 94.

5. Aristotle, *Nicomachean Ethics* V.7.

6. Thomas Hobbes, *Leviathan*, chapters 14 and 15.

14

TROELTSCH AND CHRISTENDOM

It is obviously impossible in a few short pages to do full justice to the richness and subtlety of Ernst Troeltsch's treatment of "medieval Catholicism" in the work for which he is best known, *The Social Teachings of the Christian Churches*.[1] I shall recall in a few words the salient points of that treatment for the benefit of those whose memories need to be refreshed and then make some comments of a more or less general nature about them. In so doing the only advantage I can rightfully claim, and it is at best a dubious one, is that I am not intimately acquainted with all of Troeltsch's works and can therefore approach them with a certain protective naiveté which, by reason of their competence, more experienced Troeltschean scholars presumably no longer possess to the same high degree.

I

Medieval Catholicism represents for Troeltsch the second great period in the social development of the gospel. The interest that it holds for us derives from its being one of only two instances in the entire history of Christianity in which Christian social philosophy managed to attain comprehensive historical significance and influence. Generally speaking, it is characterized by a deeper insertion of Christian ideas into the fabric of society and hence a more complete integration of the realms of the sacred and the profane than had hitherto been possible. The early Church was faced with the necessity of coming to terms with, and accommodating itself to, an already well established social order that it could not hope to transform substantively and whose way of life often conflicted openly with the gospel's ethic of radical individualism and universal love. The necessary "compromise" was effected chiefly through the annexation of what Troeltsch refers to as the Stoic doctrine of absolute and relative natural law. The state was tolerated on the ground that it was willed by God as a punishment and a remedy for sin and could thus be thought to be in conformity with the prescriptions of relative natural law. The Middle Ages, on the other hand, witnessed the emergence of a new situation that made it possible to speak of a Christian unity of civilization, at least as an ideal. This is particularly true of the West, where, more perhaps by necessity than by inner compulsion, the Church was involved from the outset as a partner in the building up of a new European civilization. The same is not true of the East, where a position of stable unity was achieved through the simple interaction of the Church with the preexisting state that had succeeded in maintaining itself without interruption since the fourth century. Having said this much, Troeltsch will focus all of his attention on the Western Middle Ages.

The Christian civilization of the West is the successor of the early Church, whose problems and tensions it inherits and to which it adds a few of its own. The Stoic-Christian conception of absolute and relative natural law is preserved, but to it is now joined the dichotomy between nature and supernature. No explanation is given as to the origin of this important innovation other than that, by reason of the greater maturity and special features of medieval politico-social institutions, the adaptation of the natural law doctrine was able to go much further than it had in the early Church. Between the Church and these institutions, to the elaboration of which the Church itself had contributed in no small measure and which reflect the needs of an essentially town-centered civilization, some kind of preestablished harmony existed. The result was a more positive valuation of the law of nature and its institutions, as opposed to the mere toleration of the early Christian period.

The Christian sociological theory that gradually evolved out of the mass of complicated data associated with this phenomenon is analyzed in terms of two complementary notions, "organism" and "patriarchalism," both of which find their embryonic formulations in Paul and both of which have their common root in the idea of love.[2] "Organism" is Troeltsch's word for the hierarchical structure of medieval society, wherein each group (priests, monks, laymen) has its honor and purpose and accepts the function assigned to it for the good of the whole. To this basic ingredient is joined the "patriarchalism of love," which fosters voluntary relationships of submission and authority, and which is all the more necessary as the whole system obviously contains large elements of compulsion, non-liberty, male domination, and other such inequities. Viewed from this vantage point, medieval society appears as a great household in which familial virtues ethically hallow and glorify all the infinitely varied relationships of humanity. The patriarchalism of love also exerts a conservative and stabilizing effect on society and this serves to counteract the revolutionary tendencies implicit in the natural law theory that lies at the heart of the organic conception. This conservatism is reinforced by the theory of the divine right of kings, as it is said to be expounded, for example, in Aquinas's treatise *On Kingship.*

The new ethic likewise introduces into the sociological temper of the times the notion of "calling," which the early Church had ignored and which, while still not entirely free of irrational elements, again favored a positive incorporation of Christianity into the life of the world.[3]

For all its remarkable advances, however, the Christian social doctrine of the Middle Ages suffered from one crucial defect: like the early Church, albeit for a different reason, it never developed a program of social reform. Nor did it feel any need for such a program, except in the very restricted sense of the Church's demand for a return to the natural law and for certain mandatory safeguards against the encroachments of the temporal power. As Troeltsch puts it succinctly, "To the early Church, social reform was too difficult; to the medieval Church, it seemed superfluous."[4]

Thus, upon closer analysis, it becomes apparent that the inconsistencies of early Christianity had not been disposed of but simply maintained at a deeper level.[5] The medieval Christian unity of civilization did not, in effect, constitute an all-out reform in accordance with Christian principles. It merely reflected the acceptance of relatively favorable conditions, which, at the time, were not regarded as a fortunate accident but rather as a logically necessary result of nature.[6] When all is said and done, we are still far from a genuinely Christian polity, one that is based on and directly molded by Christian principles.[7] The radical individualism and universal love of the gospel had no doubt made considerable inroads into society, but without completely permeating its structures.

II

Such in barest outline is what I take to be the gist of Troeltsch's argument. My own remarks, which are not necessarily intended as criticisms but as queries and demands for greater clarity, fall under two general headings. There is, first of all, the question of the historical accuracy of Troeltsch's interpretation, and, secondly, the larger question of the historical method itself, particularly as regards its application to the study of the Middle Ages. I begin with the question of historical accuracy, stressing only those points which appear to have a direct bearing on Troeltsch's assessment of the period at hand.

1. The first point concerns the distinction between absolute and relative natural law, which Troeltsch repeatedly ascribes to the Stoics and, later, to the ancient and medieval Christian writers.[8] It is no exaggeration to say that the whole issue of natural law is one of the mysteries of the early Christian period. Little is known with any degree of certitude about the origin of that doctrine, inasmuch as we have access to it only through later reports, especially those of Cicero, who was not a Stoic but an academic skeptic. The irony is that Cicero, the greatest exponent of the pre-Christian natural law theory, was not himself a natural law theorist. The best information available suggests that the doctrine in its original form was a purely theoretical view, whose practical applications were never explored by the early Stoics themselves,[9] but which was either deliberately or mistakenly construed as a practical doctrine by the Christian writers. In any event, there is to my knowledge no trace whatever of the distinction between absolute and relative natural law among the Stoics, nor for that matter among the Christian writers of antiquity or the early medieval period. In that precise form, the distinction appears to be an invention of the late Middle Ages, from which it was transmitted to the writers of the Reformation and post-Reformation periods.[10] The Stoic position, to the extent to which it can be reconstructed, is that there is one and only one natural law, which is not directly applicable to society and must therefore be diluted in order to become operative. Needless to say, a crucial difference separates a natural right or a natural law that is diluted and the notion of a relative or conditional natural right or natural law. If the principles of civil society are diluted natural right, they are much less venerable than if they are regarded as relative or secondary natural right, that is, as divinely established and entailing an absolute duty for fallen man. It is significant that Augustine, for example, never speaks of civil society as it now exists as something natural. By imputing to the Christian writers a doctrine of relative natural law, one makes them sound more conservative than they actually are; and, God knows, they are conservative enough as it is!

2. Similar remarks apply to the theory of the divine right of kings,[11] which makes its appearance only toward the middle of the sixteenth century and which, properly understood, refers to the indefeasible right that attaches to the *person* of the king by virtue of his succession, rather than to the kingly *office*. One finds it clearly expressed, among other

places, in Shakespeare's *Richard II*:

> Not all the water in the rough rude sea
> Can wash the balm off from an anointed king.
> The breath of worldly men cannot depose
> The deputy elected by the Lord.
> For every man that Bolingbroke hath press'd
> To lift shrewd steel against our golden crown,
> God for his Richard hath in heavenly pay
> A glorious angel. Then, if angels fight,
> Weak men must fall; for heaven still guards the right.[12]

Whether King Richard himself believed in the doctrine of the divine right of kings is doubtful; for the "angel" that God is alleged to have placed at his side seems to be nothing other than the British gold coin stamped in the effigy of Michael the Archangel and representing the money that Richard needed for his campaigns. But the doctrine itself is clear. What is equally clear is that no such doctrine is to be found among the early Christian or medieval writers. The majority of these writers were not opposed to sedition or even tyrannicide as a matter of absolute principle. But neither, as a rule, did they think it wise to engage in a public defense of their position. They took it for granted that true conspirators know what they have to do and do not need to be instructed in the exercise of their necessary but unpleasant duties. As a consequence, their books have about them a Jane Austenian quality to which Troeltsch is not especially sensitive.

3. In this connection, one could also quarrel with the recourse to the notion of patriarchalism, which smacks of Filmer's *Patriarcha*, more than it does of Aquinas's *Summa theologiae*, and which conjures up ideas usually associated with a state of affairs quite different from the one that prevailed in the Middle Ages.

4. The same flattening of the perspective is visible in the attribution of modern social contract notions to some of the earlier Christian writers.[13] Troeltsch again appears to be reading back into history ideas of more recent vintage that make complete sense only in the context of the emergent natural rights theories of the seventeenth century. The Church Fathers in particular were not contractualists. Most of them do not raise the question at all, and the few who do inevitably reject contractualism or political conventionalism in its classical (specifically Epicurean) form.

5. Troeltsch likewise blurs the issue when he speaks of the theocratic ideas of Augustine and Thomas.[14] Neither Augustine nor Thomas was a theocrat. Each opposed theocracy in his own way and saw the alliance between Christian wisdom and political power as at best a fragile one—a

kind of marriage of reason or perhaps a shotgun wedding. Even when, as sometimes happens, the two attributes coexist in the person of the Christian ruler, they remain distinct, cooperating with each other but never merging one into the other. If such is the case, Troeltsch would again be guilty of trying to pin a tail on a vanishing donkey, to use his own image. This may be the occasion to note in passing that, while Troeltsch pays considerable lip-service to Aquinas, Aquinas himself hardly comes to sight anywhere in the long sections devoted *ex professo* to an exposé of Thomistic principles.

Related to the general issue of theocracy is the vexing question of the appeal to the state or the so-called secular arm to coerce heretics. Recent studies have shown that, contrary to what has often been asserted by Troeltsch and others, Augustine appealed to princes as *Christians* and not as princes whenever it became apparent that matters of vital interest to society as a whole could only be settled by force. It is worth noting that, even under such extreme conditions, the Church itself does not do the coercing. It usually calls upon laymen to do its dirty work and engage in acts of "pious cruelty," as its enemies have sometimes labelled them.

6.   Finally, one could argue that Troeltsch has seriously misinterpreted the distinction between medieval and modern natural law. The transition from one to the other is viewed as a more or less continuous development in the course of which the latent implications of the primitive natural law theory are worked out and gradually brought to light.[15] One discovers here the basic reason for which Troeltsch feels justified in reading back into history certain doctrines that in their final form clearly belong to a later period. Closely linked to this view is a statement occurring at the end of Section 8 to the effect that the permanent attainment of individualism was due to religious rather than secular influences.[16] In this regard, Troeltsch, like his friend Max Weber, emphasizes the importance of the Reformation for the whole of the modern development, but only at the expense of the great secular thinkers of the same period. While such a position has a certain tactical advantage from Troeltsch's point of view, insofar as it makes it easier for him to bridge the gulf between Christianity and the modern world, it nevertheless rests on an inadequate analysis of the origins of modern thought. On the basis of the evidence at hand, one might be more inclined to view the new natural law theory of Hobbes and his followers as a radical transformation rather than a simple extension of the classic Christian theory.

In fairness, it should be pointed out that Troeltsch's views on these and allied matters did not originate with him. In most cases they can be traced back to contemporary sources, and in particular to A. J. and R. W. Carlyle's *History of Mediaeval Political Theory in the West*.[17] Interestingly enough, Troeltsch admits elsewhere that he may have been too strongly influenced by the brothers Carlyle on certain points. The question, then, is why a scholar of Troeltsch's stature should have relied so much on secondary sources at the risk of being misled by them. To answer that question, one must take a closer look at Troeltsch's method, the so-called "historical" method.

Far from being arbitrary, Troeltsch's neglect of primary sources is a direct consequence of his espousal of the historical method as he understood it. That method is predicated on the premise that human thought is at the deepest level historically conditioned, and hence, that information derived from external sources (institutions, contemporary intellectual climate, social environment, and the like) is more important for the understanding of an author than the information supplied by the author's own works. Once this is admitted, however, one fails to see how Troeltsch is able to exempt himself from his own verdict. Should not the historical principle be applied to itself? More clearly, is not the principle that all human thought is historically conditioned itself historically conditioned and hence relative?

That precise difficulty is faced in an earlier essay on "Historical and Dogmatic Method in Theology," in which Troeltsch tries to steer clear of the relativism that threatens his position by means of a partial return to a nonhistorical idealist premise.[18] Whether or not the maneuver is successful remains doubtful. As Troeltsch states elsewhere in the essay, "Give the historical method an inch and it will take a mile."[19] Yet he himself was apparently unwilling to go the full distance. Quite apart from that, a grave doubt subsists as to whether the systematic application of the historical method does not rule out a priori any possibility of understanding the older authors as they understood themselves, and hence of any genuine recovery of the thought of the past, thereby closing off the one avenue by which we might be able to free ourselves from the most powerful prejudices of our own age.

Troeltsch's reply to this objection is that all interpretations of the past are at best only plausible. Any value that they may have lies less in their demonstrable historical accuracy than in the divinatory insight they provide into the nonrepeatable character of any previous historical synthesis, and hence, into our own inescapably historical situation.[20] At the deepest level, they are motivated by the imperative demand for a new Christian ethic and are thus intended first and foremost as a call to creative action in the present rather than a form of petrified and sterile contemplation of the past.

If that is indeed the case, Troeltsch cannot reasonably be faulted for failing to do something that he had no intention of doing in the first place and that he did not believe possible anyway. To criticize him on the score of historical accuracy would again be trying to pin a tail on a vanishing donkey. I am hoping that this is so, for I might then be more easily forgiven for offering what could never be more than a subjective interpretation of Troeltsch's subjective interpretation of the Middle Ages. There is some consolation, if not much glory, in thinking that one has merely followed a noble precedent.

NOTES

1. Ernst Troeltsch, *The Social Teachings of the Christian Churches*, Torchbook edition in two volumes (New York and Evanston: Harper and Row, 1960).—This hitherto unpublished paper was written in the fall of 1971 for the bimonthly meeting of the Andover Newton-Boston College Joint Doctoral Colloquium. I am indebted to Dr. James Luther Adams, who conducted the Colloquium that year, for much of my information concerning Troeltsch. My understanding of Troeltsch may, however, differ from his in a number of unspecified ways. He is therefore not to be held responsible for any errors of interpretation on my part.

2. *Social Teachings*, 284f.

3. Ibid., 293f.

4. Ibid., 303.

5. Ibid., cf. 270.

6. Ibid., cf. 306.

7. Ibid., cf. 309.

8. Ibid., 199, 153f, 201, et passim. See also *Gesammelte Schriften*, vol. IV, 175: "Sie bildete daher die Unterscheidung eines absoluten Naturgesetzes und eines relativen Naturgesetzes aus." Ibid., 176: "Dann aber konnte die Christen auch die stoische Lehre vom relativen Naturgesetz sich aneignen."

9. Cicero, *Laws* III.6.14: "For though the older Stoics also discussed the State, and with keen insight, their discussions were purely theoretical and not intended, as mine is, to be useful to nations and citizens. The other school led by Plato provides most of our present material."

10. Cf. Ockham, *Dialogus*, Part III, Treatise ii, book III, chapter 6. Hooker, *The Laws of Ecclesiastical Polity* I.10.13.

11. Cf. *Social Teachings*, 156, 219, 291, 298, et passim.

12. Shakespeare, *Richard II*, act III, scene 2, lines 55-62.

13. Cf. *Social Teachings*, 155.

14. Ibid., 229f.

15. Cf. also *Gesammelte Schriften*, vol. IV, 188.

16. *Social Teachings*, 328.

17. A. J. Carlyle and R. W. Carlyle, *A History of Mediaeval Political Theory in the West*, 6 volumes (New York: Barnes and Noble, and London: Blackwood, 1950).

18. E. Troeltsch, "Historical and Dogmatic Method in Theology," in *Religion in History*, Essays translated by J. L. Adams and W. F. Bense (Minneapolis: Fortress Press, 1991), 11-32.

19. Ibid., 16.

20. *Social Teachings*, 1003-4 and 207, quoting Ranke's "deep and suggestive" statement that every historical epoch is "immediate to God." The meaning is that no historical epoch can claim superiority over any other epoch. The expression "immediate to God" contains an implicit reference to German constitutional law, according to which some local rulers were under the direct authority of the Emperor rather than that of some prince.

15

GADAMER ON STRAUSS: AN INTERVIEW

The following interview was held at Boston College on December 11, 1981. It was conducted and subsequently edited by Prof. Ernest L. Fortin. Prof. Gadamer has read and approved the edited version. The recording was done by Prof. Betty T. Rahv and Mr. John Walters. Prof. Frederick G. Lawrence provided valuable editorial assistance.

FORTIN: There are many philosophers and political theorists in this country who would like to know more about your lifelong relationship with the late Leo Strauss. Perhaps you could begin by describing the atmosphere at the School of Marburg in the early 1920s. That was obviously an exciting period, possibly the most exciting period in twentieth-century intellectual history. Was there a sense of that excitement among the students?

GADAMER: We were living in an age of great political change. Everyone was aware of the impact of the new parliamentary democracy in a country that was not prepared for it. The general feeling was one of disorientation. One day—I was only a youngster then—a number of us got together and asked: "What should we do?," "How can the world be reconstructed?" The answers were very different. Some thought we ought to follow Max Weber; others, Otto von Gierke; others still, Rabindranath Tagore, who was the most popular poet in Germany immediately after World War I, thanks to some moving translations of his plays. (He was a good friend of Paul Natorp and occasionally came to Germany. I saw him once: an enormous figure with the face of a prophet. Fantastic! Natorp himself was a giant in the guise of a dwarf.) These concerns were shared by the young Leo Strauss as well. He, too, was looking around in search of some orientation. He had studied under Cassirer at Hamburg but had little sympathy for his political views.

FORTIN: When did you first meet Strauss?

GADAMER: In 1920 or thereabouts. He himself never studied at Marburg, but his home town (Kirchhain) was only a few miles away and he sometimes used our library, of which I was the so-called "administrator," that is to say, the person in charge of procuring the books requested by students. Our budget was not very large but the library was a good one. Those initial encounters still stand out in my memory. He was short and I was tall. I especially recall that little look of his: furtive, suspicious, ironic, and always slightly amused. We had a common friend, Jacob Klein, who alerted me to the fact that Strauss harbored certain misgivings about me. Not that I had anything against Jews—I doubt whether he ever thought that—but he must have sensed in me the typical arrogance of a young student who is proud of his success. He was probably right. After that I was very careful not to offend him, knowing how sensitive he was. We were on good terms and talked now and then but otherwise had few relations with each other.

   Our first real acquaintance came much later, in 1933, when I availed myself of the opportunity to travel abroad. Germany was undergoing another radical change and no one was allowed to take more than 300 marks with him. For me that was a small fortune and, to that extent, hardly a restriction. But it was nevertheless a warning. I was bright enough to see that before long we would not be allowed a single penny for such purposes. I went to Paris. Strauss was there on a Rockefeller grant and we spent a very pleasant ten days together. Among other things, he introduced me to Kojève and took me to a Jewish restaurant. We talked a good deal about the situation in Germany and the French reaction to it prior to Hitler's coming to power. One day we went to the movies. The newsreel contained a segment entitled, "German Nudism," which turned out to be a report on a recent athletic event. The "nudism" referred to was that of the athletes clad in sports attire! The event had the aspect of a military parade—as you know, we are masters of organization—and the participants looked a bit like robots. The French, who were still unaccustomed to these things, found it ludicrous that human beings should be so completely regimented. The whole theatre immediately burst into laughter.[1] All of this was totally new to me who, as a young teacher with no traveling allowance, had never been outside of Germany.

Afterwards we stayed in fairly regular contact. Strauss sent me his books. The one on Hobbes I found to be of particular interest since it was related to my own research on the political thought of the Sophists. That happened to be one of my great concerns at the time, although I was forced to abandon it when it became too dangerous to discuss political matters in Germany. One could not talk about the Sophists without alluding to Carl Schmitt, one of the leading theorists of the Nazi party. So I turned to more neutral subjects, such as Aristotle's *Physics*.

After the war, Strauss came to Germany and I invited him to give a lecture (at Heidelberg, in 1954). As I recall, he spoke on Socrates. Alexander Rustow, who attended the lecture, disagreed with what he said but was utterly captivated by his charm, his wit, and the elegance of his presentation. Rustow, then in his late sixties, was a man of considerable stature. He had been a pupil of Max Weber and had succeeded him in the chair at Heidelberg. He was a twentieth-century Voltairian of sorts, who wrote some fine books on industrial society but was also an excellent classical scholar.

Strauss and I spent the rest of the day together. My wife marveled at the way in which he kept coming back to the same problems, especially when we talked about Plato. Some of these problems recurred in our published correspondence.[2] They revealed the strange overlapping of our positions along with a number of important divergences. The main divergence had to do with the question of the Ancients and the Moderns: to what extent could this famous seventeenth-century quarrel be reopened in the twentieth century and was it still possible to side with the Ancients against the Moderns? I argued that this kind of debate was necessary, that it challenged the modern period to find its own evidence, but that the choice was not really an open one. I tried to convince Strauss that one could recognize the superiority of Plato and Aristotle without being committed to the view that their thought was immediately recoverable and that, even though we have to take seriously the challenge which they present to our own prejudices, we are never spared the hermeneutical effort of finding a bridge to them.

I forgot to mention that much earlier, in the late twenties, I wrote a paper on *phronesis* in Aristotle for my classics teacher, Paul

Friedländer.[3] Friedländer was a Platonist who did not have much use
for Aristotle. I was intrigued by the way Strauss handled the problem
of the tension between Plato and Aristotle but had never heard a real
answer to that question. So I sent him a copy of the article. He wrote
me a letter (destroyed during the war) in which he praised it but
objected to my using certain modern terms, such as "sedimentation,"
to elucidate Aristotle's thought. That was exactly the point on which
we disagreed. To go into the meaning of a text does not require us to
speak its language. One cannot speak the language of another epoch.
I later wrote a critical essay on this, inspired by Hans Rose's book,
Klassik als künstlerische Denkform des Abendlands (Munich, 1937).[4]
Rose was an art historian who consistently tried to avoid modern
terminology in describing the classics. This still did not prevent him
from entitling one of his chapters "Die Persönlichkeit" ("Personality"),
which is obviously not a classical word.

FORTIN: To come back to Marburg for a moment, who was the leader
of the school in the 1920s? Natorp?

GADAMER: Yes, he was. But, you know, for the younger generation
the leader is always the one who has not yet been discovered, and that
was not Natorp; it was Nicolai Hartmann, no question. For us, he was
the great attraction. Marburg also had an outstanding faculty of ro-
mance literature with Curtius, a good friend of mine, followed by Leo
Spitzer, Erich Auerbach, and Auerbach's successor, Werner Krauss—
four distinguished scholars. Curtius's predecessor had been Eduard
Wechssler, who later moved to Berlin.

FORTIN: What made Hartmann different from the others?

GADAMER: Under the influence of Scheler, he had begun to move
away from the transcendental idealism of Cohen and Natorp. He had
been a pupil of both and above all of Natorp, but he was especially
impressed by Cohen, our most shaman-like figure. When one reads
Cohen's books today, one finds them in a way empty. They are
written in a stern, fragmentary, and dictatorial style. There is hardly
any argumentation in them. But he had a strong personality. Strauss
also had a high regard for him. He died in 1918. We never met him.
The story that Strauss told me about him came from Franz
Rosenzweig. Rosenzweig visited Cohen in Marburg one day and
asked him how he could be so taken up with modern science and still
hold to the biblical doctrine of creation; at which point Cohen began
to hedge. As for Hartmann, he was a typical Baltic man with the
Russian student's habit of drinking tea from the late morning to the
following morning. He always worked well into the night. This
prompted Heidegger to remark jokingly that when Hartmann's light
went out, his went on. Heidegger, who gave his lectures at 7:00 a.m.,

started his day very early, rising at four or five o'clock, which was about the time Hartmann went to bed.

FORTIN: Strauss used to say that the atmosphere at Marburg was very provincial.

GADAMER: Yes, in the sense that we lived in an ivory tower, absorbed in philosophy and paying little attention to the rest of the world. That continued to be the case after Heidegger's arrival—a very exciting situation. But in those years Strauss was hardly ever in Marburg.

FORTIN: When did Heidegger first start teaching there and what did he lecture on?

GADAMER: In 1923. I do not recall the exact title of his first course, but it dealt with the origins of modern philosophy. He concentrated on Descartes and developed a series of twenty-three questions. Everything was very dramatic and well organized. Hartmann, who came to the first lecture *honoris causa*, told me afterwards that not since Cohen had he seen such a powerful teacher. Twenty-three questions, that was typical of Heidegger. I doubt whether he ever got beyond the fifth one. And then there was this peculiar radicalism of his, I mean the habit of radicalizing questions almost *ad infinitum*. Some of his followers are living caricatures of him, forever asking empty questions which, through being radicalized, lose all contact with their deeper roots.

FORTIN: What about the students and student life?

GADAMER: There were close relations between Marburg and Freiburg. Students went from one place to the other, as was the custom in Germany. There was an acute housing shortage after the war and the biggest problem was to find living accommodations. I changed universities only once, when I went to Munich, but only because one of my friends had offered me a room. Munich was not an important philosophic center. The dominant trend there was phenomenology, with Pfänder and Geiger. Heidelberg was well-known because of the shadow of Max Weber and the presence of Karl Jaspers and Karl Mannheim. Jaspers enjoyed an outstanding reputation as the leader of a seminar. His star was already high when I was a student. Hamburg, originally founded as a maritime institute, had only recently grown into a full university. The city, which was wealthy, poured a lot of money into it. It had Bruno Snell and Cassirer, the greatest scholar to come from the School of Marburg. Cassirer was a voracious reader with a phenomenal memory. He was elegant, reserved, and very kind, but one would hardly describe him as a powerful personality. He had neither Heidegger's dramatic quality nor Hartmann's talent for reaching young people. As for Frankfurt, it had not yet come into its own. The university was founded in the 1920s but it was not long before it began to attract attention. Riezler, who became its president, developed it. It eventually acquired its established scholars in people like Horkheimer, Adorno, and Tillich.

FORTIN: Your discussion of Strauss in *Truth and Method* opens with the remark that his teaching at Chicago was "one of the encouraging features of our world."[5] What did you mean by that?

GADAMER: Oh, that's easy. My impression is that he attracted students by his courage to proclaim what no one else would have dared to say. Although Chicago was a citadel of progressivism, he had the guts to answer "No" to the question of whether one should believe in the progress of the human mind. It was clear to me that the University of Chicago was an unusual place. I had met Hutchins in Frankfurt in 1947 and found him to be a very open and farsighted man. I met Adler. I also met McKeon, who was a real boss. So I could imagine some of the things I had heard about Strauss: how he, too, was ambitious and tried to profile himself against McKeon. Later on, when I started coming to America, I was able to observe at first hand the dedication of so many of his students in various parts of the country: you, Allan Bloom, Richard Kennington, Werner Dannhauser, Hilail Gildin, Stanley Rosen, and others. I was frequently asked to speak at places about which I had never heard and where I knew of no one who might be acquainted with my work. Whenever that happened, I could be sure that the invitation came from a Straussian. They were always kind and open because Strauss had said some nice things about me and about our 1954 meeting in Heidelberg, to which he often referred as one of the most profitable conversations he had had in a long time.

FORTIN: Do you think Strauss would have been better off in Germany as a teacher? Would he have been able to do as much there? More perhaps?

GADAMER: No, his success was independent of such matters for the simple reason that there was nothing phony about it. You know better than I do how he drew good students, cared for them, and stayed in touch with them. I can only see the effect, not the way it was produced. My feeling is that if he had been in Germany, he would likewise have founded a real school. I did not realize until you told me how large his classes were. From his description in the 1950s, I thought he never had more than six or eight students.

FORTIN: What would you identify as his major contribution? You spoke a while ago about his having revived the old quarrel between the Ancients and the Moderns. Does that have something to do with it?

GADAMER: Yes, although I personally learned a great deal from his book on Hobbes. For the first time somebody was attempting to see Hobbes not only as a British counterpart of the new foundation of the epistemology of the sciences but as a moralist whose relationship to the Sophists could be explained by means of an analysis of his views on civil society. That made a deep impression on me. I realize that this is now a much debated question and that Strauss himself had second thoughts about his book. But that was not my field and to read something in this style was a revelation. There was also something very personal in his image of Hobbes as a man who hated the English political system and suffered greatly at the hands of British society. There is a good deal of Strauss in the Hobbes book.

The other book that I would single out is *Persecution and the Art of Writing*, where one can see both the positive and the negative or dangerous consequences of persecution for the hermeneutical problematic. The question that it raises is an enormously important one: how can one convey and express thoughts that run counter to contemporary trends or the commonly accepted opinions of one's society? The question was particularly relevant to my own studies in Plato, where the issue of public opinion and censorship comes up in even more acute fashion. It took the life of Socrates. There is always the possibility that anything worth saying will arouse opposition. One cannot be a thinker without exposing oneself to it. I pretty much agree with Strauss on that point.

FORTIN: In *Truth and Method* you also refer to his rediscovery of the esoteric mode of writing or what you call "conscious distortion, camouflage, and concealment."[6]

GADAMER: I was thinking mainly of Spinoza. He, too, had a special significance for me as a precursor of the modern historical consciousness. I was struck by the way Strauss treated the *Theologico-Political Treatise* and in particular by his analysis of Spinoza's attempt to explain miracles in terms of the cultural agenda. I studied Strauss's essays on Spinoza and Maimonides very closely. My feeling was that he was right as far as Maimonides was concerned but that the same method did not apply equally well to Spinoza. There is always the possibility that the inconsistencies uncovered in the works of an author are due to some confusion on his part. Maybe this only reflects the confusion in my own mind. As I see it, the hermeneutical experience is the experience of the difficulty that we encounter when we try to follow a book, a play, or a work of art step by step, in such a way as to allow it to obsess us and lead us beyond our own horizon. It is by no means certain that we can ever recapture and integrate the original experiences encapsulated in those works. Still, taking them seriously involves a challenge to our thinking and preserves us from the danger of agnosticism or relativism. Strauss was willing to take seriously the texts that he confronted. I resented as much as he did the assumed superiority of the scholar who thinks he can improve Plato's logic, as if Plato had not been able to think logically. On that score we were in complete agreement.

Needless to say, Strauss's attention to the external or dramatic elements of Plato's and Xenophon's works was very congenial to me. In this, I followed Friedländer to some extent but tried to go beyond him. I learned something from Hildebrandt's book on Plato, for whom Hildebrandt had a sensitive ear.[7] He was not a philosopher but a well educated psychiatrist who had a good feel for young people. This enabled him to see things in the Platonic dialogues that no one else could see.

FORTIN: Strauss credited Klein with having rediscovered the importance of the dramatic features of the Platonic dialogues. To what extent is this true?

GADAMER: There was a certain symbiosis between Klein and me. Klein had already left Marburg when I began to study the classics with Friedländer, but he often came back; so there was a genuine exchange. Friedländer did not influence Klein directly, although he did so through me. I would hesitate to say that Klein was the only one responsible for the rediscovery. However, he had a better knowledge of philosophy than Friedländer, and so did I. Together we had the merit of relating the dramatic elements of the dialogues to the philosophical problems with which they deal. I gave some courses on Plato's dialectics in which I treated the *Sophist* and the *Theaetetus*. From the center of my own studies, I tried to demonstrate that even in these late dialogues there is a certain living communication and, hence, that they contain more than is explicitly stated in the text. We were both struck by the fact that a proper attention to their dramatic component was crucial to an understanding of Plato's thought. That was the import of Klein's and Friedländer's discovery. Strauss extended this to the area of political theory. It is amazing to see how great the impact of Friedländer's book has been even on the college level, here as well as in Germany.

The only thing I would add is that in Germany philosophy is more at the forefront of Platonic studies. As a result, there is less of a tendency to overemphasize the dramatic setting of the dialogues than there is among Klein's and Strauss's second and third generation followers. I sometimes receive papers from them which abound in all sorts of clever but unfounded interpretations. Just yesterday, I had a conversation with a young student who tried to establish a connection between the circular and somewhat comical dialectic of the *Par-*

menides and the fact that the meeting with Parmenides takes place on the occasion of the Panathenaic games. I pointed out that that was all very nice but that he had to find some support for his assertion, that its relevance had to be demonstrated from the text itself, and that so far we knew no more than that it *might* be warranted.

Klein himself did not always avoid that trap. Recently, somebody showed me a copy of his lecture on the *Phaedo*, in which he says some crazy things. He points out that at the death of Socrates, fourteen persons were present. So far, so good. But he then proceeds to make a detailed comparison between these fourteen characters and the fourteen hostages Theseus had once rescued from the Minotaur with the ship that was still sent on an annual mission to Delos for the purpose of commemorating this event. That is Talmud in the wrong place.

FORTIN: That method of reading texts has often been described as "talmudic" or "rabbinical." Is that the right way to talk about it?

GADAMER: There are elements of that, at least in Strauss, just as there are in Salomon Maimon (1754-1800), one of the first Jewish philosophers of the Kantian era. Maimon wrote a very interesting autobiography in which he traces the impact of the Jewish school system on his own thinking. The book is revealing because we have a parallel here, particularly as regards the experience of suppression. Hesse, the province from which Strauss hailed, was known for its anti-Semitism in the early decades of this century.

FORTIN: In his correspondence with you, Strauss takes issue with some of your statements concerning the "relativity of all human values."[8] You certainly do not consider yourself a relativist. If I understand you correctly, you are reacting in your own way against relativism. Strauss was apparently not convinced that you had succeeded in overcoming it. Do you take his criticism to be a serious one?

GADAMER: I replied to his letter but he broke off the correspondence. I tried indirectly to challenge him in an appendix to the second edition,[9] but he did not reply to that either. We met again afterwards and I saw that he was very cordial. One day in the course of a discussion I referred to an article of mine and he said: "But you never sent it to me!" I told him it would have been pointless to send along everything I wrote since much of it was foreign to his interests. He replied, "Oh, no. I am always interested in what you write." I found that very touching. I mention it not because it reflects on my own worth but only to suggest that we were good friends. On top of that, there was the overwhelming resonance that I found among his former students. All kinds of doors were open to me when I came to this country. That also says something about his loyalty. I am not suggesting that these people demanded full agreement from me.

FORTIN: They would have been disappointed! Strauss seems to have attached more importance than you do to the crisis of our time, to what Heidegger calls the "darkening of the world," to the cataclysmic crash of all horizons of meaning and value.[10] According to him, this is the situation out of which the new hermeneutics arises, one that is characterized by the total lack of agreement about fundamental issues and in which the groundlessness of all hitherto commonly accepted notions is disclosed. You seem to make light of that.

GADAMER: That is a crucial question for me as well. The radicalism to which you allude is related to Strauss's remark about the fact that I take my cue from Dilthey, whereas Heidegger takes his from Nietzsche. That is in a way true. Of course, Dilthey is more of a contemporary of Nietzsche and is especially useful as the mediator of German Idealism, Hegel, Schleiermacher, and the romantic feeling. But behind this difference lies the central issue of the place of conceptual thinking as such. I think that without some agreement, some basic agreement, no disagreement is possible. In my opinion, the primacy of disagreement is a prejudice. This is what Heidegger called *die Sorge für die erkannte Erkenntnis*; that is, the preoccupation with "cognized cognition," the commitment to certitude, the primacy of epistemology, the monologue of the scientists. My own perspective is always the hermeneutics of the whole world. We have to become aware of the limitations of the methodology of the sciences or the epistemology of the monologue. Beneath the structures of the opinion-making technology on which our society is based one finds a more basic experience of communication involving some agreement. That is why I have always emphasized the role of friendship in Greek ethics. I allude to this in my discussion with Strauss.[11] My inaugural lecture, that is, the public lecture with which one begins one's teaching career, was on this subject.[12] My point was that what fills two books in Aristotle's *Ethics* occupies no more than a page in Kant. I was twenty-eight years old then and not yet mature enough to grasp the full implications of that fact; but I anticipated them somehow and one of my deepest insights (if I may say so) had to do with what I described as the tension between the thinker and society—one of Strauss's topics.

Here again, however, one should not lose sight of the dual nature of the relationship. Hence my insistence on the positive side of Socrates's conformism. I do not believe one can call Socrates an atheist, as Bloom does. Both Socrates and Plato maintained a certain distantiated conformism with the cult, but behind it lurks the conviction that there is the divine, something we are never able to conceive. That, in my view, is what underlies the *Phaedrus* and the other dialogues. Strauss might agree with me, but I doubt whether Bloom would, or so I gather from the discussion we had about the *Ion* and, later, about the *Euthyphro*, where the conflict between us was even sharper. Bloom took the position that Euthyphro acted in a spirit of genuine piety, as opposed to Socrates, who was emancipated from the religious tradition. I disagreed completely. I said, "No, No! That borders on sophistry, conventionalism, hypocrisy." Socrates is the really pious one. He argues on grounds of piety when he maintains that one should always respect one's father. Euthyphro's denunciation of his father illustrates the noble conflict that is typical of all of the Socratic dialogues. Someone claims a special competence; he is then convicted by means of a logical argument based on the real figure of Socrates, to whom we are always led back. Bloom defended the opposite view, arguing that Euthyphro was the pious one and Socrates the atheist. I think that is completely wrong. So we had a fierce but friendly altercation.

I never discussed these matters with Strauss or Klein at any great length. Strauss avoided them. He was very amicable and I took great pleasure in listening to him, but whenever philosophical issues came up, he shied away from them.

FORTIN: What do you think of the idea that hermeneutical ontology belongs to a transitional period, one which coincides precisely with the shattering of all horizons? Doesn't Heidegger himself look forward to the emergence of a new consensus, to the appearance of new gods, for whom we can only wait? Strauss's point is that we shall then find ourselves in a posthermeneutical situation, just as we were in a prehermeneutical situation when German Idealism was still dominant.

GADAMER: There I disagree not only with Strauss but with Heidegger as well. The point that you raise is closely connected with Strauss's remark to the effect that I work from Dilthey rather than from Nietzsche. That I regard as a fair statement. What it means is that for me the tradition remains a living tradition. I am a Platonist. I agree with Plato, who said that there is no city in the world in which the ideal city is not present in some ultimate sense. You also know the famous statement about the gang of robbers whose members need some sense of justice in order to get along with one another.[13] Well, that is indeed my perhaps overly conservative position. As you know, we are formed between the ages of fourteen and eighteen. Academic teachers always come too late. In the best instance, they can train young scholars, but their function is not to build up character. After the war, I was invited to give a lecture in Frankfurt on what the German professor thinks of his role as an educator. The point that I made was that professors have no role to play in that regard. Implied in the question at hand is a certain overestimation of the possible impact of the theoretical man. That is the thought behind my attitude. I do not follow Heidegger at all when he talks about new gods and similar things. I follow him only in what he does with the empty or extreme situation. This is his only point of agreement with Nietzsche, who likewise anticipated an extreme position of nothingness. Of course, he ended in self-contradiction.

Heidegger was not a Nietzschean in that sense. When he first started coming out with his mysterious allusions to the return of the gods, we were really shocked. I contacted him again and saw that that was not what he had in mind. It was a *façon de parler*. Even his famous statement, *Nur ein Gott kann uns retten*,[14] means only that calculating politics is not what will save us from the impending catastrophe. Nevertheless, I would criticize that too. Heidegger sometimes says more than he can cover, as he does, for example, when he looks ahead to the emergence of a new world. So I would deny that it makes any sense to speak of a posthermeneutical epoch. That would be something like the recaptured immediacy of the speculative ideas, which I cannot admit. In my opinion, it involves a confusion or a categorical fallacy. It is at best a metaphorical way of speaking and is meant to suggest only that, if we go on in this manner, technology will be enshrined as a terminal state, a final world government will come into being, and everything will be regulated by an omnipotent bureaucracy. That is the ultimate or extreme situation; and, of course, self-destruction can occur on the way to it. I do not believe in this extreme elaborated by Nietzsche. Heidegger's intention was merely to bring to light the one-sidedness of this Western way, culminating in our present-day technological society.

In one of my latest articles on Heidegger, I try to show that Heidegger was very far from any sectarian stance.[15] He did not believe in Confucius and other such exotic novelties. He was only suggesting that there exist in the Far East certain remnants of culture from which we, who have glimpsed the impasse of Western civilization, could possibly benefit. On the other hand, when he discusses the work of art and maintains that there is something beyond conceptual thinking which can claim to be true, he has my wholehearted approval. That seems basic to me and here I share his position completely.

FORTIN: You seem to regard hermeneutical philosophy as the whole of philosophy.

GADAMER: It is universal.

FORTIN: Its universality implies a certain infinity; yet you insist a great deal on human finitude.

GADAMER: They go together. Finitude corresponds to Hegel's "bad infinity." What I mean is that the "good infinity," that is, the self-articulation of the concept, the self-regulation of the system, or whatever it may be, seems to me to be an anticipation of a new immediacy. That I cannot go along with. The emphasis on finitude is just another way of saying that there is always one step more. Bad infinity in the Hegelian sense belongs to finitude. As I once wrote, this bad infinity is not as bad as it sounds.

FORTIN: You have done a lot of fine work on Aristotle and especially on his notion of *phronesis*. What troubles some people is that you seem to stress *phronesis* at the expense of *episteme*. Wasn't science or *episteme* equally important for Aristotle and doesn't one have to come to grips with that notion as well?

GADAMER: Aristotle's main point—and it is also Plato's—is that science, like the *technai*, like any form of skill or craftsmanship, is knowledge that has to be integrated into the good life of the society by means of *phronesis*. The ideal of a political science that is not based on the lived experience of *phronesis* would be sophistic from Aristotle's point of view. I do not deny that the clarification of the apodictic or demonstrative dimension exemplified by mathematics and especially by the theoretical mode of Euclidean mathematics is a great achievement in the eyes of Aristotle. But the idea of the good lies beyond the scope of any science. That is very clear in Plato. We cannot conceptualize the idea of the good.

FORTIN: Strauss once said that as a young man he had two interests—God and politics. He also said on a number of occasions that the greatest philosophers of the twentieth century—Bergson, Husserl, James, Heidegger, Whitehead—differed from their predecessors by reason of the virtual absence of any political dimension from their thought. Their philosophies may have had grave political implications but they themselves never dealt thematically with political issues. Moreover, Strauss tends to see politics as the cultural matrix of the historical consciousness. When we speak of an historian without qualification, we generally mean a political historian. You mentioned at the beginning of our conversation that you were once interested in the political thought of the Sophists but had to abandon that pursuit because of the situation in Germany. Do you still recognize the over-arching importance of politics?

GADAMER: This is the other side of the same problem, that of the place of the theoretical man in society. All is not negative here inasmuch as the theoretical man remains subordinated to *phronesis*. One of my recent articles, which has been in the press for years—it is being published in Greece and Greece needs years—deals with the problem of the theoretical and the practical life in Aristotle's *Ethics*. In it I try to show that it is always a mistake to stress the tension between these two lives or to say that, on the basis of his premises, Aristotle had to prefer the political life and defended the primacy of the theoretical life only out of deference to Plato. The article demonstrates the absurdity of that view. We are mortals and not gods. If we were gods, the question could be posed as an alternative. Unfortunately, we do not have that choice. When we speak of *eudaimonia*, the ultimate achievement of human life, we have to take both lives into account. The characterization of the practical life as the second best life in the Aristotelian scheme means only that the theoretical life would be fine if we were gods; but we are not. We remain embedded in the social structures and the normative perspectives in which we were reared and must recognize that we are part of a development that always proceeds on the basis of some preshaped view. Ours is a fundamentally and inescapably hermeneutical situation with which we have to come to terms via a mediation of the practical problems of politics and society with the theoretical life.

FORTIN: More than sixteen years have elapsed since the publication of your discussion of Strauss in the second edition of *Truth and Method* (1965). You met Strauss a number of times between 1965 and 1973, the year of his death. Do you still stand by what you said then?

GADAMER: Yes, and I hope he would agree. He was very modest and, as I mentioned earlier, he did not like to discuss his disagreements with me. I have always regretted that the dialogue was not pursued. I had made a new overture and he knew that a further discussion, though perhaps not a definitive one, was possible.

FORTIN: Are there any other survivors from the period of the early 1920s?

GADAMER: Helmut Kuhn. He was in Berlin then and now lives in Munich. He was a Protestant of Jewish extraction and had a strong religious bent. As is the case with so many other religious intellectuals, the experiences of the Third Reich prompted him to convert to Catholicism. He found a new home in the Catholic Church and became extremely conservative.

FORTIN: Litt, in the book to which you refer in *Truth and Method*,[16] describes the opposition to history as being very dogmatic. Would you not agree that the defense of history can be equally dogmatic?

GADAMER: Oh, certainly. Strauss makes that point in his letter to Kuhn.[17]

FORTIN: It was most kind of you to give us so much of your time on this, the last day of your stay in this country at least for this year. We are all very grateful to you.

NOTES

1. See Gadamer's account of this and related incidents in his *Philosophische Lehrjahre* (Frankfurt am Main: Klostermann, 1977), 50-51.

2. Cf. L. Strauss and H. G. Gadamer, "Correspondence Concerning *Wahrheit und Methode*," *The Independent Journal of Philosophy* 2 (1978): 5-12.

3. The paper was never published but an application of its results is to be found in "Der aristotelische *Protreptikos* und die entwicklungsgeschichtliche Betrachtung der aristotelischen Ethik," *Hermes* 63 (1927): 138-64.

4. See Gadamer's review of Rose's book in *Gnomon* (1940): 431-36.

5. Hans Georg Gadamer, *Truth and Method* (New York: Seabury Press, 1975), 482.

6. Ibid., 488.

7. Kurt Hildebrandt, *Platon: Der Kampf des Geistes um die Macht* (Berlin: G. Bondi, 1933).

8. *Truth and Method*, 53.

9. Ibid., 482-91.

10. See, for example, M. Heidegger, *An Introduction to Metaphysics*, translated by R. Manheim (New Haven: Yale University Press, 1959), 38.

11. *Truth and Method*, 485.

12. The lecture, delivered in 1929, was never published.

13. Cf. Plato, *Republic* 351c.

14. See the interview with Heidegger published in the 31 May 1976 issue of *Der Spiegel*, shortly after Heidegger's death. An English translation of the interview appears in *Philosophy Today* 20, no. 4 (Winter 1976): 267-84.

15. H. G. Gadamer, "The Religious Dimension in Heidegger," in *Transcendence and the Sacred*, edited by L. Rouner and A. Olson (Notre Dame: University of Notre Dame Press, 1981), 193-207. Cf. "Sein, Geist, Gott," in Gadamer, *Kleine Schriften IV*, (Tubingen: Mohr, 1977), 74-85.

16. *Truth and Method*, 490.

17. Cf. L. Strauss, "Letter to Helmut Kuhn," *The Independent Journal of Philosophy* 2 (1978): 23-26.

IV

PAPAL SOCIAL THOUGHT,

VIRTUE, AND LIBERALISM

SACRED AND INVIOLABLE:
RERUM NOVARUM AND NATURAL RIGHTS

Leo XIII's *Rerum novarum*, whose centenary was observed in 1991, is the first in a long series of official documents by means of which the Roman Catholic Church has attempted to deal with what the nineteenth century called the "social question," that is, the scandal provoked by the impoverishment of large segments of the working class in a world that was daily becoming more opulent at its expense. It is also, in the opinion of many, the most theoretical and most elegant of these documents. Its great merit is to have defined the terms in which the moral problems spawned by the industrialization of Western society would henceforth be debated in Catholic circles. In it one finds a key to the understanding of the many statements that Rome has since issued on this subject, all of which stress their continuity with it, build upon it, seek to refine it, or extend its teaching into areas not touched upon by Leo himself. In retrospect, it can be said to have affixed a canonical seal of approval to what, in the wake of Luigi Taparelli d'Azeglio's landmark treatise on natural right,[1] had emerged as an autonomous or semi-autonomous discipline now known as Catholic social ethics.[2] Pius XI called it an "immortal document," the "Magna Carta" of the new theological discipline.[3] Few papal pronouncements have had a more profound and lasting impact on Catholic life and thought.

On the assumption that the text of *Rerum novarum* is familiar or readily accessible to most people, I shall forgo any detailed analysis of its contents and focus instead on certain points that seem more problematic to us than they did to Leo and his collaborators or that call for further reflection in the light of what we have learned in the meantime about the origin and nature of modern social thought. My thesis is twofold. I shall claim that the encyclical represents an attempt to synthesize or fuse into a single whole elements derived from two independent and largely antithetical traditions, one rooted in a teleological and the other in a nonteleological view of nature. The first of these, represented preeminently in the Christian West by Aristotle and his medieval disciples, I shall refer to as the "premodern" tradition. The other, which originated with Machiavelli in the sixteenth century and achieved its most popular form in the political philosophy of John Locke a century and a half later, I shall refer to as the "modern" tradition. Secondly, I shall claim that the two components of the proposed synthesis coexist only in an uneasy tension with each other.

My argument rests on the premise that with the emergence of modern thought we come to a crossroad in the intellectual history of the West, one of those rare moments at which, through the agency of a handful of seminal thinkers, human consciousness underwent a radical transformation and the Western world was summoned to make a fundamental change in direction. Implied in this statement is the view that modern thought is not a simple derivative of premodern thought but represents a decisive break with it, defines itself in opposition to it, and on the level of its highest principles remains profoundly at odds with it. This opposition became known in the second half of the seventeenth century as the "quarrel between the Ancients and the Moderns" and was still discussed under that name by its last great witnesses, Swift and Lessing, in the eighteenth century. At stake was nothing less than the issue as to which of the two "wisdoms," that of the Ancients or that of the Moderns, was superior to the other. To this day, Swift's *Gulliver's Travels*, with its portrayal of the Ancients as giants and of the Moderns as dwarfs, stands as a towering monument to the memory of the epic battle in which these formidable adversaries were once locked for the minds and hearts of their contemporaries, to say nothing of future generations.[4]

Numerous attempts were later made to downplay the significance of that renowned quarrel by reducing it to a petty squabble between two rival groups who, lacking the necessary perspective, were blind to the implications of their respective positions. The conventional wisdom of our day is that modern thought was already latent in premodern thought, that it was gradually educed from it by a process of logical inference, and that in consequence one cannot legitimately speak of a hiatus or breach of continuity in the Western intellectual tradition from its inception in fifth-century Greece or thereabouts down to modern times. Far from heralding a break with the past, the modern development would be nothing but an effort to refurbish, update, or tailor to the needs of later ages the principles that first came to light at the dawn of the philosophic tradition.[5] Thus, it became fashionable to read Locke, the father of capitalism, as a pious seventeenth-century Aristotelian and to interpret the American founding, despite its entanglement with Enlightenment rationalism, as a spiritual off-spring of medieval theology.[6]

This revisionist account of the evolution of Western thought was never accepted by everyone. Most of the modern Popes resisted it, as is evident from even a superficial glance at the documents of the pre-Leonine period. By way of example, one has only to recall the eightieth and last of the "errors" condemned by Pius IX in the Syllabus of 1864, which is that "the Roman Pontiff can and ought to reconcile himself with progress, liberalism, and modern civilization." It has likewise been called into question by a number of eminent twentieth-century scholars—Eric Voegelin, Leo Strauss, Karl Löwith, Michael Oakeshott, Jacob Klein, and Hannah Arendt among them—who have argued that, regardless of where one's sympathies may ultimately lie, the safest and best way to make sense of the current intellectual scene is to analyze it in terms of the dichotomy between the premodern and modern modes of thought.[7] Such, at any rate, is the perspective within which I propose to examine the teaching of *Rerum novarum*.

THE CRISIS OF MODERN SOCIETY

It has often been remarked that the papacy was slow in speaking out against the injustices to which the industrial revolution and the triumph of capitalism had directly or indirectly given rise. Its intervention in this matter followed by many years the formation of a powerful social movement to which assorted nonreligious or antireligious theorists, historians, and men of letters had been lending their voices since the middle of the nineteenth century or earlier.[8] There is even a touch of irony in the fact that, by the time Rome joined the debate, the condition of the modern worker had already begun to improve.[9]

The delay, explained in part by the fact that the industrial revolution was less advanced in the traditionally Catholic countries of southern Europe than in the predominantly Protestant countries of northern Europe, may have been providential, for it gave the Church enough time to equip itself with the intellectual tools needed to tackle a problem of this magnitude. Only in the last quarter of the century, thanks to the revival of Thomism, to which Leo XIII had long been committed both as bishop of

Perugia and as Pope, would it be in a position to undertake such a task and bring it to a happy conclusion. The result was a far more probing analysis of the problems at hand than would have been possible at an earlier date.

Other critics, from Rousseau to Nietzsche, had attacked modern society for its small-mindedness, its bourgeois mediocrity, its lack of elevation, or its inability to produce human beings of noble character. *Rerum novarum* proceeds on different and more narrowly moral grounds. It criticizes contemporary Western society for its injustices and traces these injustices to their roots in the two great political systems of modern times, liberalism and socialism, both of which it subjects to a rigorous scrutiny guided in the main by the newly rediscovered Thomistic principles. It denounces with vigor the abuses to which liberalism and the economic system in which it finds its classic expression, capitalism, lend themselves, and it alerts its readers with even greater vigor to the threat to human freedom posed by the spread of socialism. Liberalism is rebuked for its excessive individualism; socialism, for its dangerous collectivism. The former exalts the individual at the expense of the community; the latter sacrifices him to the collectivity. Both systems are destructive of human dignity and detrimental to the health of society.

The bill of particulars in each case is too well known to warrant more than a brief reminder. Liberalism exploits the worker and treats him as chattel. It "grinds him down" with excessive labor, "stupefies" his mind, "wears out his body," and repays him by robbing him of the fruit of his industry (no. 42). It imposes inhumane conditions on him and, by obliging him to bargain for his salary, leaves him entirely at the mercy of his employers. Its laissez-faire economics sanctions the unlimited acquisition of wealth, emancipates greed, and breeds flagrant inequities. It deprives the poor of the necessities of life and multiplies the wealth of those who already have more than they need. It denies that labor and commodities have any intrinsic value and allows their price to be fixed exclusively by the free interplay of market forces or the law of supply and demand. It is insensitive to the workers' spiritual needs. It destroys the structure of the home by condoning child labor or by making it necessary for both parents to work. And the list goes on.

Socialism, Leo's *bête noire* and the only one of the morally defective systems to be dealt with in a separate section (nos. 4-15), is responsible for even greater evils. Its eagerness to find a cure for the cupidity stimulated by the free enterprise system is laudable, but the proposed remedy, namely, the total suppression of private property, is worse than the disease itself. It throws open the door to "envy, mutual invective, and discord." It removes the necessary incentives to ingenuity and causes the fountains of wealth to dry up. With nothing that he can call his own, the individual is left with a freedom that has no way of expressing itself and therefore remains purely abstract. Whereas capitalistic society reduces one social class, the workers, to a condition of slavery, the socialist regime produces "uniform wretchedness and meanness for all" (no. 15). It violates everyone's rights: those of lawful owners, those of workers, and those of parents, for whom it would substitute itself. To make matters worse, it is strictly utopian. Its "absurd equality," *absurda aequabilitas* (no. 38), runs counter to the nature not only of the individual but of society, which requires for its proper functioning the services of human beings with diverse and unequal talents (no. 17). Whatever the socialists may think, these inequalities are both natural and necessary. Society would be worse off without them. Again, the list goes on.

To its credit, the encyclical never forgets that, different as they may be from each other, liberalism and socialism thus understood have much in common, grounded as they are in a materialistic and mechanistic understanding of human existence. Both share the same animosity toward premodern thought, the same obsession with economic factors, the same "scientific" or nonteleological conception of the universe, and the same view of the human being as a being who sets for himself only such goals as are geared to the satisfaction of his bodily needs. Seen in this perspective, the two movements are not polar opposites. Socialism comes into being by way of a radicalization rather than a repudiation of the basic premises of modern liberalism.

The strength of *Rerum novarum* is that it boldly attacks this specifically modern view of life, not, as others had been doing, in a spirit of romantic sentimentality or Traditionalist conservatism, but by means of theological and rationally defensible principles. It stresses man's natural sociability, defends private property without omitting to tell us that it must be used for the good of all, dwells on such long-forgotten themes as fair prices and just wages, underscores the importance of the family, encourages the formation of private trade unions as long as their goals do not conflict with those of the larger society, rejects the Marxist notion of class conflict, demands decent working conditions for laborers, reminds society of its special obligations toward the "lowly and the destitute" (no. 37), waxes eloquent when it evokes the peculiar dignity of human beings, urges respect for religion, and pleads for the moral regeneration of society through the cultivation of virtue and the pursuit of the common good.

In a nutshell, what the encyclical calls for is nothing short of a wholesale return to a premodern and, by and large, Thomistic understanding of the nature and goals of civil society, whose insights are brought to bear on the new situation created by the rise of capitalism and the socialist reaction to it. That alone would be enough to set *Rerum novarum* apart as a document of unique theoretical and historical importance. One thing is certain: its publication marks the spectacular reentry of the Church into an arena from which it had been excluded as a major player by the great intellectual and political events of the Enlightenment and its aftermath.

The other side of the story, and it is no less fascinating than the first, is that in elaborating its program for the reform of society *Rerum Novarum* resorts to a number of categories that are proper to modern thought and not easily squared with its basic Thomism. Two of these merit special consideration: the overwhelming emphasis on the naturalness of private property and the doctrine of natural rights.

RERUM NOVARUM AND PRIVATE PROPERTY

That the encyclical should be outspoken in its support of private property is understandable insofar as much of what it says was intended as a challenge to the growing power of the socialist movement, whose stated aim was to concentrate all property in the hands of the state. Yet the position that it takes is anything but a mere restatement of what the Catholic tradition had previously taught on this subject. While the Church had long been a strong advocate of private property, its defense of it was usually couched in rather more moderate terms. The prevailing view was the one expounded by Thomas Aquinas in the *Summa theologiae* II-II, qu. 66, a. 2, which raises the question of whether it is lawful for someone to possess a thing as his own. The question receives an affirmative answer supported by three arguments taken from Aristotle's *Politics*. The first is that human beings bestow greater care upon things that belong to them individually than they do upon common things. Secondly, they are less confused than they tend to be when everyone is responsible for everything indiscriminately. Finally, they have a better chance of living at peace with one another if they know beforehand what belongs to whom. Simply put, private property is a good idea. Although not an absolute demand of natural right, it is entirely in accord with it and ought to be favored whenever possible.

The encyclical refers with approval to Thomas's article (no. 22) but proceeds to give the right of private property a much firmer grounding in nature. As was just noted, Thomas left it at saying that it is "lawful" and, in practice, necessary for human beings to possess certain things as their own: *licitum est quod homo propria possideat*. The encyclical for its part presents private property not only as superior to other possible arrangements and eminently desirable for that reason, but as a "stable and perpetual" (no. 6), inviolable,[10] and indeed "sacred" right, *ius sanctum* (no. 46).[11] This right has its source in nature itself. All the laws, the divine as well as the natural and the civil, sanction it (no. 11). Since it is not conferred by civil society, it cannot be taken away by it and must be protected by it. It matters little that the argument forms part of a critique of socialism's "pernicious tendency" to invade people's possessions in the name of some "absurd equality" (no. 38), for it is still rooted in what claims to be a universal principle.

Astonishingly, no one at the time seems to have noticed that this notion of private property as a natural and imprescriptible right had only recently been imported into Catholic theology, in all probability by Luigi Taparelli d'Azeglio, the biggest name in nineteenth-century Catholic social thought and, incidentally, the man who had been appointed rector of the Roman College when it reopened in 1824, the year Gioacchino Pecci, the future Leo XIII, enrolled there as a student at the age of fourteen.[12] Prior to that time, it was understood that according to natural law the earth originally belonged to everyone and that its subsequent division, dictated in large measure by reasons of expediency, was a matter of human or positive law. That older view is summarized as follows in Gratian's *Decree*: "The division of property and slavery belong to the 'right of nations' (*ius gentium*); . . . by the right of nature all things are common and everyone is free."[13] It is the view that Thomas himself sets forth in *Summa theologiae* I-II, qu. 95, a. 4, where in like manner the division of property is assigned, not to the natural law simply, but to the "right of nations," defined as that part of the positive law (*ius positivum*) whose principles are derived from the natural law as conclusions from premises.[14]

A telltale sign of the encyclical's departure from the Church's long-standing teaching on this point is its failure to include any reference to the key notion of *ius gentium*. Since the latter had been given due prominence in the first and third drafts, both of them prepared mainly by the Jesuit Matteo Liberatore, its omission from the final text cannot have been an oversight.[15] Why was it left out? A plausible answer is supplied by Liberatore's *Principles of Political Economy*, first published in 1889 and thus very close in time to *Rerum novarum* From this work we learn that other contemporary theorists were turning Thomas's categorization of the *ius gentium* as part of positive law into an argument against private property. If the latter is a creation of the civil authority to begin with, it can be abrogated by it. As his spirited response shows, Liberatore was confident that even on Thomas's account of the *ius gentium* a strong case could be made for the naturalness of private property. Thomas's *ius gentium* is not simply a matter of positive law. It stands "midway between natural right and civil right." Its principles are deduced from strict natural law principles and share in their necessity. The proof is that for Thomas these principles are binding quite apart from any human enactment.[16]

All well and good, save that the Pope, who participated regularly and actively in the deliberations of the drafting committee, was not about to take any chances. It would not be at all surprising if the initiative to strike from the text any reference to the *ius gentium* had come from him. Not that Liberatore was any less committed than Leo to the defense of private property. Far from it, for he appears to be the one who injected the notion of the sacredness of property into the debate when, in the first draft, he wrote with breathtaking candor: *la proprietà privata è sacra*—"private property is sacred."[17]

What the encyclical and much of nineteenth-century Catholic theology took to be a teaching coeval with Christianity was in fact a startling innovation ultimately traceable to John Locke and propagated by his numerous disciples throughout the eighteenth century and beyond.[18] Locke himself stopped short of calling private property "sacred," although, in view of the preeminence accorded to it as the cornerstone of his political system, he might have. That honor fell to later writers and in particular to Adam Smith, a certified Lockean, who hails the right of property as the "most sacred and inviolable" of rights.[19] The expression caught on and was enshrined soon afterwards in Article XVII of the French Constituent Assembly's Declaration of the Rights of Man and Citizen (August 26, 1789), which states unequivocally that "Property being an *inviolable* and *sacred* right, no one can be deprived of it" save in cases of extreme public necessity.[20] Seventy-five years later, J. S. Mill could still write: "The sacredness of property is connected, in my mind, with feelings of the greatest respect."[21] It is to this emphatically modern tradition that, consciously or unconsciously, the encyclical owes its extraordinary doctrine of the sanctity of private property. To borrow an image from Sherlock Holmes, Locke, or perhaps Adam Smith, is the dog that should have barked but curiously did not bark in *Rerum novarum*. That everyone should have assumed that Taparelli and his intellectual progeny were merely echoing a time-honored Christian teaching shows how little was then known about ancient and medieval theories of property. It would obviously be unfair to accuse Leo and his collaborators of deifying mammon by declaring private property intrinsically holy. What they meant and had every right to insist on is that workers should not be defrauded of their earnings through excessive taxation, unfair wages, the seizure of property by eminent domain without proper indemnification, and above all, total expropriation through the abolition of private property. In their minds, "sacred" was synonymous with "inviolable," the term with which it is often coupled and by which it is translated in the official Italian version of the encyclical.[22] Yet this alone would not make its application to material objects other than those reserved for the cult any less incongruous were it not for the interesting historical background of the new usage. It so happens that the tandem "sacred" and "inviolable" is indigenous to another context where its significance is more easily grasped, namely, that of the notion of sacred kingship as it had developed in the

West from the Hellenistic period onward, eventually reaching its culmination in the seventeenth-century theory of the divine right of kings. If the king is the anointed of God, if he rules by right of hereditary succession, if that right is indefeasible, if he is entitled to the total submission of his subjects, and if his actions are not to be judged by anyone save God, it makes some sense to refer to his person as sacred and inviolable. Any attempt on his life, challenge to his prerogatives, or resistance to his rule, even in the name of religion, becomes a sacrilege.[23]

The same tandem shows up among other places in *The Federalist Papers* no. 69 (Hamilton), which states expressly:

> The person of the king of Great Britain is *sacred* and *inviolable*; there is no constitutional tribunal to which he is amenable; no punishment to which he can be subjected without involving the crisis of a national revolution (emphasis added).[24]

Such a view becomes doubly interesting when we recall that it is the one Locke attacks in the *First Treatise of Civil Government* and to which he goes on to oppose his own teaching, that of the *Second Treatise*, where private property is characterized, if not literally at least in equivalent terms, as sacred.[25] One can say that in the course of the seventeenth and eighteenth centuries the divine right of kings was replaced by the sacred right of capitalists.[26] The wonder in all of this is that the drafters of *Rerum novarum* should have been so eager to adopt a terminology of which, as sworn opponents of modern liberalism, they had ample reason to be wary.

Further confirmation of the Lockean ancestry of the encyclical's view of private property is to be found in its endorsement of the concept of labor as the sole source of property (nos. 8-10). In Locke's words, which the encyclical echoes almost textually, it is by mixing one's labor with an object that one gains possession of it.[27] To my knowledge, nobody prior to the seventeenth century had claimed that labor was of itself a title to property, let alone the only title to it. All that was ever said is that the laborer deserves a just wage or a material reward commensurate with his efforts.[28] Nor, unlike the encyclical, did anyone divorce completely "just ownership" (*iusta possessio*) from "just use" (*iustus usus*).[29] The commonly accepted view was that, in strictest justice, wealth or property belongs to the person who has the wisdom and moral rectitude necessary to insure its proper use and make it serve the common good. True, well-ordered societies will usually allow actual proprietors to keep as their own all legally acquired goods regardless of the use that they make of them. If they do so, however, it is not because private property is a perfectly just solution to the problem at hand but because under normal circumstances it is the practically best solution to that problem. Any other course of action invariably results in either chaos or greater injustices. There are natural limits to what even the wisest of rulers can accomplish in this regard. Not only the profound analyses of Plato's *Republic* and Aristotle's *Ethics* but some of the most painful experiences of our own century are there to remind us of the dangers involved in any fanatical attempt to impose standards of absolute justice on society at large.

A well-known passage from Xenophon's *Education of Cyrus* I.3.16 will clarify the point. Say that a big man owns a coat that is too small for him and that his neighbor, a small man, owns a coat that is too large for him. Reason would seem to dictate that they exchange coats, but should they refuse, the law will most often respect their wishes, if only to avoid the innumerable disputes in which it would otherwise become embroiled. On the level of everyday life, such a practice has much to recommend it. Still, there is nothing sacrosanct about it. In order to recognize and, circumstances permitting, redress some of the graver injustices imbedded in a particular political structure or legal system, one needs a more truly natural criterion. Because it pays insufficient attention to this higher criterion, the encyclical ends up by being more conservative than it has to be. Its final teaching is that, since ownership is not only distinct from use but independent of it, it cannot be forfeited through misuse or for any other motive. Not only is the worker entitled to an adequate salary, he has a "true and perfect right" to dispose of it as he sees fit: *ius verum perfectumque . . . non modo exigendae mercedis sed et collocandae uti velit* (no. 5).

To many a thoughtful reader, this virtual absolutization of the right of private property has always smacked of something like fin-de-siècle capitalism or come across as a one-sided response to the anxieties stirred up among the upper classes by the rising tide of socialism, in which Leo could see nothing but a worldwide conspiracy to destroy the Catholic Church.[30] The Pope's position was not only hard to defend theoretically, except on Lockean terms, it had its practical shortcomings as well. By adopting it, the Church was again laying itself open to criticism. Just as it had once been assailed for its support of the Old Regime against the bourgeoisie, so now it would be accused of allying itself with the bourgeoisie against the "proletariat," the term by which *Rerum novarum*, along with much of the nineteenth century, designates the new working class.

There is no denying that Leo's primary concern was with the modern laborer, whose lot he was eager to improve, but neither is it possible to deny that his proclamation of the sacredness of private property benefitted the rich as much as, if not more than, it did the poor. To that extent at least, the mentality that informs the encyclical is not far removed from the one that Anatole France so effectively satirized when he spoke of the "majesty of the laws, which forbid rich and poor alike to sleep under bridges, to beg in the streets, and to steal their bread."[31] It is unlikely that France wrote this with *Rerum novarum* in mind, even though he was not above poking gentle fun at papal documents once in a while. He was not thinking of the Pope but of the French Revolution and the bourgeois society to which it had given birth. He knew that for most of his contemporaries property meant everything: "In any civilized state, wealth is a sacred thing; in democracies it is the only sacred thing."[32] He was well aware of the modern arguments in favor of private property: "The poor live from the wealth of the rich; that is why this wealth is sacred. Do not touch it; that would be an act of wanton cruelty."[33] And he had a keen sense of the transformation that the overthrow of the old aristocracy had effected in his country: "Money has become honorable. It is our sole nobility, the most oppressive, the most insolent, and the most powerful of them all."[34] Truly, the sacred right of property had superseded the divine right of kings. The fact was undeniable. By the end of the nineteenth century, even the papacy was ready to acknowledge it.

Quadragesimo anno made a valiant attempt to redress the balance by insisting, more than *Rerum novarum* had done, on the social purposes of property. It corrected some of the inaccuracies that had crept into Leo's text, such as the assertion that labor is the only source of property. It introduced a few new elements into the discussion, most notably the now famous "principle of subsidiarity" and the Taparellian notion of "social justice,"[35] which Leo had rejected because of its non-Thomistic pedigree. It likewise protested vehemently against the accusation—the "calumny"— that the Church had "taken the side of the rich against the non-owning workers."[36] But while it reinforced *Rerum novarum*'s defense of workers by stressing their "sacred rights," *iura sacra* (no. 28),[37] it did not alter its basic position on private property. On the contrary, it enshrined it by repeating Leo's statement to the effect that the division of property is a "sacred right"; it adopted Leo's sharp distinction between "ownership" and "use"; and, like Leo, it transferred "use" from the sphere of justice and legislation to that of Christian charity.[38] It thus left us once again with a juxtaposition rather than a bona fide integration of two originally antithetical views, one that takes its bearings from the notion of duty and the other from the notion of rights. The roots of the tension generated by this juxtaposition will come into sharper focus if we take a closer look at *Rerum novarum*'s stand on the issue of rights versus duties.

NATURAL RIGHT AND NATURAL RIGHTS

Leo's teaching on private property forms part and parcel of a larger view of civil society in which the notion of "natural rights" comes close to playing a dominant role. The encyclical states clearly and loudly that these rights are inherent in each human being—*in hominibus insunt singulis* (no. 12)—and that they are "inviolable."[39] Since they are given to us by nature, they antedate civil society and exist independently of it: *minime [pendent] a republica* (no. 12). Human beings are endowed with them prior to their entry into that society and retain them once they have become members of it. The encyclical takes it for granted that human beings first existed in a prepolitical state and that civil society later came into being as a means of protecting the rights they enjoyed in that state:

> Nature must have given the human being a stable and ever present source to which he might look for a constant supply of goods that nothing save nature with its abundant fruits can provide. There is no need here to bring in civil society. The human being precedes civil society and, prior to the formation of any civil society, must have by nature the right to sustain his life and care for his body (no. 7).

Never before had this notion of natural rights figured so prominently and so massively in a pontifical document. The phenomenon is the more remarkable as natural rights are totally foreign to the literature of the premodern period. The Bible certainly knows nothing of them. The term "rights," *iura*, does not appear even once in the Vulgate, for centuries the standard version of the Bible in the West. *Ius* in the singular occurs approximately thirty times, but always to designate some legally sanctioned arrangement. Genesis 23:4 speaks in this sense of a *ius sepulchri* or right of burial apropos of Abraham, who negotiates with the Hittites the purchase of a tomb for Sarah. No attempt is made to define the nature of this or any other right. Since the Hebrew Bible has no word for "nature" and in any event does not engage in philosophic speculation, it can scarcely be expected to furnish us with a full-blown theory of "natural" rights.

Neither can the natural rights doctrine be said to play a significant role in medieval thought. Thomas Aquinas, to refer once again to the authority from whom the encyclical purportedly draws the bulk of its theological inspiration, either had never heard of them or did not deem it necessary to incorporate them into his scheme. According to Busa's exhaustive *Index Thomisticus*, the word *iura* occurs a total of fifty-four times in Thomas's voluminous corpus, but never in the sense of *natural* rights. In all cases, the reference is to canonical or civil rights, or to the ancient as distinguished from the new codes of law, or to the laws governing warfare and the like.[40]

What is true of Thomas holds for the rest of the Middle Ages, both before and after him. There is little textual support for the view, defended most energetically by the distinguished French legal historian, Michel Villey, that the father of the rights theory as we know it is William of

Ockham. For Villey, everything hinges on the distinction between objective right ("the right thing," "one's due," Ulpian's *suum ius cuique tribuere*) and subjective right, by which is meant a moral power or faculty (*facultas*) inhering in individual human beings. The difference that sets the two notions apart becomes plain when we recall that "right" in the first sense does not necessarily work to the advantage of the individual whose right it is. In Rome, the right of a parricide was to be stuffed in a bag filled with vipers and thrown into the Tiber.[41] Ockham, the villain of Villey's story, is the man who consummated the break with the premodern tradition by accrediting the notion of subjective rights or rights that individuals possess as opposed to rights by which, so to speak, they are possessed. His is the work that marks the "Copernican moment" in the history of legal science.[42] Others, such as Richard Tuck and Brian Tierney, have lately argued against Villey that subjective rights or rights understood as faculties are older than Ockham, that they are essentially a twelfth-century invention bequeathed to us by the canon and civil lawyers of that period.[43] In Tierney's words:

> The doctrine of individual rights was not a late medieval aberration from an earlier tradition of objective right or of natural moral law. Still less was it a seventeenth-century invention of Suarez or Hobbes or Locke. Rather, it was a characteristic product of the great age of creative jurisprudence that, in the twelfth and thirteenth centuries, established the foundations of the Western legal tradition.

Tierney's point is well taken. The surge of interest in legal theory, stimulated by the need to resolve one of the most pressing problems of the Middle Ages, that of the relationship between the spiritual and the temporal authorities, seemed to dictate an approach to moral matters that focused to an unprecedented degree on rights and duties. Yet it is doubtful whether the definition of rights as faculties suffices by itself to separate the modern from the ancient concepts of right. From the absence of any explicit distinction between objective and subjective right in their works,[44] one cannot infer that the Greek philosophers would have objected to the designation of rights as powers or faculties, for such they must somehow be if by reason of them human beings have the ability to do or refrain from doing certain things. The crucial issue concerns the status of these rights, which must be either natural or legal.

If the texts cited by Tuck, Tierney, and Villey prove anything, it is that rights as the Middle Ages understood them were subordinated to an antecedent law that circumscribes and relativizes them. For Ockham, a "right" was a "lawful power," *licita potestas*.[45] For his contemporary, Johannes Monachus, it was a "virtuous power," *virtuosa potestas*.[46] As the adjectives used to qualify them indicate, these rights were by no means unconditional. They were contingent on the performance of prior duties and hence forfeitable. Anyone who failed to abide by the law that guaranteed them could be deprived of everything to which he was previously entitled: his freedom, his property, and in extreme cases his life. Not so with rights in the modern sense, which are variously described as absolute, inviolable, imprescriptible, unconditional, inalienable, or sacred. In Hobbes's version of that doctrine, which, whatever else may be said about it, has the merit of consistency, a criminal who has been justly sentenced to death retains his right of self-preservation and may kill his hangman if he has the opportunity to do so.[47]

It may be objected that, since rights are already implied in the notion of duty—if I have a duty to do something, I must have the right to do it—there is no point in opposing them to each other. What they represent would be nothing other than the two sides of a single coin. Thus, as Tierney observes, the precept "Honor thy father and thy mother" is not only a commandment; it also means that parents have a subjective right to the respect of their children.[48] Fine, but this leaves open the question as to which of the two comes first and is to be given the right of way in the event of a conflict between them. It is often when they sound most alike that premodern and modern writers are furthest apart. A case in point is the notion of self-preservation, which for the medievals is first and foremost a duty—one is forbidden to commit suicide or do anything that would impair one's health—and for the Moderns a right. The point is well made by Godfrey of Fontaines, who writes:

> Each one is obliged by the right of nature to sustain his life, which cannot be done without exterior goods; therefore also by the right of nature (*iure naturae*) each has dominion and a certain right (*ius*) in the common exterior goods of this world, which right also cannot be renounced.[49]

All of this is to say that premodern Christian ethics was primarily an ethics of duty and not an ethics of rights. Its great representatives wrote books on law, *De legibus*, or on duties, *De officiis*, and stressed one's obligations toward God and neighbor.[50] The notion of natural rights was according to all appearances unknown to them. Not so with the Moderns, who come out resolutely on the side of rights. The pivotal text on this score is the programmatic statement of *Leviathan*, chapter 14, where, breaking with a two thousand year-old tradition, Hobbes begins with rights and boldly asserts their priority over duties:

> The right of nature, which writers commonly call *ius naturale*, is the liberty

each man has to use his own power as he will himself for the preservation of his own nature—that is to say, of his own life—and consequently of doing anything which, in his own judgment and reason, he shall conceive to be the aptest means thereunto.

From this fundamental right of self-preservation Hobbes goes on to infer the various "natural laws" to which human beings would do well to bind themselves if, once civil society has come into being, they wish to enjoy any measure of freedom. To repeat, what distinguishes the new notion from the old is not the understanding of rights as powers but the concentration on rights rather than duties or law as the absolute moral phenomenon.

I do not wish to imply that Hobbes's revolutionary view of morality is the one that *Rerum novarum* was trying to pass off as authentic Christian doctrine. On the contrary, the encyclical is at pains to Christianize the rights theory by inserting it into a properly moral framework. It does this in the only way possible by reversing Hobbes's procedure. Just as Hobbes derives the law of nature from a fundamental *right* of nature, namely, the right of self-preservation, so the encyclical attempts to derive natural rights from a fundamental *law* of nature. Its clearest statement to this effect is the one that presents self-preservation as a duty from which stems the right to procure the necessities of life for oneself and one's family (no. 44). The concern evinced for the family in this context and elsewhere is already a clear sign of the encyclical's desire to work within a premodern framework. No such concern is present in modern liberalism, which has two main themes: the atomic individual possessed of prepolitical rights, and the contractual society into which he "enters" for the protection of these rights.

Numerous other features of the text point in the same direction. By often speaking of rights in conjunction with duties, the encyclical indicates that it has no intention of separating the two, let alone opposing them to each other. Duties are both stressed and properly listed (e.g., no. 19). The law that imposes them is even referred to as "most sacred," *sanctissima* (no. 13). We are further reminded that some of these duties, such as the observance of the day of the Lord, have God as their object and must therefore be "religiously" (*sancte*) observed (no. 40).

The trouble is that the encyclical speaks in the same way of rights, which, as we have seen, it likewise labels "sacred"[51] and to which in other instances it seems to accord a certain priority over duties. It mentions rights roughly twice as often as it does duties and usually ahead of duties when the two appear together.[52] Elsewhere, it asserts that rights become "stronger" (*validiora*) when considered in connection with duties, thereby implying that they are already strong apart from any relation to duties (no.

12). It calls self-preservation a duty in one place and a natural right in another, again without specifying whether it is first a duty and then a right or vice versa (nos. 7 and 44). On one occasion, what the Bible expresses in the form of a commandment it inexplicably translates into the language of rights. Thus, the injunction to "increase and multiply" (Genesis 1:28) becomes the "natural and primeval right to marry," *ius coniugii naturale ac primigenum* (no. 12). In a text devoted to moral matters, details of this sort are not without significance. Wittingly or unwittingly, the message conveyed is that at the very least rights are to be placed on more or less the same footing as duties. It is not an unimportant message.

One runs into similar ambiguities when one tries to combine pre-political rights with the notion of the common good. The modern rights doctrine in its original and still most powerful form amounts to nothing less than a proclamation of the sovereignty of the monadic individual. The common good, on the other hand, presupposes the subordination of the individual to the community to the extent that, lacking self-sufficiency, he is dependent on it for the attainment of his end or perfection. If in some respects he transcends civil society, it is not qua individual but qua member of another society, called by Augustine the "city of God," whose good surpasses that of any temporal society. Thomas Aquinas's often repeated dictum still holds: the good of the whole takes precedence over the good of the part.

Leo had good reason to decry the individualism of the age, but in the long run his case against it was bound to be weakened by his acquiescence, however cautious, in the principle that anchors modern individualism at its deepest level, the inviolability of natural rights. What we are left with is a diluted version of rights as well as a diluted version of the common good, which will soon be conceived as nothing more than the sum of the conditions required to insure the free exercise of one's individual rights. The common good ceases to be the proper (albeit not the private) good of the individual members of society, as Thomas thought it was and as it must be if it is to be truly common; it becomes an alien good, for which there is no natural inclination to sacrifice oneself. The logical outcome is the characteristically modern phenomenon known since Rousseau as the "bourgeois," the man who lives for himself in the midst of people on whom he depends for his well-being and in whom he must therefore pretend to be interested—in other words, the man who distinguishes his own good from the common good, as opposed to the citizen, who identifies the two.[53]

The difference between the two positions comes out most clearly when one considers the encyclical's assertion that the individual is "older," *senior*, than civil society and endowed by nature with the right to life and the protection of his body "prior to his entry into any civil society."[54] Such is not the view put forward by Thomas, who saw no reason to disagree with Aristotle's statement that civil society is prior to the individual.[55] For him, there was never a moment when the human being was not subject to a higher authority and hence, in principle, a member of a community governed by that higher authority. The Garden of Eden bears no similarity whatsoever to what the early modern political theorists called the "state of nature," by which they meant essentially the Hobbesian "war of every man against every man" or, in Locke's polite reformulation of the same doctrine, a state in which every individual, having the "executive power of the law of nature," was free to take the law in his own hands and do whatever he personally deemed necessary to insure his self-preservation.[56] The original natural rights theory is of a piece with this teaching and unintelligible without it.

Differently and more cogently expressed, the state of nature is not a fresh or updated version of the biblical account of human beginnings: it is a substitute for it. If anything, its implications are profoundly atheistic. Leo XIII may have had no choice but to fall back on the rights theory as a bulwark against the inroads of socialism since there was little on the intellectual horizon that could have served his purpose equally well, but the strategy had its drawbacks. There are better ways of guarding against modern totalitarianism, its deification of the state, and its threat to freedom. After all, Aristotle and Thomas were not totalitarians. They never talked of civil society as if it were not subject to the higher norm of reason, nature, or the natural law. Individual human beings and civil societies, they thought, were to be judged not by what they are absolutely but by what constitutes their good, their end, or their perfection.

One could go a step further and argue that, paradoxically, modern liberalism, the regime that prides itself on its ability to secure for each human being as much freedom as is compatible with everyone else's freedom, paves the way for a new kind of tyranny, the tyranny of each individual over every other individual. Such is the situation that is apt to develop, for example, when in the name of sacred rights known criminals are granted the same immunities as all decent citizens. The anomaly is precisely the one that the older tradition was careful to avoid. To quote Thomas himself, no human being is ordered to another human being as to his end: *creatura rationalis, quantum est de se, non ordinatur ut ad finem ad aliam, ut homo ad hominem. Si hoc fiat, non erit nisi inquantum homo propter peccatum irrationalibus creaturis comparatur.*[57]

THE NEW SYNTHESIS

I began by suggesting that the teaching and the language of *Rerum novarum* stem from two distinct traditions, one premodern and the other modern. The first is teleological and stresses duties. It holds that human beings are naturally political and directed to some preestablished end in the attainment of which they find their perfection or happiness. The second is nonteleological and stresses rights. It denies that there is any supreme good to which human beings are ordered by nature and views them from the standpoint of their beginning or the passions by which most of them are habitually moved, namely, the desire for security, comfort, pleasure, and the various amenities of life. For the same reason, it denies that they are natural parts of a larger whole whose common good is superior to the private good of its individual parts.[58] In the course of the discussion, I pointed to some of the difficulties involved in any attempt to blend the two approaches. At this juncture, two alternatives come to sight. The combination can take the form of an eclectic compromise that remains on the plane of the original positions and splits the difference between them, or it can take the form of a genuine synthesis, effected on the basis of a principle that transcends the plane of the original positions. Can the encyclical be said to have achieved a genuine synthesis?

The challenge that it faced was rendered particularly acute by the radical heterogeneity of the positions whose amalgamation was being sought. The heart of the modern project was from the beginning and has remained ever since its so-called "realism." Its originators had concluded on the basis of experience that premodern ethical and political thought had failed because it made impossibly high demands on people. It studied human nature in the light of its noblest possibilities and was thus compelled to speak about an ideal that is seldom if ever achieved among human beings. In a word, it was utopian or, to use Descartes's image, quixotic.[59]

The best way to remedy the situation, it was decided, was to lower the standards of human behavior in order to increase their effectiveness or propose an ideal that was more easily attainable because less exacting. This led in due course to a recasting of all basic moral principles in terms of rights. Most people would rather be informed of their rights than reminded of their duties. Edmund Burke knew whereof he spoke when he said, "The little catechism of the rights of men is soon learned; and the inferences are in the passions."[60]

The beauty of the new scheme is that it supposedly made it possible for everyone to enjoy the benefits of virtue without having to acquire it, that is, without undergoing a painful conversion from a premoral concern with worldly goods to a concern for the goodness of the soul. The intelligent pursuit of one's selfish interest would do more for the well-being of society than any concerted attempt to promote the common good. Mandeville, another bona fide Lockean, captured the spirit of the new enterprise as well as anyone else when, in *The Fable of the Bees: Private Vices, Publick Benefits* (1705), he argued that the day bees started worrying about moral virtue the hive would be ruined and that it would recover its prosperity only when each one returned to its vices. This, to a higher degree than it perhaps imagined, is the position that the encyclical would synthesize with its basic Thomistic outlook.

Syntheses, it has been aptly said, produce miracles. The miracle in this case consisted in using the categories of modern thought to restore something like the lofty morality they were originally calculated to replace. That miracle has yet to be attested. Catholics heard more about rights from Leo than they had from any other pope, and, thanks in large part to him, they were destined to hear more and more about them as time went on. Yet opinion continues to be divided as to whether this new emphasis has led to the spiritual renewal that Leo hoped to foster. Suffice it to say that in recent years the Church has had its hands full trying to resist the changes that are being urged upon it by some of its members in the name of the very rights it now promotes. In the best instance, the rights for which Leo pleaded had to do with the freedom to worship God and fulfill the rest of one's God-given duties. Once these rights were granted independent status, however, it was only a matter of time before they were parlayed into a moral argument to criticize the law rather than to justify one's observance of it.

This brings me to my last question, which is whether the drafters of the encyclical—Liberatore, Zigliara, Mazzella, Boccali, Pope Leo himself, and others—were fully aware of their dependence on modern modes of thought. The simplest answer is Yes. The Pope may have felt that the time had come to abandon the intransigence of his immediate predecessors and temper their inflexible principles with a more flexible policy, even if this meant adopting a terminology that is neither native nor congenial to the older Catholic tradition. Without capitulating to modernity, Catholics had to develop attitudes that were appropriate to living in the modern world. After all, there was much to be said for liberal or constitutional democracy, which, of the two viable alternatives on the contemporary horizon, is the one that came closest to what Christianity had always recommended. Moreover, something could be done to correct its defects, whereas the defects of socialism were seemingly irremediable. Then, too, it was always possible to live in one's time without sharing the principles of that time. This much was evident from the examples of Swift and Lessing, not to mention others.

There is nonetheless reason to suspect that Leo and his mentors were invincibly blind to the theoretical implications of some of their statements. The nineteenth century is unfortunately not the most auspicious period in the history of Catholic theology. University life had been severely disrupted by the French Revolution and the Napoleonic wars. The leading centers of theological learning were shut down at the end of the eighteenth century and remained closed for at least three decades. The Roman College was not reopened until 1824; Louvain, until 1830. Their reorganization would take a minimum of two generations, for the core of new teachers needed for this gigantic undertaking would first have to be trained before they could train others. As for the French universities, they were stripped of their theological faculties and never regained them. The slack was eventually taken up by the new "Instituts Catholiques," but not until the last decades of the century. More importantly, an age-old tradition had been broken, and we know from past experience that the restoration of a lost tradition is an uncommonly difficult task.

It is true that Thomas Aquinas was beginning to attract attention after close to two centuries of virtual neglect, but relatively little was known about him. Upon his arrival in Rome in 1846, the year that followed his conversion to Roman Catholicism, John Henry Newman, whom Leo XIII would elevate to the rank of cardinal thirty-three years later (in 1879), could complain in a letter to an English friend about the "prevalent depreciation of St. Thomas" in Italy.[61] Another letter to the same correspondent paints a gloomy picture of Catholic intellectual life in that country. A person with whom he was living had just informed him that

> we should find very little theology here, and a talk we had yesterday with one of the Jesuit fathers shows we shall find little philosophy. It arose from our talking of the Greek studies of the Propaganda and asking whether the youths learned Aristotle. O no—he said—Aristotle is in no favor here—no, not in Rome—not St. Thomas. I have read Aristotle and St. Thomas and owe a great deal to them, but they are out of favor here and throughout Italy. St. Thomas is a great saint—people don't dare to speak against him—they profess to reverence him and put him aside. I asked what philosophy they did adopt. He said *none*. Odds and ends—whatever seems to them best—like St. Clement's *Stromata*. They have no philosophy. Facts are the great things, and nothing else. Exegesis but not doctrine. He went on to say that many privately were sorry for this, many Jesuits, he said; but no one dared oppose the fashion.[62]

The picture had brightened somewhat in the forty-five years that separate Newman's stay in Rome from the publication of *Rerum novarum*. There was greater sympathy for Thomas in Italy and Germany, and the opposition to Thomism in influential Traditionalist circles had abated. Yet, even as late as 1879, the year of *Aeterni patris*, the encyclical that established Thomas as the Church's foremost authority in matters of philosophy and theology, Thomism was far from having won universal approval among Catholics. Reservations about its Aristotelian component persisted among Traditionalists, as well they might, given Traditionalism's historicist premises, and the Thomistic school itself was torn by divisions between Suarezians, whom Leo disliked, and non-Suarezians. The other crucial factor to be taken into account is the pervasive influence of modern liberalism, especially in the form that it took in the philosophy of John Locke, the dominant political theorist of the modern age.

A prime example of the syncretism to which the confluence of these variegated streams of thought could lead is Taparelli d'Azeglio, whose *Saggio teoretico di dritto naturale* became the prototype of the manuals of Catholic moral theology that followed one upon the other throughout the ensuing century. In a letter to his Jesuit Provincial, Taparelli admits that he knew next to nothing about natural right when he began to write on it at the age of fifty, and, moreover, that whatever thoughts he did have came mainly from Locke and the other modern authors on whom he had been weaned. This is what prompts L. de Sousberghe to write:

> As we can see from his letters, Taparelli became an innovator without any conscious intention of modifying the tradition and simply by applying a Christian good sense, freed from all tradition, to the modern philosophy of his age. He thus excerpts from the heritage of the eighteenth century (where he no longer even recognizes the elements of Scholastic origin) whatever he deems capable of being assimilated by Christian teaching.[63]

Although one of Thomism's most ardent promoters,[64] Taparelli made surprisingly little use of Thomas in his own works. Recent studies have shown that his immediate authority was Christian Wolff, whose eclectic but on the whole conservative views had gained enormous popularity among Catholics in Germany, Austria, and Italy during the second half of the eighteenth century. Wolff's appeal to Neo-Thomists was all the greater as he himself claimed to have borrowed more from Thomas than from Leibniz, the other great source on whom he drew copiously.[65] Little wonder that Leo's theologians, some of whom had been disciples and colleagues of Taparelli, should have had so much trouble separating the wheat from the chaff or distinguishing between what was or was not compatible with standard Catholic doctrine.

To sum up, the problem with *Rerum novarum* is that it lives in two worlds between which it cannot choose and which it is unable to harmonize completely. The surprising fact is not that it fell short of its stated goal, assuming that it did, but that its authors were able to accomplish so much with the somewhat meager resources at their disposal. Nothing that I have said, more for the sake of clarity than by way of criticism, should diminish our admiration for their work.

The enterprise was a daunting one. It has never been easy to explain how Christianity, the only essentially nonpolitical religion known to history, should relate to the ambient world. Since, unlike the Hebrew scriptures, the New Testament does not supply us with a detailed code of laws by which to live—as Cardinal Caetani is made to say in Ignazio Silone's *The Story of a Humble Christian*, "You can't govern with the Pater Noster"[66]—Catholic theology has no choice but to look to philosophy for guidance in this matter. In the Middle Ages, Aristotle became its best ally. Needless to say, our own world is very different from that of our medieval forebears. Specifically, the intellectual forces of which it is the product may not be as easily reconcilable with the moral teachings of the Bible as was Aristotelian philosophy. Still, it is the only world in which we have been made to live. The task of Christian leaders is to understand it and find new ways of causing the light of the gospel to shine within it. *Rerum novarum*'s pioneering achievement is to have convinced them of the urgency of that task and made them aware of the nature of the obstacles that stand in the way of its realization. True fidelity to its spirit requires that anyone who would follow in its footsteps do what its authors wanted to do and probably would have done had they had access to the vastly more adequate, albeit still largely untapped, resources that have since become available to us.

NOTES

The present chapter is a revised and enlarged version of a paper presented at the International Conference on *Rerum novarum* held at the Lateran University in Rome on May 6-9, 1991. The original text, entitled *"Rerum novarum* and Modern Political Thought," appeared in the Proceedings of the conference, *Rerum novarum: L'uomo centro della Società e via della Chièsa*, Rome: Libreria Editrice Vaticana, 1992, 81-109.

1. Luigi Taparelli d'Azeglio, *Saggio teoretico di dritto naturale appoggiato sul fatto* (Palermo, 1840-43 and Rome: La Civiltà Cattolica, 1855). The book, an unexpected success, was soon reprinted in Naples, Livorno, and Rome, sometimes without the author's knowledge. A French translation, supervised by Taparelli himself, appeared in 1857. The book was also translated into German (1845) and into Spanish (1866-68). Cf. R. Jacquin, *Taparelli* (Paris: P. Lethielleux, 1943), 175.

2. In *Centesimus annus* (1991), Pope John Paul II refers to it as the Church's "social doctrine," "social teaching," or "social magisterium" (no. 2; cf. nos. 13, 54, 56).

3. *Quadragesimo anno*, nos. 147 and 39. The paragraph numbers for this encyclical as well as for *Rerum novarum* are those of the English translation as it appears in *The Papal Encyclicals*, 5 vols., edited by Claudia Carlen, IHM (Wilmington, NC: McGrath Publishing Co., 1981).

4. For an interpretation of *Gulliver's Travels* along these lines, see A. Bloom, "Giants and Dwarfs: An Outline of *Gulliver's Travels*," in *Giants and Dwarfs: Essays, 1960-1990* (New York: Simon and Schuster, 1990), 35-54.

5. The great philosophic authority behind this view is Hegel, who was followed with a variety of modifications by numerous historians, among them R. W. and A. J. Carlyle; see their monumental *History of Mediaeval Political Theory in the West*, vol. I, 2d edition (Edinburgh and London: W. Blackwood and Sons, Ltd., 1927), 2: "There are no doubt profound differences between the ancient mode of thought and the modern; . . . but just as it is now recognized that modern civilization has grown out of the ancient, even so we think it will be found that modern political theory has arisen by a slow process of development out of the political theory of the ancient world—that, at least from the lawyers of the second century to the theorists of the French Revolution, the history of political thought is continuous, changing in form, modified in content, but still the same in its fundamental conceptions."

6. For a recent survey of the debates surrounding this complex issue, cf. T. L. Pangle, *The Spirit of Modern Republicanism: The Moral Vision of the American Founders and the Philosophy of Locke* (Chicago and London: The University of Chicago Press, 1988), 7-39. Also P. A. Rahe, "John Locke's Philosophical Partisanship," *The Political Science Reviewer* 20 (1991): 1-43.

7. For a clear and penetrating statement of the problem, see *inter alia* Leo Strauss, "Progress or Return? The Contemporary Crisis in Western Civilization," in *An Introduction to Political Philosophy: Ten Essays by Leo Strauss*, edited by H. Gildin (Detroit: Wayne State University Press, 1989), 249-310. The text is also reprinted in *The Rebirth of Classical Political Rationalism: Essays and Lectures by Leo Strauss*, edited by T. L. Pangle (Chicago and London: University of Chicago Press, 1989), 227-70. The impetus for reconsidering this issue originally came from Heidegger, who, in an attempt to "uproot" a philosophic tradition that had supposedly gone astray almost from the beginning, was led to study Aristotle with a thoroughness that had not been seen since the thirteenth century. To a good number of his disciples, the exhumed Aristotle proved more convincing that Heidegger's critique of him.

8. The trend is well illustrated not only by Marx and the other nineteenth-century socialists but, on a more popular level, by such widely read books as Victor Hugo's *Les Misérables* and Michelet's *Le Peuple*. Another work by Hugo, entitled "Ascension Humaine," casts the problem in the sharpest possible light by depicting a downtrodden and evanescent humanity rising to new heights, magnificently (and, one might add, blasphemously) accomplishing by its natural powers a deed that the Christian tradition reserved for Christ. Virtue was to be discovered by plunging into the depths of society and a new "religion of humanity" would soon replace Christianity as the instrument of mankind's salvation. See on this subject the recent account by R. Emmet Kennedy, Jr., "The French Revolution and the Genesis of a Religion of Man," in *Modernity and Religion*, edited by R. McInerny (Notre Dame: University of Notre Dame Press, 1994).

9. This is not to say that on a local level numerous efforts had not already been made to come to grips with the problem. For these earlier developments in France, see J. B. Duroselle, *Les débuts du catholicisme social en France* 1822-1870 (Paris: Presses universitaires de France, 1951). R. Talmy, *Aux sources du Catholicisme social: L'Ecole de La Tour du Pin* (Tournai: Désclée, 1963).

10. No. 9 states explicitly that private property is a right that no one is ever allowed to violate: *nec ullo modo . . . violare cuiquam licet.* See also no. 15: *privatae possessiones inviolate servandae [sunt].*

11. For similar uses of *sanctus* in its adjectival or adverbial form, cf. no. 22: *res proletariorum, quo exilior, hoc sanctior habenda; no. 7: iura quidem . . . sancte seranda sunt.*

12. Cf. P. Thibault, *Savoir et pouvoir: Philosophie thomiste et politique cléricale au XIXe siècle* (Quebec: Presses de L'université de Laval, 1972), 123; ibid., 49. Taparelli's treatment of private property occurs in book 2, ch. 4, of his *Saggio teoretico,* 2d editition (Rome: La Civiltà Cattolica, 1855), vol. I, 252-61.

13. *Decretum,* D. I, cap. 7: "*Iure gentium est distinctio possessionum et servitus; iure naturae est communis omnium possessio et omnium una libertas.*" The text is taken from Isidore, *Etym. Libr.* XX.5.4.

14. See also, ibid., II-II, 57, 3. On the nineteenth-century conceptions of private property among Catholic theologians, see the important article by L. de Sousberghe, "Propriété de droit naturel: thèse néo-scolastique et tradition scolastique," *Nouvelle Revue Théologique* 72/6 (June 1950): 580-607.

15. The pertinent text from the first draft reads as follows: "Essa [la proprietà privata] discende per immediato discorso da' primi principii della ragione; e pero è constituita universalmente ed appellata: *De iure gentium.*" *L'Enciclica RERUM NOVARUM: Testo autentico e redazioni prepatorie dai documenti originali,* edited by G. Antonazzi (Roma, 1957), 40, line 37. For the text of the third draft, cf. ibid., 95, lines 351-60. A second edition of this work was published by Edizioni di Storia e Letteratura of Rome in 1991.

16. *Summa theologiae* II-II, qu. 67, a.3, ad 3. Cf. M. Liberatore, *Principles of Political Economy,* translated by E. H. Hering (New York: Benziger, 1891), 117-28.

17. Antonazzi, *Testo autentico,* 46, line 265. The notion of sacred rights already appears in Liberatore's *Institutiones Philosophicae,* vol. III: *Ethica et Ius Naturae,* 10th editition (Rome, 1857), 166: *Qui viribus caret, is omni iure destitueretur, et sola praepotentia ius gigneret; cum contra quo magis imbellis est et sine viribus, eo sanctiora putari debeant iura quae possidet ac maiori veneratione colenda.*

18. For a perceptive analysis of Locke's controverted theory of property, see in particular L. Strauss, *Natural Right and History* (Chicago: University of Chicago Press, 1952), 234-46.

19. Adam Smith, *The Wealth of Nations* I.10.2: "The property which every man has in his own labour, as it is the original foundation of all other property, so it is the most *sacred* and *inviolable.*" The "sacred rights of property" are mentioned again in

I.11. 2. The formula had already been employed by J.-J. Rousseau, *Emile*, bk. V, translated by Allan Bloom (New York: Basic Books, 1979), 461: "The right of property is inviolable and sacred for the sovereign authority as long as it remains a particular and individual right." See also Rousseau, *Discours sur l'économie politique, Oeuvres complètes*, Vol. III (Paris: Gallimard, 1964), 263: "Le droit de propriété est le plus sacré de tous les droits et plus imporatnt à certains égards que la liberté même."

20. R. Schlatter notes that this statement was intended "as a weapon that would destroy feudal property and absolutism with one blow," *Private Property: The History of an Idea* (London: Allen and Unwin, 1951), 222. See also the *Déclaration* passed by the National Convention, June 23, 1793: "Le peuple français, convaincu que l'oubli et le mépris des droits naturels de l'homme sont les seules causes des malheurs du monde, a résolu d'exposer dans une déclaration solennelle ces droits *sacrés et inaliénables* . . . ARTICLE 1: Le but de la société est le bonheur commun. Le gouvernement est institué pour garantir à l'homme la jouissance de ses droits naturels et imprescriptibles. ARTICLE 2: Ces droits sont l'égalité, la liberté, la sûreté, la propriété."

21. *Morning Star* (13 March 1868). By that time, the notion that private property might be sacred had long been under attack from socialists. See, *inter alia*, Proudhon's famous essay, *Qu'est-ce que la propriété?* (Paris: Garnier-Flammarion, 1966, c. 1840), chapter 2: "Car, si la propriété est de droit naturel, . . . tout ce qui m'appartient en vertu de ce droit est aussi *sacré* que ma personne." Chapter 1 of the same work had begun by declaring that "Property is theft." The fifth letter of Schiller's *On the Aesthetic Education of Man* alludes in ironic fashion to the "sacred rights" of society.

22. See the text in Antonazzi, *L'enciclica* 227: *[L']inviolabilità* del dritto di proprietà è indispensabile per la soluzione pratica ed efficace della questione operaia." Leo's earlier encyclical, *Quod Apostolici Muneris* (1878), where the subject of private property is already taken up, speaks of its "inviolabilty" but not of its "sacredness": *ius proprietatis ac dominii, ab ipsa natura profectum, intactum cuilibet et inviolatum esse [Ecclesia] iubet. Acta Sanctae Sedis* 11 (Rome, 1916), 377.

23. The classic treatments of this subject are J. N. Figgis, *The Divine Right of Kings*, 2d edition (Cambridge: Cambridge University Press, 1914); Fritz Kern, *Kingship and Law in the Middle Ages* (Oxford: B. Blackwell, 1939), Part I: *The Divine Right of Kings*. Also, for a summary account, G. H. Sabine, *A History of Political Theory*, 3rd edition (New York: Rhinehart and Winston, 1961), 391-95. The doctrine is well expressed by Shakespeare in *Richard II*, act III, scene 2, lines 55-62 (quoted on p. 169). Shakespeare's Richard is not as pious as he sounds. The "glorious angel" that God has placed at his side is probably nothing more than the gold coin stamped in the effigy of Michael the Archangel with which he was able to pay his soldiers. See the frequent allusions to the "sacredness" of the person of the king elsewhere in Shakespeare, e.g., *Winter's Tale*, act II, scene 3, line 85; act V, scene 1, line 172; *King John*, act III, scene 1, line 148; *Richard II*, act I, scene 1, line 119; act

I, scene 2, lines 12 and 17; act III, scene 3, line 9; act V, scene 2, line 30; *Henry VIII*, act II, scene 4, line 41; act III, scene 2, line 173.

24. See also, toward the end of the same paper: "The President of the United States would be an officer elected by the people for four years; the king of Great Britain is a perpetual and hereditary prince. The one would be amenable to personal punishment and disgrace; the person of the other is *sacred and inviolable*."—The remote origins of this hierocratic conception of political rule are to be sought in the Old Testament custom of anointing kings, beginning with Saul; cf. I Sam. 10:1. As the anointed of God, the person of the king, bad as he might have become, was inviolable; cf. 2 Sam. 1:14, where David says to the Amalekite who, in obedience to Saul's command, had put an end to the dying king's life: "How is it you were not afraid to destroy the Lord's anointed?" David himself had previously refused to kill Saul when the opportunity presented itself on the ground that "no one can put forth his hand against the Lord's anointed and be guiltless" (I Sam. 26:9). As for the use of the terms *sacer* and *sanctus* in connection with these matters, they seem to have made their appearance in the course of the twelfth century for the purpose of securing the independence of the Empire from the papacy. Cf. F. Kern, *Kingship in the Middle Ages*, 66-67. The theory of sacred kingship suffered a severe setback in the second half of the thirteenth century with the rediscovery of Aristotle's *Politics* and its account of the natural foundations of political rule. It was later revived in England as a means of consolidating the authority of the British crown after the break with Rome.

25. On the individual's absolute right to property, see, for example, *Second Treatise of Civil Government*, no. 138: "The supreme power cannot take from any man part of his property without his own consent," for the preservation of property is "the end of government and that for which men enter into society . . . Men, therefore, in society having property, they have such right to the goods which by the law of the community are theirs, that nobody hath a right to take their substance or any part of it from them without their own consent; without this, they have no property at all."

26. The parallel is suggested by Locke himself, who alludes at least twice to the "sacred" character of the monarch who rules by divine right. Thus, *Second Treatise*, no. 205: "[I]n some countries, the person of the prince by the law is *sacred*; and so whatever he commands or does, his person is still free from all question of violence, not liable to force, or any judicial censure or condemnation." *First Treatise*, no. 107: "This designation of the person our author is more than ordinarily obliged to take care of, because he, affirming that the 'assignment of civil power is by divine institution', hath made the conveyance as well as the power itself *sacred*; so that no consideration, no act or art of man, can divert it from that person to whom by this divine right, it is assigned, no necessity or contrivance can substitute another person in his room."

27. *Second Treatise*, no. 27: "Though the earth, and all inferior creatures be common to all men, yet every man has a property in his own person; this nobody has any right to but himself. The labour of his body and the work of his hands, we may say, are properly his. Whatsoever then he removes out of the state that nature hath

provided and left it in, he hath mixed his labour with, and joined to it something that is his own, and thereby makes it his property . . . For this labour being the unquestionable property of the labourer, no man but he can have a right to what that is once joined to, at least where there is enough and as good left in common for others." Cf. *Rerum novarum*, no. 9: "Truly, that which is required for the preservation of life, and for life's well-being, is produced in great abundance from the soil, but not until man has brought it into cultivation and expended upon it his solicitude and skill. Now, when man thus turns the activity of his mind and the strength of his body toward procuring the fruits of nature, by such an act he makes his own that portion of nature's field which he cultivates—that portion on which he leaves, as it were, the impress of his personality; and it cannot be but just that he should possess that portion as his very own, and have right to hold it without anyone being justified in violating that right." No. 10: "For the soil which is tilled and cultivated with toil and skill utterly changes its condition; it was wild before, now it is fruitful; it was barren, but now it brings forth in abundance. That which has thus altered and improved the land becomes so truly part of itself as to be in great measure indistinguishable and inseparable from it . . . As effects follow from their cause, so it is just and right that the results of labor should belong to those who have bestowed their labor."

   28. For a concise statement of the contrast between the older view and the new one, cf. P. Larkin, *Property in the Eighteenth Century with Special Reference to England and Locke* (London and New York: Longmans, Green, 1930), 1-20. On the development of the modern notion of property, see, most recently, T. A. Horne, *Property Rights and Poverty: Political Argument in Britain, 1605-1834* (Chapel Hill: University of North Carolina Press, 1990).

   29. Cf. *Rerum novarum*, no. 22: "It is one thing to have a right to the possession of wealth and another to have a right to use wealth as one wills. Private ownership, as we have seen, is the natural right of man, and to exercise that right, especially as members of society, is not only lawful but absolutely necessary . . . But if the question be asked: 'How must one's possessions be used?' the Church replies without hesitation . . . . 'Man should not consider his material possessions as his own, but as common to all, so as to share them without hesitation when others are in need'." For the distinction between ownership and use, cf. Aristotle, *Politics* 1263a26ff.: "Property ought to be in a certain sense common, but, generally speaking, private . . . . Clearly, it is better for property to be owned privately but made common as to its use."

   30. Cf. P. Thibault, *Savoir et pouvoir*, 176-77. For further parallels between *Rerum novarum* and Locke's theory of property, see, in addition to L. de Sousberghe, loc. cit., R.L. Camp, *The Papal Ideology of Social Reform* (Leiden: F. J. Brill, 1969), 54-56. Also, for a general assessment of Leo's theory and the tacit correction of some aspects of that theory in *Quadragesimo anno*, see Camp, ibid., 66-67, 84-87, 98-99, and 146-47. With unnecessary bluntness, R. Schlatter remarks apropos of *Rerum novarum*: "One of the last upholders of the conservative interpretation of the natural right of property was the Church of Rome. That the Church should accept a doctrine born of the Enlightenment and promulgated by the French Revolution is surprising."

Private Property: The History of an Idea (London: Allen and Unwin, 1951), 278.

31. Anatole France, *Le lys rouge* (*The Red Lily*), ch. 7 (Paris: Calmann-Lévy, 1960).

32. Anatole France, *L'Île des pingouins* (*Penguin Island*), bk. 6, ch. 2 (Paris: Calmann-Lévy, 1973).

33. Ibid., bk. 2, ch. 4.

34. Anatole France, *Le Mannequin d'osier*, in *Histoire contemporaire*, ch. 5 (Paris: Calmann-Lévy, 1981).

35. See Taparelli's *Saggio teoretico di dritto naturale*, vol. I, bk. 2, ch. 3, entitled: "Nozioni del dritto e della giustizia sociale." Taparelli is for all practical purposes the inventor of that celebrated, if somewhat vague, notion. One can say without undue simplification that social justice is the object of what was being considered more and more as a separate discipline, namely, social ethics.

36. *Quadragesimo anno*, no. 44. O. von Nell-Breuning, who drafted the new encyclical, attributes this "calumny" to bad translations or false interpretations of *Rerum novarum* and wonders why these were never challenged: "*Rerum novarum:* A General Appraisal," *Social Justice Review* 82 (May-June 1991), 76 (translated from Nell-Breuning's *Soziallehre der Kirche*, 3rd edition, Vienna, 1983). This may not be the whole story, however. Nell-Breuning himself admits that, like much of the teaching of *RN*, "the doctrine of private property bears unmistakable traces of the then prevailing spirit of the age." See also Nell-Breuning's more detailed commentary on *Quadragesimo anno's* doctrine of private property in his *Reorganization of Social Economy* (New York: The Bruce Publishing Company, 1936), 91-121. For the opposition to some aspects of Leo's teaching on the part of La Tour du Pin and other members of the French Oeuvre des Cercles, cf. Talmy, *Aux sources du catholicisme social* (Tournai, 1963), 103-50. Much useful information regarding these matters is to be found in G. Jarlot, S.J., *Doctrine pontificale et histoire*, vol. I (Rome: Presses de l'université Grégorienne, 1964), 202-25; vol. II (Rome, 1973), 247-79.

37. *Quadragesimo anno*, no. 28; Latin text in *Acta Apost. Sedis* 23 (1931), 185. The text notes that this whole matter had by then become the object of a new branch of legal science, one that was unknown to previous ages: *Ex hoc autem continenti atque indefesso labore nova iuris disciplinae sectio superiori aetate prorsus ignota orta est, quae sacra opificum iura ab hominis christianique dignitate profluentia fortiter tuetur.*

38. Cf. *Rerum novarum*, no. 22, which *Quadragesimo anno* correctly interprets as follows: "In order to place definite limits on the controversies that have arisen over ownership and its inherent duties, there must be first laid down as a foundation a principle established by Leo XIII: the right of property is distinct from its use. That justice called commutative commands *sacred* respect for the division of possessions and forbids invasion of others' rights through the exceeding of the limits of one's own property; but the duty of owners to use their property only in a right way does not come under this type of justice, but under other virtues, obligations of which 'cannot

be enforced by legal action'. Therefore they are in error who assert that ownership and its right use are limited by the same boundaries; and it is much further still from the truth to hold that *a right to property is destroyed or lost by reason of abuse or non-use*" (no. 47, emphasis added). Cf. Denzinger, *Enchiridion Symbolorum*, no. 2255. Nell-Breuning later stated expressly that the intention of the new encyclical was not to modify Leo's teaching but merely to complement it by means of a greater emphasis on the social function of property. For his overall assessment of the situation, see "*Rerum novarum*: A General Appraisal," loc. cit., 72-77. Cf. also on this matter the useful bibliographical references assembled by M. Habiger, O.S.B., *Papal Teaching on Private Property, 1891-1981* (Lanham: University Press of America, 1990), 90ff.

39. Cf. *supra*, note 11.

40. It should not surprise us that the best book on "natural rights" in Thomas, F. Rousseau's *La croissance solidaire des droits de l'homme* (Paris: Desclée and Montreal: Bellarmin, 1982), is unable to come up with a single Thomistic text in which such rights are mentioned.

41. M. Villey, "Suum ius cuique tribuere," *Studi in onore di Pietro de Francisci*, vol. 2 (Milan: Giuffre, 1956), 364.

42. M. Villey, *La formation de la pensée juridique moderne*, 4th edition (Paris: Montchrestien, 1975), 261.

43. R. Tuck, *Natural Rights Theories: Their Origin and Development* (Cambridge: Cambridge University Press, 1979). B. Tierney, "Villey, Ockham and the Origin of Individual Rights," in *The Weightier Matters of the Law: Essays on Law and Religion*, edited by J. Witte, Jr., and F. S. Alexander (The American Academy of Religion, 1988). See also Tierney's richly documented follow-up article on the same subject, "Origins of Natural Rights Language: Texts and Contexts, 1150-1250," *History of Political Thought* 10 (1989): 616-46, and, by the same author, "Marsilius on Rights," *Journal of the History of Ideas* 52 (1991): 3-17.

44. To illustrate: it is not clear whether, in defending himself against his accusers, Socrates is exercising a right or fulfilling a duty. All he says is that it is right or just (*dikaios*) for him to do so; cf. *Apol.*, 18a.

45. Cf. Villey, "La genèse du droit subjectif chez Guillaume d'Occam," *Archives de philosophie du droit 9* (1964): 117.

46. Johannes Monachus, *Glossa aurea* (Paris, 1535), fol. xcir, *Glossa ad sext.* I.6.16. In Gerson and his pupil, Jacques Alamain, "right" (*ius*) is a power or faculty subordinated to the "rule of right reason," *dictamen rectae rationis*; cf. Tuck, *Natural Rights Theories*, 28.

47. Cf. Hobbes, *Leviathan*, ch. 21: "Therefore, if the sovereign command a man, though justly condemned, . . . not to resist those that assault him, . . . yet has that man the liberty to disobey." Ibid., ch. 28: "For by that which has been said before, no man is supposed bound by covenant not to resist violence, and consequently it cannot be intended that he gave any right to another to lay violent hands upon his person. In the making of a commonwealth, every man gives away the right of

defending another, but not of defending himself." For a clear statement of the traditional position on this subject, see Thomas Aquinas, *Summa theologiae* II-II, qu. 69, a. 4: "It is not permissible for a justly condemned man to defend himself." Thomas adds that, while the condemned party is not allowed to kill his executioner, he is not morally obliged to stay in jail and may escape should the occasion present itself.

48. Loc. cit., 20. A third possibility, which would have occurred more naturally to a medieval thinker, is that it is objectively right for children to respect their parents. The same could be said of Tierney's second example, to the effect that the natural law precept "Thou shall not steal" implies that others have a natural right to acquire property. Again, this is not quite the same thing as saying that it is right for some people to acquire and own property.

49. *Quodlibet* 8. qu. 11, *Philosophes Belges*, vol. 4 (1921), 105.

50. "Duty," like "rights," has no exact equivalent in Greek. It is the word commonly used to render the Latin *officium*, which in Cicero combines the meanings of two Greek words, *katorthûma* ("that which has been done rightly," "right action") and *kathêkon* ("the fitting"). Cf. Cicero, *De officiis* I.3.8.

51. Cf. *supra*, note 11.

52. Cf. nos. 2, 12, and 58: *iura et officia*. In no. 25, *officia* precedes *iura*.

53. See, for a penetrating analysis of this remarkable phenomenon, A. Bloom, Introduction to J.-J. Rousseau, *Emile or On Education* (New York: Basic Books, 1979), 5.

54. *Rerum novarum*, no. 7: *est enim homo, quam respublica, senior*. Cf. no. 12, where *antiquior* is substituted for *senior* without any apparent change of meaning. See, for a strong statement of the same position, Liberatore, *Principles of Political Economy*, 130: "[T]he right of property arises in us as an individual and domestic right, and therefore as substantially prior to civil society and independent of it; just as the human person and the family are prior to civil society and independent of it. The State has authority over the rights that come from itself. It has no authority over rights that come from nature—rights that precede the State in history and in reason."

55. Thomas Aquinas, *Sententia libri politicorum, opera omnia*, vol. XLVIII (Rome, 1971), 79, *ad* 1253a19.

56. John Locke, *Second Treatise of Government*, no. 13. The fact that, unlike Hobbes, Locke begins by speaking of a "law" rather than of a "right" does not separate him from Hobbes on this point, for in the state of nature this law is to be interpreted in accordance with the dictates of a fundamental passion, the desire for self-preservation. Not without reason does Locke refer to his teaching as "very strange" or novel; ibid., nos. 9 and 13. For a statement of the premodern position, according to which the meting out of sanctions is the prerogative of rulers and not a right to be exercised by private individuals, cf. Thomas Aquinas, *Summa theologiae* II-II, qu. 64, a. 3: "[T]he execution of a criminal is licit insofar as it is ordered to the welfare of the whole community. Hence it is reserved for the one to whom the care

of preserving the community has been entrusted . . . Now the care of the common good is entrusted to rulers invested with public authority. Therefore, only they, and no private persons, are permitted to put a criminal to death."

57. *II Sent.* d. 44, qu. 1, a. 3, c.

58. See on this point L. Strauss, *Natural Right and History* (Chicago: University of Chicago Press, 1952), 177-86.

59. René Descartes, *Discourse on Method*, Part I: "Thus it happens that those who regulate their behavior by the examples they find in books are apt to fall into the extravagances of the knights of romances and undertake projects which it is beyond their ability to complete or hope for things beyond their destiny . . . I compared the ethical writings of the ancient pagans to very superb and magnificent palaces built only on mud and sand: they laud the virtues and make them appear more desirable than anything else in the world; but they give no adequate criterion of virtue, and often what they call by such a name is nothing but cruelty, apathy, parricide, pride or despair." See in a similar vein Francis Bacon, *The Advancement of Learning*, book 2, c. fin.: "As for the philosophers, they make imaginary laws for imaginary commonwealths, and their discourses are as the stars, which give little light because they are so high." The argument was first made by Machiavelli, who writes in ch. 15 of *The Prince,* the manifesto of the new movement and a book that Descartes held in high esteem (see his précis of it in a letter of September, 1646, to Elizabeth of Bohemia), "[H]ow we live is so far removed from how we ought to live that he who abandons what is done for what ought to be done will rather learn to bring about his own ruin than his preservation." Spinoza, a disciple of Machiavelli in more ways than one, was of the same opinion, convinced as he was that the older philosophers had all made the mistake of conceiving of human beings "not as they are but as they themselves would like them to be" (*Political Treatise*, ch. 1, Introd.).

60. Edmund Burke, *Thoughts on French Affairs*, in *Works of Edmund Burke*, vol. III.377.

61. J. H. Newman, *The Letters and Diaries*, vol. XI, edited by Ch. St. Dessain (London: T. Nelson, 1961), 303.

62. Ibid., 279.

63. Loc. cit., 594. On Taparelli in general, see R. Jacquin, *Taparelli* (Paris, 1943), and, for a rapid overview, M. Prélot, "Taparelli d'Azeglio et la renaissance du droit naturel au XIXe siècle," *Archives de Philosophie Politique* 3 (1959): 191-203. Because of the inauspicious moment at which it appeared, Jacquin's brilliant study never received the recognition that it deserves. Nor for that matter has Taparelli himself, who was held in highest esteem well into this century. In footnote 33 to his encyclical on *The Christian Education of Youth* (*Divini Illius Magistri*, 1929), Pius XI speaks of his *Saggio* as "a work never sufficiently praised and recommended to university students." In an allocution pronounced by the same Pope on December 18, 1927, and referred to in the encyclical, Taparelli is all but placed on a par with Thomas Aquinas: "Ma oltre le opere dell'Aquinate, altre ve ne sono che hanno non

meno eccellenza e freschezza di dottrina, che possono essere studiate e consultate in ogni tempo. Una di queste opere è certamente il *Saggio teoretico di dritto naturale* del P. Taparelli d'Azeglio." Cf. Jacquin, *Taparelli*, 159 and 343, n. 7. Jacquin adds that he personally heard Pius XI praise Taparelli in similar terms on the occasion of a papal audience held at Castel-Gandolfo in August 1936; ibid., 157.

64. See, for a statement of his Thomistic convictions, Taparelli's programmatic essay, "Del progresso filosofico possibile nel tempo presente," *Civiltà Cattolica*, Serie seconda, vol. 3 (1853), 265-87. Also on this subject R. Jacquin, "Le P. Taparelli d'Azeglio et le renouveau de la philosophie thomiste," *Aquinas* 5/3 (1962): 427-30.

65. Cf. M. Thomann, "Christian Wolff et le droit subjectif," *Archives de Philosophie du Droit* 9 (1964): 163. For further details and the appropriate bibliographical information, see Thomann's introductions to Wolff's *Jus Naturae* and *Jus Gentium, Gesammelte Werke*, Abt. II, vol. 17 (Hildesheim, 1972), XII-XXIII; vol. 25 (1969), XLVI-LI; and vol. 26 (1969), V-XLII. Taparelli's dependence on Wolff was first pointed out by Thomann in the above-mentioned article, "Christian Wolff et le droit subjectif," 174. The dependence is especially noticeable in regard to the twin pillars on which social ethics as a more or less autonomous discipline may be said to rest: the priority of the individual to civil society and the emphasis on prepolitical rights. The complete title of Taparelli's main work, *Theoretical Essay on Natural Right Based on Facts*, reflects the influence of Wolff, whose ambition was to combine two rival methods of scientific inquiry, the rational or deductive and the empirical or inductive. Whatever Taparelli's intention may have been, his own method is far more deductive than inductive. The "facts" with which he is concerned are not those of modern experimental science. They are roughly synonymous with common sense or prescientific knowledge, that is, with such knowledge as is likely to be part of universal human consciousness. Wolff's influence on Taparelli appears to have been reinforced by Wolff's follower, Burlamaqui, whose immensely popular manual on natural right Taparelli had begun by using before writing his own. On Burlamaqui and Wolff, cf. P. Meylan, *Jean Barbeyrac et les débuts de l'enseignement du droit dans l'ancienne académie de Lausanne* (Lausanne: F. Rouge, 1937).

66. Ignazio Silone, *Avventura d'un povero cristiano* (*The Story of a Humble Christian*), translated by William Weaver (New York: Harper and Row, 1970).

17

FROM *RERUM NOVARUM* TO *CENTESIMUS ANNUS*: CONTINUITY OR DISCONTINUITY?

Centesimus annus, coming as it does on the heels of the dramatic collapse of socialism in Europe and throughout the world, has indicated in a spectacular way *Rerum novarum*'s prophetic (if a bit one-sided) stand on private property, but not without adding a good deal of its own to what Leo had said. I shall limit myself to a few observations about the points on which it seems to me to go beyond *Rerum novarum*, depart from it, or otherwise improve upon it.

The most obvious of these concerns, the emphasis on the "universal destination of material goods," the issue on which, if my interpretation is correct, Leo's encyclical was most vulnerable. Leo had made it plain that material goods, although privately owned, were intended for everyone's benefit, but by shifting their use from the sphere of justice to that of charity, he deprived that principle of all legal force and left its largely unspecified application to the initiative of individual Christians or groups of Christians.

On that score, *Centesimus annus* has a lot more to offer. It devotes a long chapter to this matter and goes into considerable detail about such timely issues as aid, not only to the poor and the marginalized among us, but to poorer nations, the protection of the environment, the maximization of human as well as of material resources, and the like. It also prudently refrains from speaking of private property as a sacred right, thus tacitly repudiating a key teaching of *Rerum novarum* or at least rephrasing it in such a way as to make it conform more closely to standard Church teaching. The change was all the more proper as the rapidly evolving political situation in Eastern Europe had done away with the need to trumpet the inviolability of private property as a bulwark against the rising tide of socialism. When Pope John Paul II affirms that for Leo private property was not an "absolute value" (no. 6) or an "absolute right" (no. 30), he appears to be offering what used to be called a "benevolent interpretation" of his predecessor's teaching. Granted, Leo's vigorous defense of private property is hedged about with all kinds of admonitions regarding its charitable use, but the gist of his argument is precisely that, contrary to what the Church and virtually everyone else had previously held, one's ownership of legitimately acquired goods is not forfeited by one's misuse of these goods. I might add that John Paul II's reinterpretation is benevolent in more ways than one, for there is a good chance that, were he alive today, Leo would be among the first to acquiesce in it.

If *Centesimus annus* takes a less extreme view of private property than *Rerum novarum,* in other respects it shows itself far more open to modern modes of thought. To measure the distance that separates the two encyclicals, one has only to check how often Thomas Aquinas is quoted in each of them—several times in Leo's encyclical; not once to my knowledge in *Centesimus annus.* This is not to suggest that Thomas is totally absent from the latter but only that its tone is demonstrably less Thomistic than that of *Rerum novarum.*

Centesimus annus notes with good reason that in discussing the organization of society "the Church has no models to present," since truly effective models can only arise within the framework of different "historical situations" (no. 43). It likewise freely admits that the Church "is not entitled to express preferences for this or that institutional or constitutional solution," the devising of which is a task usually best carried out by people on the spot.

These disclaimers notwithstanding, it would be hard to deny that the encyclical offers us at least a general model comprising three distinct elements. First, a liberal democratic structure that acknowledges the freedom and transcendent dignity of the person as a being endowed with prepolitical rights that neither the state nor anyone else is permitted to curtail. Second, a free-market mechanism that encourages entrepreneurial initiative as a means of stimulating economic growth and enabling workers to build a better future for themselves and their families. For the first time in a document of this kind, the encyclical speaks in positive terms of "capitalism," a word long held in suspicion by theologians because of its tainted origins but now redefined in keeping with the norms of Christian morality (cf. no. 42). The third element is the common destination of material goods, which we have already mentioned and which is expressly introduced as a safeguard against the glaring inequities that frequently if not habitually arise when the economic life is controlled by market forces alone (cf. no. 19). History is there to remind us that the founders of modern liberalism were wrong in thinking that, in and of itself, the free interplay of these market forces can be trusted to bring about a reasonably just distribution of goods within a particular society. It is simply not the case that enlightened self-interest is the sole key to the success of our communal endeavors. Without a generous measure of civic virtue, no society, liberal or otherwise, is likely to endure, let alone thrive. Hence the encyclical's concern to reintroduce into the debate such cardinal but long-forgotten notions as the "common good" or the "spirit of cooperation and solidarity" (cf. no. 61 et passim).

As Pope Leo had already argued, the liberal democratic system just outlined has enormous practical advantages over its great modern rival, Marxist socialism. It does a better job of producing the goods that we require for our subsistence and well-being. It makes these goods readily available to a much larger number of people. By doing away with the "bureaucratic oppression" associated with socialism, it makes it possible for human beings to exercise the freedom on which their personal dignity depends. Finally, it is capable of self-correction in a way in which socialism is not. These are the reasons that justify both *Centesimus annus*'s ringing endorsement of modern liberalism and the enthusiasm with which that endorsement had been greeted by thoughtful analysts here and abroad. Liberal democracy is after all the modern regime that comes closest to what the Christian tradition had always recommended. Such was Leo's conclusion and such also is the conclusion at which, on the basis of far greater empirical evidence—Leo and his generation had not yet seen socialism in action—the new encyclical arrives.

What is equally important but more easily overlooked in the midst of the euphoria generated in some circles by the publication of *Centesimus annus* are the severe warnings that accompany its defense of liberalism. The least that can be said about the encyclical is that it is anything but naive about the state of our liberal democratic affairs or blind to the dangers of our free-market economy. It does not suffice to say that the Western democracies have thus far failed to live up to their own principles. The problem lies much deeper, in the principles themselves, which do not lead to a high level of morality or foster the kind of spiritual life that the encyclical calls for.

The term most often used by the Pope to describe this morally less attractive side of American life is "consumerism," the modern version of hedonism, that is, the excessive attachment to material goods and the unrestrained commitment to their pursuit. The problem, although more visible today than ever before, has been with us for a long time. Americans, Tocqueville once suggested, had performed the extraordinary feat of elevating "egoism" to the level of a philosophical principle. He spoke not as a critic but as a true friend of liberal democracy, one who was willing to administer the bitter medicine it needed if its promises were to be fulfilled. *Centesimus annus* does not argue differently. One of its fears is that the Western countries will interpret the demise of socialism as a one-sided victory that dispenses them from making "necessary corrections" in their own system (no. 56).

And there is much to be corrected. The list of evils has seemingly grown longer rather than shorter with the passage of time. One has only to think of the breakdown of the family, the abortion plague and the push for the legalization of euthanasia, the sexual revolution, the principled defense of pornography in the name of freedom of speech, the scandal of poverty and homelessness in the midst of great opulence, the squalor of our ghettos and devastated neighborhoods, the growth of the drug culture, the proliferation of blue-collar and white-collar crime, the scams and rip-offs of which we are the often unconscious victims, the pitiable state of American education at all levels, the trivialization of the arts and the media, and with that the list is far from complete.

One can always argue that the liberal democratic system is not itself responsible for these evils, that it is an essentially benevolent system whose aim is to provide decent living conditions for as many of us as possible, and that the amount of crime, licentiousness, and vulgarity with which we have to put up is the price to be paid for the freedom we enjoy. While the argument is not without merit, it nevertheless fails to take into account some important facets of the problem. In particular, it overlooks the fact that by and large the needs which our market economy strives to satisfy are not natural but artificial needs, stimulated by the market itself and calculated to increase its profitability. Instead of trying to inculcate habits of self-restraint in the minds and hearts of its citizens, it attaches them ever more firmly to the material goods in terms of which it teaches them to measure their standard of living. What it confronts us with is not the simple hedonism to which common human nature is prone but a dynamic hedonism that feeds upon itself, endlessly creating the new desires that will fuel our commercial enterprises.

The fact of the matter is that modern liberalism has always been better at taking care of our bodies than of our souls. Nor has it ever tried to do anything else. This is not to deny that the system occasionally succeeds in producing its own brand of morality. The redefined capitalism of which the Pope speaks—"capitalism properly understood," as Michael Novak calls it in a phrase reminiscent of Tocqueville's "self-interest rightly understood," is not amoral. It, too, has its virtues, but they are of a different order. They are the instrumental virtues of bourgeois society, virtues that are more concerned with the proper functioning of the system than with the perfect order of the soul. As Pope John Paul II writes:

> Important virtues are involved in this process, such as diligence, industriousness, prudence in undertaking reasonable risks, reliability and fidelity in interpersonal relationships, as well as courage in carving out decisions which are difficult and painful but necessary, both for the overall working of a business and in meeting possible setbacks (no. 32).

For such virtues we can be grateful, but it is doubtful whether they will give us all what we could and should have as human beings and Christians.

Professor Buttiglione goes straight to the heart of the matter when he says that the core of *Centesimus annus* is its attempt to replace the alliance between libertinism and a free market economy by the alliance between a free market and solidarity. What remains unclear is how concretely this alliance is to be brought about and what will hold it together once it has been forged, especially since its two poles originate in different parts of the soul and tend to pull us in opposite directions. Unless something is done to cement it, the new alliance is liable to prove much less stable than the first. According to the encyclical, the solution to the problem is to be sought in religion, morality, law, education, and culture, all of them "values" to which it grants, and urges the state to grant, autonomous status (cf. no. 19). All well and good, at least until such time as one begins to wonder about the grounding of these autonomous values.

Following Thomas Aquinas, Leo XIII looked to nature for that grounding. The assumption was that human beings are ordered to certain preexisting ends, such as the knowledge of the truth and the achievement of moral excellence, to which they are inclined by nature itself. It was the task of education to build upon, strengthen, and purify these natural inclinations through the acquisition of the virtues, both dianoetic or intellectual and moral. One learned to look upon the good of the whole of which one was a part, not as an alien good, but as one's own good. Solidarity or action in concert with others for the good of the whole became, as was said, "connatural." What reason prescribed or pointed to, nature supported. The whole idea was to harmonize duty and inclination in such a way as to overcome our alienations and recover the wholeness of which our fallen condition so often deprives us.

One can understand Pope John Paul II's reluctance to adopt this line of thought, based as it is on a teleological conception of nature that has supposedly been destroyed by modern science. But there may also be other reasons for his apparent lack of enthusiasm for Thomism, one of them being that the Thomistic school had long been under attack for its failure to come to grips with the problems of the modern age. To make matters worse, twentieth-century Thomism has been further discredited by the involvement of some of its leaders in right-wing political movements on the continent before and during World War II. Pastoral concerns, if nothing else, likewise dictated that the Pope's message be delivered in a language that is more congenial to most of our contemporaries, for whom the categories of medieval Scholasticism are now mostly unintelligible.

Accordingly, *Centesimus annus* makes abundant use of what Professor Buttiglione calls a "personalistic metaphysics" the object of which is to establish by means of a phenomenological analysis, not indeed of nature, but of the acting person the requirements of a just and full human life. The strategic advantage of such an approach is that it abstracts from natural inclinations and is thus immune to the criticisms leveled at teleology by modern science and philosophy. An added advantage is that it presents us with a loftier conception of morality than the one associated with modern liberalism in its primitive or Lockean form.

One must nonetheless ask how, without further support from nature, this moral ideal will prevail over the powerful forces ranged against it— how, humanly speaking, the "ought" of practical reason will be able to hold its own against the "is" of the passions that resist it. In plain terms, there is strong support on the part of nature for economic entrepreneurship and its monetary rewards. The passions that it calls into play are actual at all times and do not require a painful conversion from a concern for material goods to a concern for the good of the soul. The same cannot be said of the moral "ought," which gets a better billing in the new scheme, but not necessarily better results. What we end up with most of the time is the dreary spectacle of people who talk as if the "ought" were the only thing that matters and act as if there were nothing but the "is." Commerce and industry prevail over morality and culture and a more or less intelligent selfishness determines the course of their impoverished lives.

As the title of this chapter suggests, the problem that I set out to explore is whether the content of *Centesimus annus* is continuous or discontinuous with that of *Rerum novarum* or, since the answer is apt to be "both," whether one of these elements outweighs the other. Professor Buttiglione, who is well aware of the issue, resolves it in an eminently sensible way by distinguishing between Christian social teaching, which is immutable, and Christian social doctrine, which develops as a result of the application of that teaching to changing historical situations and is therefore mutable. As a living tradition, Christianity implies more than faithfulness to the letter of the heritage it seeks to preserve; it demands faithfulness to the creative impulse to which it owes its greatness and which, one presumes, is still operative in it. The decisive question in that case is not whether there are changes to be made but whether the proposed changes are compatible with the principles underlying the basic teaching. I am reasonably certain, for example, that Leo XIII would have had no trouble accepting John Paul II's emended version of his stance on private property. Would he have been equally comfortable with the many other novel features of his doctrine?

The question is not an easy one to answer, but we can make a beginning of sorts by adverting to John Paul II's unprecedented insistence on the more or less Kantian notion of the "dignity" that is said to accrue to the human being, not because of any actual conformity with the moral law, but for no other reason than that he is an "autonomous subject of moral decision" (no. 13). The more usual view, which Kant was rejecting, is that one's dignity as a rational and free being is contingent on the fulfillment of prior duties. That dignity could be forfeited and was so forfeited by the criminal who had no respect for and no desire to abide by the moral law. One's goodness or dignity was not something given once and for all; it was meant to be achieved. Its measure was one's success in attaining the end or ends to which one was ordered by nature. The Rousseauean and Kantian notion of the sovereign or sacred individual had yet to make its appearance. To be and to be good were two different things.

The matter would obviously require a much more careful examination than any that we have time for, but the little that I have said about it may help cast the problem in its proper light. Just as *Rerum novarum* bears traces of the transition from late medieval to early modern thought, that is, from the divine right of kings to the sacred right of private property, so *Centesimus annus* bears traces of the transition from early modernity to late modernity, that is, from the Lockean notion of the sacredness of private property to the eighteenth-century notion of the sacredness of the sovereign individual. My question is whether this rather remarkable change in the Church's "Doctrine" might not be indicative of a more than inconsequential change in its "Teaching."

V

PAGAN AND CHRISTIAN VIRTUE

18

THE SAGA OF SPIRITEDNESS:
CHRISTIAN SAINTS AND PAGAN HEROES

Christianity is not the kind of religion to which one turns in the first instance for examples of what the classical tradition meant by heroes and heroism. While it recognizes the need for military valor, it cannot be said to extol it, and since it is not essentially a political religion, it does not prize the virtues of the magnanimous statesman as highly as they tend to be prized elsewhere. Few soldiers adorn its calendar of saints. Martin of Tours is the most famous one and he is remembered more for giving up his arms than for having borne them. There is no evidence that he ever used them. One does find a number of kings, queens, and princes in that calendar—Louis IX of France, the German emperor Henry II, Stephen of Hungary, and others—all of them included in it on account of what they did for the Church rather than of what they did for their countries.

In the East, Constantine the Great, the first Christian emperor, was considered by many to have been a saint. The unofficial canonization was never recognized by the universal Church, perhaps with good reason. The year in which Constantine convoked the Council of Nicea was also the one in which he had his son and his sister's son murdered, needless to say, for political motives. These were not his only crimes. Reconciling the demands of the faith with the duties of statesmanship has never been an easy task, as we know from *Henry V*, Shakespeare's account of the advent of the Christian empire, as could be shown from the numerous allusions to the reign of Constantine in the play. In the *City of God*, Constantine and the other famous early Christian emperor, Theodosius, are praised for their private rather than their public virtues: Constantine because he legitimized Christianity and restored to the Church the properties that had been confiscated by his predecessors; Theodosius because he submitted to the will of Ambrose, the bishop of Milan, and sat in ashes.

The Middle Ages made Trajan into a Christian saint, again not because he had given the Roman Empire its largest expansion through the conquest of the Parthians and Dacians, but because according to popular legend he had been brought back to life by Pope Gregory the Great and baptized before his final destiny was sealed. One day, on the occasion of his triumphal return to Rome after a successful campaign, he supposedly stopped to console the bereaved mother of one of the Roman soldiers killed in battle. The scene was duly portrayed on a marble column commemorating the victory and placed next to the forum, where it still stands. Struck by that scene, Pope Gregory decided that any pagan who showed that much concern for the lowly and the suffering deserved a better fate. Accordingly, he resurrected him and conferred upon him the baptism that secured his entry into heaven. No modern historian vouches for the authenticity or the theological orthodoxy of the story, and I am not sure that anyone ever did. But it is an edifying story, one that has every right to be included in the *Golden Legend*, that incomparable treasury of medieval *mirabilia*. Besides, some points of doctrine regarding the state of souls after death were still in a state of flux. The *Divine Comedy* had yet to be written and the topography of the next world lacked clarity.

Nothing that I have said thus far should come as a great surprise. Christ, from whom Christians take their name and whom they are supposed to imitate, did not distinguish himself by virtues that qualify as heroic in the usual sense of the term. He never ruled—according to the temptation scene in the gospel, he turned down the devil's offer to exercise dominion over all the nations of the earth. He does not appear to have been motivated by the desire for glory, which is what heroes strive for above all else. He rebuked his disciples for aspiring to honors higher than those granted to others. And he encouraged everybody to seek the last places. Forbearance, gentleness, the forgiveness of offenses, and the other so-called "passive virtues" were to be the hallmark of his followers, who had to learn from him that he was "gentle and humble of heart" (Mt. 11:29). The point was made with the utmost clarity in the Sermon on the Mount, the charter of the new religion, which declares that the "poor in spirit" are the ones who will inherit the kingdom of heaven. The idea was not entirely new, for these "poor in spirit," *ptokoi tô pneumati*, are the remote descendants of the *anawim* celebrated by the prophet Zephaniah, the poor who are closer to God because their material poverty makes it easier for them to acknowledge their total dependence upon him. To quote Zephaniah himself,

> For I will leave in the midst of you a people humble and lowly. They shall seek refuge in the name of the Lord, those who are left in Israel. They shall do no wrong and utter no lies, nor shall there be found in their mouth a deceitful tongue.[1]

Christ himself would suffer an ignominious death on the cross, a punishment reserved for common criminals. No religion ever held up such an image by which to inspire its followers. The pagans could not think of their gods as being anything but beautiful. A "suffering servant" was beyond their ken and provoked nothing but revulsion. None of them would have regarded humility as a form of human excellence and elevated it to the rank of a virtue.

One puts it mildly when one says that in its official teaching Christianity was never high on spiritedness. The Church Fathers often compared Christ to Odysseus, who had himself tied to the mast of his ship so as not to succumb to the song of the Sirens. The scene, often represented in ancient art, reminded them of Christ with his arms outstretched on the cross. They never compared Christ to Achilles, the best and most spirited of the Achaeans. In his famous *Address to Young Men on Reading Greek Literature*, Basil the Great lists all of the virtues that Christians might look for in the works of the pagan classics.[2] One virtue is conspicuous by its absence: courage, the virtue most closely allied to spiritedness.

Perhaps it was felt that by that time there was no longer any real need for it. Roman expansionism had long come to an end and consolidation was the order of the day. In this endeavor, Christianity, whose universality paralleled that of the Roman Empire, had a unique contribution to make. It stressed the virtues of harmony and cooperation among all human beings regardless of ethnic origin, language, or national customs and could thus be pressed into service to counteract the forces of disunity at work within Roman society. The cost on the other side of the political ledger was, all things considered, minimal. Even the disparagement of military valor, which at other moments and under different circumstances, could only be thought of as a liability, had suddenly turned into an asset, favoring the ends to which in its self-interest imperial policy was now committed. By the same token, emperors had little to fear from a religion that derived the institution of civil government from the will of God, frowned upon sedition, and discouraged worldly ambition with as much vigor as it extolled the virtue of obedience to one's divinely sanctioned rulers. If even under the reign of Nero St. Paul had preached the duty of unconditional submission to tyrants, how much more willing would Christians be to acquiesce in the rule of a prince who was at the same time a patron and a defender? It is no accident that Christianity eventually made its bed in the Roman Empire and has been, so to speak, wedded to it ever since. To this day, its geographical parameters coincide by and large with those of the Rome of old, its colonies, and the colonies of its colonies, with the notable exception of Russia. Nowhere else, it seems, has Christianity been able to strike deep roots or make substantial inroads among the native populations.

One could go a step further and argue that Christianity not only does not encourage heroism but destroys its very possibility. The point would require a much more extensive treatment than any that can be accorded to it here, but one cannot help wondering what happens to the tragic hero once it is understood that there is a full personal life, together with the prospect of eternal rewards or punishments, after death. Tragic heroism is, of course, not the only form of heroism, but in some way it incorporates all the others and functions as a mirror in which their merits are reflected. What gives to the tragic deed its supreme worth is that by engaging in it the hero risks the permanent loss of everything that is dearest to him—a situation which can never be that of a Christian who takes his faith seriously since by forfeiting his life on earth he only gains a better one in heaven.

There were other reasons to think that the old-style heroism was in for a hard time, most of them neatly laid out in books I-V of the *City of God*, Augustine's critique of Rome and of the pagan notion of glory. That critique anticipates by twelve centuries the one that will be developed by Descartes, who argues in the *Discourse on Method* that what the Ancients called by the name of virtue is "nothing but cruelty and apathy, parricide, pride or despair."[3] Contrary to popular belief, Augustine never said that the virtues of the pagan were "splendid vices," but he came mighty close to doing so. On that score, Harnack, who attributes the expression to him, guessed right, especially if one gives due prominence to the implications of the word "splendid."

There are two sides to the tale of Rome's greatness, one bad and one good. Everyone admitted that Rome had been guilty of the most horrendous crimes. Its empire was a gigantic larceny, as is made clear by the story of Alexander and the pirate in the *City of God* IV.4, the shortest and perhaps most devastating chapter of the whole work. In many ways, Roman glory was nothing but a mask for the endless slaughter of kinsmen and allies by a group of people who believed that "the greatest renown resides in the greatest empire."[4] The critique is not entirely original. In elaborating it, Augustine was piggybacking on Cicero, Rome's first philosopher and greatest demythologizer. The difference is that he is more out-spoken than Cicero, boldly lifting the veil with which the latter had attempted to cover Rome's atrocities. Yet Augustine, an ambitious man himself, realized that there is a distinction to be made between the love of human glory, *cupiditas humanae gloriae*, and the love of domination, *cupiditas dominandi*. Ambition may be a vice, but, unlike greed or the sheer lust for power, it comes close to being a virtue.[5] It enabled the Romans to live for something greater than their individual selves and thus kept them from falling into base vices. Even in his later years, Augustine could still feast his soul on the lives of Regulus, Torquatus, Fabricius, Lucius Valerius, and other noble *maiores*.[6] True, they had sought glory, but they lived simply, were wholly dedicated to the good of their fellow citizens, returned all that they earned to the city, and died in self-imposed poverty.

This being the case, one is sometimes tempted to think that with a nudge here and there a suitable compromise between pagan and Christian virtue might be reached or that some *tertium quid* might be called in to mediate their differences. The illusion must be resisted. The bottom line is that the pagan ways, resplendent as at their best they undoubtedly were, no longer constituted a worthy object of imitation. They had been super-seded by gentler ways, which would henceforth prevail, or so the old bishop hoped. The hope was not unfounded. Centuries later, Machiavelli was forced to admit that "not the slightest trace of the ancient virtue re-mains." He added that the majority of those who now read the ancient histories "take pleasure only in the variety of events which they relate, without even thinking of imitating the noble actions, deeming that not only difficult but impossible."[7]

The context of the late Roman Empire to which Christianity originally belonged, the *imperium sine fine* sung by Virgil, was not, in fact, destined to last forever. There would be moments in the history of the West that called for a resurgence of the heroic spirit. Most prominent among them is the one that witnessed the armed conflict against Islam, which necessi-tated the mobilization of all of Christendom's military resources. The holy places, captured by the infidel, had to be retaken and there were other Christian lands from which the invader had to be expelled. At one mo-ment, most of the Middle East, all of North Africa and Spain, and half of France were in enemy hands. The long struggle—it lasted close to a thou-sand years—gave rise to a novel form of heroism that goes under the name of chivalry, a complex phenomenon about which for present pur-poses little needs to be said except that it is distinguished from older forms of heroism by two main features: its expressly religious dimension and the role that it assigns to women in the elaboration of the chivalric ideal.

Spiritedness is too much a part of the human soul to be limited to the resistance to external enemies. In the absence of such enemies, human beings inevitably seek new objects for it and new opportunities to exercise it. The early Christians did not have to look far for them. They came face to face with them in the person of the emperor, who decided early on that their presence posed a threat to Roman life. In those days, to become a Christian was to declare oneself a candidate for persecution. The situation called for a different kind of courage, the one displayed by the martyr, the "witness" par excellence to his faith. The comparison was drawn by Augustine, who announced that "if the ordinary language of the Church allowed it, one might more elegantly call those who gave their lives for the faith our heroes."[8]

Barely more than three centuries had elapsed before the persecutions abated and Christians found themselves in positions of power within an empire that was or would soon become officially Christian. Since human beings cannot live without certain distinctions that set those who excel among them apart from everyone else, heroism had to undergo a metamorphosis. Some Christians came to the conclusion that, now that the world had stopped rejecting them, they had to reject it. A new age had dawned, that of the Desert Fathers, whose prototypical hero was the ascetic. The goal was, if not the actual annihilation of one's natural impulses, at least the total mastery over them. The pursuit of that goal occasionally led to practices best described as exotic. A number of ascetics became stylites, living for as many as thirty-five years on narrow platforms at the tops of pillars. On the theory that the devil made his abode in the desert, others withdrew into desolate places, as Christ himself had done at the beginning of his public ministry. The idea was to beard the lion in his den. Not surprisingly, the biographies of these hermits abound in vivid accounts of their daily encounters with him.

Competition among the monks became increasingly fierce, inasmuch as each one felt obliged to match the austerities indulged in by others. If one of them managed to survive for a month on two loaves of bread, his neighbor, not to be outdone, had to get by on twenty cabbage leaves. The controlling metaphor was supplied by St. Paul, who speaks in the Letter to the Philippians of the runner who strains for the finish line with every ounce of energy at his disposal. The key word, which is the one Paul had used, is *epecteinô*, to "stretch out" or "press forward." The underlying notion turned out to be quite Aristotelian. Virtue, defined as the limit of one's capacity, is demonstrated in its employment. One knows what one can do when one does it. The results can be surprising.

A similar concept of spiritual perfection was at work in the lives of the Irish monks, such as Columban and Boniface, who would soon be preaching the gospel to the tribes of northern Europe. One assumes that such endeavors were inspired by a desire to convert others to the faith, but the impression is misleading. Behind them, it seems, lay an ideal of detachment akin to the one sought by the Desert Fathers. Theirs was the vocation of Abraham, who had been summoned by God to leave his country, his kindred, and the comforts of his father's house and go off into an unknown and possibly dangerous land.

The missionary enterprise paid off in other ways as well. It prepared the ground for the emergence of a Christian civilization on the European continent. Like its predecessors, the new civilization had to have its heroes. Their definition was invented by Thomas Aquinas, who spoke for his age with almost as much authority as Augustine had done for his. As so often happened at that time, the terminology was borrowed from Aristotle, who in the *Nicomachean Ethics* (VII.1) speaks of certain virtues which, by reason of their superhuman character, deserve the title of heroic or divine. A Christian equivalent was found in the seven "gifts of the Holy Spirit": understanding, counsel, science, wisdom, piety, courage, and the fear of the Lord. These were the Christian counterparts of Aristotle's "heroic or divine" virtues:[9] *Virtutes heroicae sunt dona Spiritus Sancti.* The clue came from Isaiah, who was transformed into a Scholastic theologian for the occasion and made to sound like a medieval Aristotelian, fully apprized of the distinction between the speculative and the practical intellect, along with all of their different operations. A new order of greatness had been established and codified, the "order of charity," as Pascal would call it, which was as incommensurable with the order of reason as was the latter with the material order.

In the process, the Christian hero was given what had earlier been and still is his most popular name. He was called a saint, the term that expresses in the most comprehensive way the ideal of charity for which all Christians are urged to strive even if only a few of them come close to it. That ideal is summed up in the paradoxical commandment of love, which demands the greatest sacrifices and must indeed take the form of a commandment if the love in question has as its object all of one's fellow human beings, the ugliest as well as the most attractive.

The trouble is that, by distinguishing between saints and ordinary Christians, one runs the risk of reintroducing into the Christian scheme an order of rank that is alien to the spirit of the Gospel. There is always a danger that the aspirant to personal holiness will undermine his accomplishments by taking pride in them, an insidious temptation against which it is almost impossible to guard. As Augustine, who was speaking from experience, remarks, what distinguishes pride from the other vices is that it has the knack of insinuating itself even into one's good deeds, thereby vitiating them. This, more than anything else, is what accounts for the overwhelming emphasis on humility in Christian spirituality. The New

Testament reveals its awareness of the problem by issuing seemingly contradictory injunctions in regard to the display of virtuous behavior. It tells its readers to let their light shine before men so that they may see their good works (Matt. 5:16), and in the same breath it reminds them that when they fast, they should powder their faces so that their fasting may not be seen by men but only by the Father, who sees in secret (Matt. 6:17). How the proper balance between these conflicting demands is achieved is a question that cannot be answered in the abstract and must be left to the discretion of each individual.

I began by saying that Christianity depreciates spiritedness by subordinating it to the universal love and service of God and neighbor. But love, as we know, is a problematic passion, one that is prone by nature to all sorts of excesses. Its object is the end on which one's heart is set, and the end, as distinguished from the means, is pursued without any limit. The love that can be reckoned is something other than true love. There is "beggary" in it, as we learn from *Antony and Cleopatra*. The consequences are frightening. Peter Chrysologus, a fifth-century Church Father, speaks not only for himself but for a host of others when he writes:

> The law of love is not concerned with what will be, what ought to be, what can be. Love does not reflect. It is unreasonable and knows no moderation. It refuses to be consoled when its goal proves impossible and despises all hindrances to the attainment of its object. Love destroys the lover if he cannot obtain what he loves. Love follows its own promptings. It does not think of right and wrong. It inflames desire and is impelled by it toward forbidden things. But why continue?[10]

Why indeed? Experience shows that what characterizes the Christian tradition is neither its spiritedness nor the lack thereof, the two contradictory tendencies for which it is most often blamed, but its unpredictability. To many an impartial observer, historical Christianity comes across as a loose cannon. The remedy, as the Church Fathers and their medieval disciples discovered, was to inject into it a solid dose of Platonic or Aristotelian wisdom. The results varied in accordance with circumstances, but they are generally well illustrated by Dante's *Comedy*, an epic poem that stands at a considerable distance from the old epic tradition insofar as its hero is not, strictly speaking, an epic hero. Two themes, neither of them essentially heroic, dominate it: justice and philosophy. This is not to imply that the book lacks passion. On the contrary, it is full of it. Above all, it is full of anger and moral indignation. My suspicion is that both of these are present for the purpose of lending an air of gravity to the project as a whole. At any rate, the serious reader might be inclined to take them more seriously if the book were not so funny. In Dante's view, what the age most needed was an intellectual nobility of the kind that he himself exemplified. The *Comedy* is the only great epic poem whose hero is its own author, and he was never canonized. Anyone who "in these Christian times," as Dante calls them, had the gall to portray himself as the savior of the human race was more likely to end up on the *Index of Prohibited Books*. Only the patriotic zeal of the Italian clergy prevented this from happening when the *Index* was established in the sixteenth century.

We are left with the classic view of the saint as the true exemplar of Christian virtue and with the problem that the cult of the saints poses for some of our contemporaries. That problem was well stated by Gibbon in the famous chapter 28 of *The Decline and Fall of the Roman Empire*, which deals with the introduction of what he misleadingly calls the worship of saints and relics among Christians. Gibbon, who takes his cue from Hume, describes the cult in question as a relapse into the kind of gross superstition that an earlier generation of Christians had left behind. The trend, he thought, was but another manifestation of what Hume had called the "flux and reflux of the human mind." Only a thin line separated the old polytheism from the new cult. As he put it:

> The imagination, which had been raised by a painful effort to the contemplation and worship of the Universal Cause, eagerly embraced such inferior objects of adoration as were more proportioned to its gross conceptions and imperfect faculties. The sublime and simple theology of the primitive Christians was gradually corrupted; and the Monarchy of heaven, already clouded by metaphysical subtleties, was degraded by the introduction of a popular mythology which tended to restore the reign of polytheism.[11]

In retrospect, the argument is less than persuasive, for it ignores completely the chasm that separates the cult of the saints from that of pagan heroes. Two points deserve special mention. The first is that, far from being a mere return to ancestral practice, the new cult established a strong rapport between the world below and the world above, thus obliterating the fault that traditionally ran across the face of the universe. The graves of the saints became privileged places where categories once taken to be distinct were merged and where the contrasted poles of heaven and earth finally met. The second point, which is not unrelated to the first, is that the Christian saint, unlike the deified pagan hero, was thought to enjoy close intimacy with God and was thus in a position to intercede with him on behalf of the living.

My intention in bringing this matter up is not to dwell on the differences that, *pace* Gibbon, set the cult of the Christian saints apart from the conventions of the pagan world but to suggest a possible reason for which such a cult was deemed necessary or useful. A religion that was universal in principle and became all but universal in fact, at least as far as the Roman Empire is concerned, faces the problem of adapting itself to vastly different and constantly changing environments or material circumstances. The cult of the local saint alleviated that problem. It tended to bridge the gulf between the universal and the particular and, thus, offered a measure of compensation for the losses incurred as a result of the absorption of once independent political entities into what purported to be an ecumenical empire.

The broader question is whether the ideal that I have tried to present can still attract many followers today. That ideal understood itself as the fulfillment, that is, the carrying to completion and therewith the transformation, of both the Mosaic law and Greek philosophy, "the Old Testament of the gentiles," as Clement of Alexandria named it. It purportedly fused their respective virtues into a genuine synthesis. It was a matter of thought as well as of deeds. It was contemplative as well as active, Socratic as well as Homeric, philosophic as well as heroic—all of this in an eminent way insofar as it encompassed everything that was eternally valid in the highest aspirations of both Athens and Jerusalem. St. Paul summed it up in one sentence when he urged his disciples to think only of what is "honorable, just, pure, lovely, gracious, excellent, and worthy of praise" (Phil. 4:8).

NOTES

1. Zeph. 3:12-13.

2. St. Basil the Great, "Address to Young Men on Reading Greek Literature" in *The Wisdom of Catholicism*, edited with an introduction and notes by Anton C. Pegis (New York: Random House, 1949), 8-26.

3. René Descartes, *Discourse on Method and Meditations*, translated, with an introduction by J. Lafleur (Indianapolis and New York: The Bobbs-Merrill Company, 1960), 7.

4. *De civitate Dei* (*The City of God*) III.14.

5. *De civ. Dei* V.12.

6. *De civ. Dei* V.18.

7. Machiavelli, *Discourses on Livy*, first book, Introduction.

8. *De civ. Dei* X.21.

9. Thomas Aquinas, *Summa theologiae* I-II, qu. 68, a. 1, ad 1.

10. Peter Chrysologus, Sermon 147 in Migne, *Patrologia Series Latina* (PL) 52, col. 595.

11. Edward Gibbon, *The Decline and Fall of the Roman Empire*, vol. II (New York: Simon and Schuster, 1962), 539-40.

19

IN DEFENSE OF SATAN:
CHRISTIAN PERSPECTIVES
ON THE PROBLEM OF EVIL

Moral evil comes in two forms, active and passive—the evil that human beings do and the evil that they suffer, or, to formulate the distinction in specifically religious terms, sin and guilt on the one hand and punishment on the other: *malum culpae* and *malum poenae*. To our understanding of this subject, whose timelessness is vouched for by the unspeakable horrors of the twentieth century, revealed religion, with its customary insistence on man's proneness to evil and the vicissitudes of the human heart, presumably has an important contribution to make, and the more so as Greek philosophy has no word that corresponds exactly to our notion of guilt. Nowhere outside of revealed religion does one find a greater emphasis on the gravity of sin, which the Bible views not as a form of madness or a simple miscalculation, but as a personal affront to a kind and solicitous God. Nothing in the literature of classical antiquity matches the forceful denunciations of evil that one encounters in the prophets of Israel; and while Christianity has its own interpretation of this phenomenon, it does not question it in a fundamental way. The New Testament all but begins with a call to repentance, reform, or conversion, three terms used interchangeably to render the Hebrew word *t'shuv,* "return," one of the leitmotifs of Hebrew prophetic literature. It tells us that Christ, the Messiah, came to call sinners, and when he is introduced by John the Baptist, it is precisely as the "Lamb of God who takes away the sin of the world." It is significant that the Christian tradition is the one that transformed Satan into an encyclopedic image of evil, equating him with the serpent in Genesis and elevating him to a preeminent position among the evil spirits that abound in the religious lore of the ancient Near East.

Revealed religion stands or falls by the notions of sin, guilt, and atonement; without sin and the recognition thereof, there is no need for or possibility of redemption. Adopting a procedure similar to that of Pascal, Cardinal Newman laid down as the "large and deep" foundation of religion, both natural and revealed, the innate sense of moral obligation and moral failing.[1] "Its many varieties," he wrote in his classic *Grammar of Assent,* "all proclaim that man is in a degraded, servile condition, and requires expiation, reconciliation, and some great change of nature." Sin is universal. No one is free from it. To quote Scripture once again, even the just man sins seven times a day. Anyone who is unaware of having done wrong is surely guilty of some hidden sin or some unconscious sin of omission.

Religion not only thrives on evil, but at times almost seems to encourage it. If sin does not exist, it must be invented. The point was made in a comical way by Eisenhower who, as a candidate for the presidency, once defied the Republican law of gravity by telling a joke. The story concerns a young Catholic by the name of Bo MacMillan, as fine a lad as the town had ever seen: a prize student, president of the senior class, star of the football team, editor of the school paper, and holder of just about every other honor to which a teenager can aspire. One quiet Saturday afternoon, thinking he was alone in the street, Bo picked up a stone and tossed it through a window. The local constable, who happened to spot him, stared in disbelief and wondered what on earth could have impelled him to do such a thing, to which the boy replied: "You see, sir, I didn't really want to do it, but I'm on my way to confession and I was a mite short of material."

As proof of the seriousness with which sin is taken in Christianity, one has only to cite the number of books written on it across the centuries. My own favorite example of the genre is the eighteenth-century Roman Catholic theologian and doctor of the Church, Alphonsus Liguori, whose mammoth treatise of moral theology, comprising over 3,100 pages in quarto, discusses every imaginable sin and a few others besides. I stopped reading it when I realized it was making me think of all sorts of things the knowledge of which did not seem indispensable to the perfection of one's intellect or the guidance of one's life. No wonder the book, written in Latin, was never translated into any modern language!

(There are still some rewards left for those who take the pains to learn that dead tongue.) Confessors were expected to be aware of all of this. As for the rest of the faithful, the less they knew about it the better. There was no point in giving them ideas.

Since our interest is in politics and since the New Testament, with its concentration on personal salvation, has practically nothing to say about politics, it is to the theological tradition that we must turn for the information that is most pertinent to us. Given the limited amount of time at our disposal, I shall restrict myself to a digestible if somewhat pedestrian synopsis of the view of evil found in that tradition, particularly as it bears on our topic. For that purpose, I shall rely heavily on St. Augustine, the first theologian to elaborate a coherent theory of evil and still Christianity's most authoritative voice in the matter, although not the one most often listened to nowadays. My thesis is that what passes for the traditional Christian position on this topic is a distortion, sometimes unwitting and sometimes deliberate, of the original Augustinian position.

It is important to note that the context within which the problem first came to the fore is ultimately religious rather than strictly political. That context was supplied by Manichaeanism, an eclectic religious philosophy the most readily graspable aspects of which can be seen as a radicalization of certain elements borrowed from classical or Platonic philosophy. The experience from which Manichaeanism starts is most familiar to us from the discussions of the best regime in the *Republic* and the *Politics* and concerns the unavoidable evils of human existence. The health of any civil society and therewith the happiness of its members is directly proportionate to the degree of justice that informs its public life. Without justice, the overarching social virtue, cities and nations are nothing but compacts of wickedness or gigantic larcenies, differing from robber bands only by the magnitude and the impunity of their crimes. Yet experience teaches that no regime is ever perfectly just and that actual societies are inherently incapable of living up to the high moral standards laid down by the philosophers and insisted upon by the Bible. The evil in question is not reducible to sporadic acts of injustice engaged in by assorted crooks and gangsters, of whom there are always plenty around, for these can usually be dealt with by legal means. Al Capone may have gotten away with it for a while, but the Feds did catch up to him in the end. Besides, a few rotten apples do not spoil the harvest. Common criminals can only do a limited amount of harm. They have no interest in bringing down the whole of society, if only because this would leave them with nothing to exploit. No, the problem has much deeper roots. It lies not with the lawbreaker but with the law itself, which is relative to the regime and shares in its imperfection. In the best instance, that regime is either a monarchy, an aristocracy, a democracy, or some more or less felicitous combination thereof; but this means that the city is in fact ruled by a part, which only claims to rule in the interest of the whole. Seldom if ever does a society distribute its honors to those who most deserve them or its wealth to those who best know how to use it. To opt for a new regime when that rare opportunity presents itself is not to replace injustice with justice but to exchange one set of injustices for another.

Worse still, even the most decent of citizens cannot help becoming a party to the inequities of his society. By the mere fact that he lives and works in it, pays his taxes to it, and perhaps holds office in it, he contributes to the enshrinement of its unjust way of life. Civil society is constituted in such a way as to compel us to choose between playing the hammer or the anvil, between victimizing others or being victimized by them. In simple Aristotelian terms, it is only under the best regime that the good citizen and the good man coincide; but the best regime is a practical impossibility and so therefore is moral goodness, despite its being demanded of everyone. Evil is built into the very structure of the universe. We live in an absurd world.

If such is the case, however, God's justice would seem to be called into question. Life on earth appears as a cruel joke, bearing witness not to the wisdom of a loving creator, but to the mischievous intent of some evil genius. Manichaeanism tried to solve the problem by attributing evil to matter and denying that matter was created by God. It saw the world as ruled by two independent and coeternal principles, to one or the other of which could be traced all that is good or bad in it. To escape evil, one had to shun as much as possible everything connected with the body—marriage, procreation, the killing of animals to obtain food, and the like—or else show one's utter contempt for the body by flouting the laws of society. Hence the extremes of asceticism and libertinism for which the members of the sect became known.

FREEDOM AND GUILT

It goes without saying that this metaphysical dualism was incompatible with biblical monotheism and the doctrine of divine omnipotence. The alternative was to locate the source of evil not in the cosmos, but in the rational creature. Human beings have themselves and no one else to blame for the evils from which they suffer and which are meted out to them in the form of a punishment for their sins. This implies, however, that man's fall into sin was a matter of choice rather than of necessity; for no one can be held to account for something he cannot avoid. Thus was born the notion of free will as a spiritual faculty that is wholly independent both of the intellect, even though it continues to be guided by it in the order of specification, and of the sensitive appetite, which can influence it extrinsically but which it has the capacity to resist. Without free will or freedom of indifference, as it is sometimes called, there is no guilt, and without guilt there is no justifiable punishment.

It hardly needs to be added that such a position goes well beyond anything that Greek philosophy had seen fit to assert. Just as it is hard to translate the term "guilt" into Greek, so it is hard to find the exact equivalent of what the Christian tradition understands by free will. This is not to suggest that the Greek philosophers subscribed to no-fault murder or denied that human beings bear any responsibility for their crimes, but only that they regarded the assessment of the degree of responsibility as fraught with much greater uncertainty. Human beings were not entirely blameless, but neither could they always be held fully accountable for what they did.

This raises the interesting question of how man could have sinned in the first place. It does not suffice to say that the very notion of freedom carries with it the possibility of misusing that freedom. If, as we must assume on the basis of divine revelation, man was created in a state of intellectual and moral perfection, one still has to account for the fact that he did go astray. From a good tree one expects only good fruit. I shall not burden you with the subtleties of the theological answer to this thorny question and leave it at saying that, once the order of creation and with it the harmony of the soul had been disrupted, evil was bound to perpetuate itself and increase as time went on. Crime begets crime, and the more prevalent it is, the greater the urge to acquiesce in it becomes. One's vision of the good is impaired and bad social habit combines with natural inclination to render its accomplishment ever more difficult. Attachment to one's own takes precedence over the attachment to the good of reason and keeps us from becoming perfect lovers of justice. Therein lies the real reason for which no human society will ever be perfect.

An omnipotent and omnicompetent God could no doubt wipe out evil altogether, especially since the redemption wrought by Christ has removed its cause; yet he has his reasons for allowing it to persist, the main ones being that it functions as a test of one's faith and a trial by which virtue is won. Anyone whose sense of justice continues to be offended at the sight of unmerited suffering can take satisfaction in the thought that the evildoer will sooner or later receive his due. Divine providence guarantees the moral consistency of the universe and sees to it that all wrongs are righted, if not in this world at least in the next.

In the meantime, the sufferer was not left wholly uncomforted. It would be unfair to suggest that the pagan philosophers and poets were indifferent to human suffering or unmoved by it; but they had no answer to it. In the presence of evils as great as the ones that had befallen Priam, even the pious Virgil had no more to say than that "there is pity for a world's distress and sympathy for short-lived humanity": *Sunt lacrimae rerum et mortalia mentem tangunt* (*Aen.*, I.462). Stoic *apatheia* was not the solution either. There are limits to human endurance, and they appear to have been severely tested by the atrocities of imperial Rome and its hateful tyrants. Not surprisingly, Stoicism, the most popular philosophy of the Hellenistic period—it reigned for five hundred years—vanished as a school of thought after the second century A.D. and in time ceased to be discussed at all in the rhetorical schools of late antiquity. Christianity had more to offer. Instead of denying suffering or rising above it, it assumed it and gave it meaning. By willingly undergoing it, one shared in the sufferings of Christ and became identified with him. The startling image of a God on the cross was there to remind us that the disciple is not above the master and that where the head is, there also the body shall be.

The irony in all of this is that the religion which is most sensitive to evil and most eager to espy it everywhere is often regarded as the one least able to resist it and most liable to inflict it. Both tendencies can be shown to have their common root in the peculiar inwardness that sets Christianity apart from its closest rivals, Judaism and Islam. Its typical response to the evils inherent in the political life was to enlist people as members of a spiritual kingdom that transcends all national boundaries and in which alone perfect justice is to be found. The unintended result was a propensity to demand either too little or too much justice in this life.

The new religion taught its adherents to look upon themselves as strangers or wayfarers on earth and to seek first the kingdom of God and its righteousness (Matt. 6:33). It told them to bear patiently the evils of this world, rejoice when persecuted, requite evil with good, and obey their rulers whoever they might be. It thereby weakened the bond of society and bred indifference to its regime. Since life here below is so quickly spent, and since all regimes are defective anyway, it matters little under which of them one lives. The Christian in the *Taming of the Shrew* is the one who falls asleep when the conversation turns to politics and who expresses his willingness to let the world "slide."[2]

Above all, it made them wary of engaging in what Nietzsche would later call the "necessary dirt of politics"[3] and fostered a reluctance to inflict harm on others, lest by doing so they should transgress the divine law. Better to forbear than risk eternal damnation. Revenge may have been in order at certain moments, but it was best left to the Lord. "Thus," says Hamlet, "conscience does make cowards of us all."[4] The same proclivity was reinforced by the doctrine of divine providence, which causes us to overlook the power of evil on the ground that victory over it is assured by God himself. Nothing is better calculated to undermine the self-sacrificing patriotism on which the city relies for its defense. The little that remains of spiritedness, which is too much a part of human nature to be destroyed once and for all, is turned inward and used to combat one's evil inclinations by means of self-renunciation and ascetical practices. Christians, Julian the Apostate observed wistfully, were better at praying for their rulers than at fighting for them.

The other side of the coin, and it is scarcely more reassuring, is that Christianity, the religion of love par excellence, breeds its own form of intolerance and has often succumbed to the very evils it ostensibly seeks to counteract. In compliance with the requirements of the gospel, the faith must be preached to all nations. Conquest, when favored by circumstances, becomes a religious obligation and pious cruelty a permanent temptation. If unbelief or apostasy—as opposed to, say, tyranny, civil war, or high treason—is the capital sin, it must be eradicated whenever possible and by force if necessary. Repressed spiritedness is unleashed and, inspired by zeal for the faith, is redirected toward the infidel. In a strange but all-too-human reversal of roles, the persecuted becomes the persecutor.

In like fashion, the extraordinary moral sacrifices that Christianity demands of its followers disposed them to make the same demands on others. These sacrifices must be justified, and there is no better way to prove to oneself that they are than by imposing them on everyone else. One's impulse is to ascribe all wrongdoing, not to ignorance or weakness, but to the malice or free will of the sinner, who invites not pity but retribution. Evil is everywhere present and everywhere to be feared. But it is impossible to think that badly of one's fellows unless one experiences the same secret desires within oneself. The monsters that populate one's own dreams end up by being projected onto the neighbor, where they can be more safely exorcised.

These two sets of objections sound as if they had been lifted straight out of the works of the great modern critics of Christianity—Machiavelli, Rousseau, and Nietzsche foremost among them. In point of fact, they come from none other than Augustine himself and define the issues that lie at the heart of the *City of God,* the dual purpose of which was to provide a much needed corrective to the extremes of religious passivity or resignation on one side and religious fanaticism or apocalypticism on the other.

The gist of Augustine's argument is that Christianity is indeed compatible with the duties of noble citizenship and that it does not disarm the city, encourage its followers to turn their backs on it, or make them any less intent on opposing injustice within it. Upon examination, that argument comes across as considerably tougher than the one to which we have lately become accustomed on the part of the religious establishment. To be sure, Augustine speaks a great deal about mercy, but the mercy that he has in mind is anything but the gentle sentiment for which it is commonly mistaken. It is a virtue, part of whose object is to moderate the excesses of one's natural aversion to suffering. As such, it has little to do with the compassion that is made so much of in contemporary religious and political discourse and which, as John Stuart Mill pointed out long ago, owes more to Rousseau's sentimental deism than it does to the gospel. Augustine was not prepared to rule out torture as a means of wresting the truth from a suspected criminal, and he had fewer reservations than do the American Catholic bishops, to say nothing of other Christian groups, about the role of war in human life. My suspicion is that he also had fewer illusions than they do about the possibility of any perfectly just war.

Furthermore, his notion of divine providence left ample room for the exercise of retributive justice. God, being all-knowing and all-powerful, can permit evil and use it to achieve his goal. Human beings are not in that situation. Since God's plan remains inscrutable, they cannot take their bearings by it but must rather be guided by the divine law, which simply forbids us to do evil and requires us to oppose it within the limits of possibility. Leibniz, and not Augustine, is the one who severed the relation between evil and retribution by stipulating that ours is the best of all possible worlds and, hence, that the use of evil is sanctioned by the grandeur of the overall end to be attained by it. There is no "theodicy" in Augustine, if by that term we mean what Leibniz meant by it. That is why I purposely avoided the expression in discussing Augustine's account of evil.

WHEAT AND TARES

As I have indicated, the *City of God* also had a second, less obvious purpose, which was to curb the fanaticism to which the lofty moral teaching of the Sermon on the Mount can give rise. The awareness that it betrays of the limits of human justice and of the ineradicability of evil comes as a timely reminder that one can only go so far in enforcing obedience to the moral law. There is a similar lesson to be drawn from the way in which God deals with sinners, frequently withholding his punishment so as to give them a chance to repent. By prematurely weeding out the garden, one runs the risk of pulling up the wheat with the tares. Moreover, the line that separates the good from the bad runs through the Church itself, which, though incarnating God's offer of mercy, never fully lives up to its divine mission and cannot simply be equated with the kingdom of God. Membership in it is no guarantee of the righteousness of one's actions. Like Noah's ark, to which it is often compared in patristic literature, the Church too contains every species of animals, pure as well as impure. As a medieval scribe noted in the margin of his manuscript of the Book of Genesis, next to this famous passage, "Except for the sight of the terrible things going on outside, the stench inside would have been unbearable."

On a deeper psychological level, one is well advised to bear in mind the fundamental ambiguity of all human actions, which must not only be good in themselves but performed for the right motive. Yet human motives remain obscure for the most part. They belong, not to the secrets of the heart, which one could reveal if one chose to do so, but to the secret intention of the heart, which is hidden from all human beings, including the agent himself. In particular, one can never be sure that pride has not insinuated itself even into one's good deeds in order to corrupt them. The remedy is humility, the most fundamental though by no means the highest virtue. Such humility is not synonymous with abjectness and does not entail any lack of dignity or nobility of taste. It is itself a conquest the essential task of which is to vanquish pride, the root cause of all evil.

The delicate balance that Augustine tried to strike was hard to maintain and frequently disrupted. For this, Augustine himself cannot be held responsible. His *City of God* was hardly read at all during most of the Middle Ages, and when people did begin to read it again from the mid-fourteenth century onward, they misunderstood it. Meanwhile, the political problem on which he put his finger did not go away. It came back to haunt the Christian kings of Spain, who found a novel way of coming to terms with it. Unable to reconcile the seemingly contradictory injunctions of the New Testament, they carried them out in succession. They spent most of their lives brutally conquering the New World and then used the gold that flowed up the Quadalquivir into the royal coffers to build the Escorial, the fabulous monastery in which, as humble penitents, they ended their days.

With the breakdown of Christendom and the emergence of the national state, the situation underwent a radical change. The Church does not exist in a vacuum and is often at the mercy of the civil authority for the freedom that it demands for itself and its followers. Sad to say, its precarious position in this regard is often such as to allow it to be manipulated by ruthless tyrants. For a variety of historical reasons, it has been lured into alliances with totalitarian regimes of the right or the left, whose rise to power it is known to have facilitated. The concordat with Nazi Germany, the involvement of clerics in revolutionary movements throughout the world, and the vagaries of the Vatican's so-called *Ostpolitik* are only three prominent manifestations of the same complex syndrome.

I began with an allusion to Satan and this may also be the best way to conclude in so far as his biography comes close to being a mirror image of the history of the Church. Our ancient forebears did not have much trouble believing in him; indeed, quite the contrary. Augustine sniffed his presence in every corner but most of all behind the gods of polytheism, the religion against which in his day Christianity was still competing. As long as the devil was there, everything was fine and Christ, who helped us in our fight against him, had his raison d'être.

With the advent of modernity, Satan lost a good deal of his credibility. The problem now was to make him plausible once again, and no one succeeded better at this than Milton. If anything, Satan became too plausible, less perhaps as an archvillain before whom one cringes in horror than as a pathetic victim to whom our hearts go out in sympathy. It would not be long before the belief in his universal power, the necessary accompaniment of a universal faith, began to turn against the religion that gave him birth. A Satan who is that powerful must be entitled to greater respect and, as we know from *Faust*, invites dangerous new experiments. Lending credence to him ups the ante, so to speak, and spices vice by adding an extra kick to it, or so the *poètes maudits* of nineteenth-century France discovered.

Then a funny thing happened to him. He disappeared, the first in a list of doctrinal casualties that was destined to grow longer as time went on. Orthodox Christians were quick to point out that this was not at all the case; that, when it comes to recognition, Satan, the Prince of Darkness, is less fussy than God, that he prefers to work behind the scenes, tricking mortals into following him without their being conscious of it. Yet he cannot remain totally hidden, for otherwise nobody would even mention him. Where then are we to look for him?

Here, I confess that I am at a bit of a loss for an answer. One might expect thoughtful Christians to point to the likes of Hitler and Stalin as suitable candidates, but such has generally not been the case. The great surprise of our century is that, after all this talk about the devil, so few people recognized him when they came face to face with him. If our modern sensibilities, shaped in part by Christianity, tell us anything, it is that nobody can be that bad.

It is nevertheless doubtful whether most human beings can dispense with the need to personalize the forces of evil. I may be wrong, but it seems to me that there is still someone abroad who fulfills this need. The modern world did not really kill Satan; it democratized him and made him into a bourgeois. His name is Dr. Moriarty, the wicked genius who capitalizes on the peculiar invisibility afforded by liberal society to extend his own empire and against whom the local police are utterly powerless. It takes a supersleuth, Sherlock Holmes, another bourgeois, to penetrate the workings of his mind and outwit him. The enterprise is more ingenious than heroic, but then a society that is unable to dedicate itself to any cause higher than the self can get along with lesser gods. All it insists on is that they be interesting. Moriarty commits his crimes, not for the sake of gain—he does not need the money—but because he is bored. The theatrical, cocaine-using Holmes solves them because he is bored. And people read the stories because they are bored.

If Nietzsche's diagnosis is correct, the last and perhaps the greatest enemy of mankind is modern liberal society, which prides itself on having laid the old superstitions and the evils attendant upon them to rest but cannot supply us with a vision of the good or satisfy our deepest longings. The only remaining question is whether this state of affairs is itself the final tragedy, as Nietzsche seems to have thought, or merely the prelude to some catastrophe of unimaginable proportions brought on by our inability to believe any more in the binding power of evil.

NOTES

1. John Henry Newman, *An Essay in Aid of a Grammar of Assent* (Oxford: Clarendon Press, 1985), 258. Cf. all of chapter 10.

2. Shakespeare, *Taming of the Shrew*. Induction, scene 1, line 6; cf. scene 2, line 146, and act I, scene 1, line 254.

3. Friedrich Nietzsche, *Beyond Good and Evil*, aphorism 61.

VI

THE AMERICAN CATHOLIC CHURCH

AND POLITICS

20

SOCIAL ACTIVISM AND
THE CHURCH'S MISSION

The newly found social activism of most of our major church bodies has again raised the age-old question of the proper relationship of the Church to the political order and brought forth a flurry of conflicting statements, the latest but not the least notable of which is Richard J. Neuhaus's programmatic essay on *Christianity and Democracy*.[1] Whether all the problems mentioned or alluded to by Neuhaus can be settled on the level of principle alone is open to considerable doubt. It is tempting to argue, as some have done, that the Church ought to steer clear of politics altogether and restrict its pronouncements to spiritual and moral matters. Unfortunately, the two domains are not as readily separable as one would like them to be. Laws pertaining to social justice, civil rights, abortion, public welfare and the like have definite moral implications to which the Christian churches cannot remain indifferent and on which they have the duty to speak out. The great theologians of the past, Augustine and Thomas Aquinas among them, were the first to acknowledge this duty, but they also warned that it should be exercised in a manner consistent with the transcendent character of the Church's mission.

The reasons for the caveat are only too obvious. Political issues are inherently controversial and divisive. They rarely admit of clear-cut solutions and leave room for reasonable disagreement on the part of decent and thoughtful persons. By taking sides on such issues, the Church inevitably commits itself to a partial view of justice and runs the risk of compromising its integrity. Alexis de Tocqueville pointed out long ago that in a society such as ours the effectiveness of organized religion as a moral force was largely contingent on its willingness to dissociate itself from political power and eschew direct political action.

Having said this much, one has to admit that it is not always easy to decide at which point a statement about political matters becomes a political statement. Prodded by a growing number of progressives within their own ranks and on their own staffs, the Roman Catholic bishops of this country have seen fit to pronounce themselves on virtually every major issue to come up for public discussion in recent years, from the ratification of the SALT Agreements and the Panama Canal treaties to nuclear disarmament, military conscription, conscientious objection, multinational corporations, capital punishment, national health insurance, federal aid to nonpublic schools, and world hunger. Since 1966 alone, when the old National Catholic Welfare Conference was replaced by the U.S. Catholic Conference and the National Conference of Catholic Bishops, they have issued more documents on social problems than they did during the first 200 years of their existence as an established hierarchy.[2] They perhaps came as close as they ever did to direct political involvement during the 1972 presidential campaign, when they ran a series of ads listing, in three parallel columns, their own positions on the leading campaign issues along with those of the two major candidates. The advertisement left little doubt as to who their preferred candidate was and, as it turned out, had little effect on the outcome of the election.

The wisdom of all such efforts to repoliticize the Church has been rightly questioned by impartial observers as well as by some of the bishops themselves. By reason of their background and calling, few Church leaders are equipped to speak with competence on the immensely complex problems of modern industrial society. The point was made succinctly a few years ago by a sympathetic critic who remarked that, if the bishops' proposals concerning multinational corporations were to be put into effect, immediate world starvation would ensue.

Contrary to what is often asserted nowadays, the Church has not been divinely mandated to restructure society. The gospel, as anyone who reflects on it soon discovers, is a notoriously apolitical document. It shows little awareness of the distinction between political regimes, does not indicate any preference for one over the others, imposes none of its own, and makes no specific recommendations for the reform of the social order. It was to be preached to all nations but was not destined to replace them or meant to compete with them on their own level. It takes it for granted that Christ's followers will continue to organize their temporal lives within the context of the society to which they happen to belong and, while it is strenuously opposed to all forms of injustice, it leaves the administration of public affairs to the authorities whom God has ordained for this purpose. Nor does it give much encouragement to political speculation. It realizes that "Christians," as they would soon be called (Acts 11:26), were liable to be dragged before governors and kings but tells them only not to worry about defending themselves. What they were to say would be given to them in that hour by the Holy Spirit (Matt. 10:18). Its own ideal, in short, is strictly transpolitical. It makes a good deal of sense to speak of a Hebrew polity, since this is what the Hebrew Scriptures are all about; it makes no sense at all to speak of a Christian polity, however frequently people may have done so in the past. Christ himself renounced all political ambitions. His kingdom was emphatically not of this world.

This is not to deny that Christianity was fraught from the outset with grave political consequences. By calling human beings to a higher destiny and reserving the best part of their existence for the service of God, it effectively destroyed the regime as a total way of life. It cultivated a passion for an elusive kingdom of God beyond history and thus tended to turn people's minds away from the only realities that reason is capable of knowing by itself. In the process, civil society was displaced as the locus of virtue and the object of one's strongest and most noble attachments. It ceased to be the sole horizon lending meaning and substance to the activities of its citizens. The love of one's own was no longer concentrated within specific borders and citizenship itself lost its fundamental significance. Even the greatest human achievements were robbed of their former splendor. One's eternal salvation mattered more than the fate of empires. In the words of Shakespeare, kingdoms were "clay" and it was "paltry to be Caesar."[3]

Such are the real roots of the opposition that Christianity encountered when it first began to spread throughout the Roman empire, and it is to this problem that its early apologists were eventually compelled to address themselves. The new religion would have gone the way of the radical sects of late antiquity had it not succeeded in making a case for the compatibility of its demands with those of the political life. Over the centuries a mutually acceptable solution was worked out which was not always observed in practice but which could be invoked when the balance tilted too far on one side or the other. The premise on which it rested was that Christianity could accommodate itself to any and all regimes but was not wedded to any one of them. They were all imperfect anyway—Augustine referred to them as "compacts of wickedness"—and they all had their inner contradictions, but they were not totally unjust, either, for otherwise they would have no chance of survival, and they could be improved. It was indeed possible to live as a Christian in an immoral society as long as one did one's best not to become a party to its manifest injustices. Even under a Nero, St. Paul had no qualms about preaching submission to the Roman rulers (Rom. 13:1-7). The alternative was to withdraw from society altogether—an enticing solution to which some extremists were occasionally drawn—but then no one would have been left to bear witness before others to the truth of the gospel. It was not up to the individual to decide when or to whom the offer of God's mercy was to be made. As Aquinas finally put it, the Christian is wholly a member of the political community, even though he is not a member of it by reason of the whole of himself. Like everyone else and perhaps more than anyone else, he is expected to work for its betterment, always with the understanding, however, that there are practical limits to what can be accomplished in any given set of circumstances. Exactly what these limits are is not something that can be determined in advance. Of necessity it ends up by being the object of a prudential judgment based on a careful consideration of actual and infinitely varied situations.

Only the wildest fanatic or the most naive optimist would maintain that the requirements of perfect justice can be met on the level of society at large. That being the case, practical decisions affecting the life of the nation are best entrusted to lay persons who can devise and implement appropriate courses of action without engaging the authority of the Church as such. As Vatican II has reminded us, the role of the hierarchy is not to substitute itself for the laity, but to prepare Christian laymen and laywomen capable of assuming public responsibilities in an area that is properly theirs. By discharging that function faithfully, it both avoids the pitfalls of clericalism and insures that the light of the gospel will be brought to bear on the problems of everyday existence.

The lesson, needless to say, has not always been heeded. It was certainly not heeded in the Middle Ages and there is evidence that it is again falling on deaf ears. Work on behalf of justice, we are now being told, is a "constitutive dimension" of the Church's ministry, on a par with the celebration of the sacraments and the preaching of the word of God.[4] One would have little reason to quarrel with this view if one were sure that the justice in question is that of the Sermon on the Mount and not the latest version of the modern egalitarian creed or what goes on under the new name of "social justice." If social justice is so central an element of the gospel message, one wonders why no one ever bothered to talk about it prior to the middle of the nineteenth century when the neologism—to be discussed in the next chapter—was first put into circulation, probably by the Italian philosopher Taparelli d'Azeglio.

Implied in that vague notion as it is currently being used is the idea that there is one and only one legitimate regime and that it is unjust to support any other. Our predecessors were of a different opinion and, for all anyone can tell, they may have been on the right track. Justice is not the monopoly of any single regime. Each regime stresses one particular aspect of it, on which it stakes its claim to serious consideration, but none embodies or can embody the whole of it. And that is why the argument goes on. One is always faced with a trade-off of some sort. What we gain at one end, we lose at the other. If this were not the case, we should have no reason to look elsewhere, beyond politics, for an adequate solution to the problem of human justice. That is also why, as Richard Neuhaus has so aptly stated, the kingdom of God is not to be equated "with any political, social, or economic order of this passing time."[5]

Still, one has the right to ask whether, by attempting to ground his own argument for liberal democracy in a theological principle, Neuhaus has not left himself open to the charge of repoliticizing the Christian faith. It is one thing to defend liberal democracy as compatible with Christianity and quite another to claim a biblical warrant for it. The Bill of Rights is undoubtedly a good idea. It has served us well over the years and we can be thankful that, after much debate, it was adopted by the Founders. But to say that it or its equivalent is "theologically imperative" is as implausible as it is useless. The New Testament approaches the problem from another angle altogether; it is completely silent on the subject of rights and, like the Hebrew Scriptures, prefers to remind people of their duties toward God and neighbor. For better or for worse, the perspective of the New Testament is notably different from, though not necessarily irreconcilable with, the one that we moderns generally find most congenial.

The difficulty at hand becomes even more apparent when one reflects on the spiritual impoverishment of modern life. The United States prides itself on being the first nation to have secularized its political life or to have institutionalized the separation between church and state. From the beginning it showed itself remarkably accommodating to all religious traditions and provided a home within its boundaries for the most diverse religious groups. In many ways the experiment has been extraordinarily successful. Yet history has demonstrated that the privatization of religion via its subordination to an independent standard of political unity is not an unqualified asset. Modern democracy may discourage intolerance and religious strife, but by the same token it is not especially favorable to an intense spiritual life or to outbursts of religious fervor. Its soil offers scant nourishment for those rare passions that rend the souls of mystics and disclose the heights and depths of authentic religious experience. It treats all religions as equal, conveniently overlooking the fact that each one looks upon itself as superior to the others and hence worthy of the noble sacrifices that it demands of its followers. As mentioned, America has not produced any Pascals. This observation by Alexis de Tocqueville is as pertinent today as it was then. In the century and a half that has elapsed since it was first made, not much has transpired that would cause us to doubt its validity.

Nor do the dynamics of the regime conduce to a high level of civic morality. Its basic principle is utility, and utility, effective as it may be as a means of rallying the consent of the multitude, is not the natural breeding ground of excellence. The decency and respectability that it fosters are those of the banker, for which we can be grateful but about which it is hard to be enthusiastic. If liberal democracy excels at anything, it is at teaching individuals to live for themselves in the midst of others. Public-spiritedness is not its strong suit and expressions of it are rarely divorced from the pursuit of private interests. Americans are apt to raise a hue and cry about the breakdown of law and order when they or their daughters are mugged but they do not like to pay their taxes. It is hardly surprising that, for all our prosperity and incredibly high standard of living, we should hear so many complaints about the quality of our common life. One has only to observe the loneliness, the feeling of alienation and the pervasive sense of lostness that mark the lives of such large numbers of our contemporaries to realize that we do not have all that we could and should have in the way of human fulfillment.

Neuhaus rightly points to the "deep human hunger for a monistic world" as the source from which the totalitarian impulse draws its force, but he does not show us how that natural hunger can be moderated. It does not suffice to say that liberal democracy is an ongoing process which, if left to itself, can be trusted to take care of the bugs in the system. Unless someone comes up with a way of compensating for the defects of the system itself by restoring a measure of wholeness to our fractured selves, modern liberalism will remain vulnerable to attack on both flanks, right and left. There are some parts of the soul that it is congenitally unable to satisfy and to which, in its own interest, it would do well to give serious attention. Since it has so much else to recommend it, it is unfortunate that no one has yet undertaken a thorough reexamination of its theoretical basis in the light of more recent developments. The last full-scale defense of it that I know of goes back to J. S. Mill, well over a century ago. Virtually all the most influential political works of the late nineteenth and twentieth centuries have been attacks on it, inspired by either Marx or Nietzsche, the two great fountainheads of modern leftist and rightist totalitarianism. Given this situation, a concerted effort to come to terms with its shortcomings would do more to help its cause than a simple rehearsal of its undeniable merits. What we need is a new Tocqueville, who could both defend it and tell us how to improve it, this time by means of a comparison with the new right and especially the new left, whose power Tocqueville, bereft as he was of the privilege of hindsight, seems to have grossly underestimated.

Meanwhile, a modest first step in the direction just indicated might be to restore political philosophy to the position of honor that it once occupied in the Christian tradition. One can only regret that this crucial discipline has been neglected for so long in American higher education. The fault does not lie with our Church leaders alone. If they themselves have so little acquaintance with it, it is because it is conspicuously absent from the best theological and philosophic literature of the twentieth century. Only in recent years, thanks to the efforts of a small group of outstanding scholars, has it once again begun to be studied with any degree of seriousness, and not everywhere at that. It is still totally lacking in most of our seminaries, where it has been replaced by narrowly conceived and often ideologically oriented courses in social ethics (the scientific study of our quasi-mythical "social justice") or in liberation theology, itself an offshoot of the latest trend in European thought, however eager it may be to pass itself off as a native Latin American product.[6] A return to the traditional sources of Christian political wisdom could conceivably help the churches to recover a true sense of their social mission and, instead of giving them a spurious relevance that they do not need, win the respect of numerous outsiders who look to them for some kind of guidance in an age in which there has perhaps never been so much confusion about fundamental issues.

NOTES

1. Richard J. Neuhaus, "Christianity and Democracy: A Statement of the Institute on Religion and Democracy," *Center Journal* 1, no. 3 (1982): 9-25.

2. For a survey and a discussion of these documents, see J. Brian Benestad, *The Pursuit of a Just Social Order: The Policy Statements of the United States Catholic Bishops, 1966-1980* (Ethics and Public Policy Center, 1982). Many of the documents themselves can be found in *Quest for Justice: A Compendium of Statements of the United States Catholic Bishops on the Political and Social Order*, edited by J. B. Benestad and F. J. Butler (Washington, D.C., 1981).

3. Shakespeare, *Antony and Cleopatra*, act I, scene 1, line 35; act V, scene 2, line 2.

4. See J. Bryan Hehir, "The Ministry of Justice," *Network Quarterly* 11, no. 3 (Summer 1974): 2.

5. Neuhaus, "Christianity and Democracy," 9.

6. See the brilliant analysis of this strange phenomenon by the late Gaston Fessard, *Chrétiens marxistes et théologie de la libération* (Paris: Lethielleux, 1978). See also Fortin's review of Fessard's book in *Classical Christianity and the Political*

Order: *Reflections on the Theologico-Political Problem*, edited by J. Brian Benestad (Lanham, MD: Rowman and Littlefield, 1996), 366-69.

21

CHURCH ACTIVISM IN THE 1980s:
POLITICS IN THE GUISE OF RELIGION?

By virtually all accounts, the most striking feature of present-day Christianity is the massive involvement of its leadership in the great social and political debates of our time. Three simple examples, chosen at random from three different continents, will illustrate the point. A few years ago, a French clergyman with whom I chanced to be living mentioned in the course of a heated discussion that, in his view, "the bourgeoisie have only one thing left to do: disappear!" The statement needed no further proof. It was self-evident. One had only to read the gospel in order to be convinced of its truth. No decent person was at liberty to disagree with it. Secondly, Denis Hurley, the Catholic archbishop of Durban, South Africa, who was indicted for his allegedly slanderous denunciations of police brutality, was cheered in his cathedral when, the day before the trial was scheduled to begin, he declared, "The gospel is political!" Lastly, on the occasion of John Paul II's first visit to South America, a missionary stationed in Peru was quoted as saying that his experience in that country had finally taught him what Christianity was all about: to defend the weak and the poor, abolish misery, and wage war on injustice. Strange as it might have sounded at other moments in history, the reforming zeal to which these assertions bear witness is shared in varying degrees by a growing number of Christians who have come to view their faith as an en-terprise dedicated to eradication of the evils that plague human existence by transforming society along more or less leftist lines.

There is reason to suspect that what my European friend, the South African archbishop, and the Peruvian missionary took to be the timeless (if only newly discovered) wisdom of the gospel is little more than the distilled "wisdom" of the sages in whom, to speak like Goethe, "the mir-rored age reveals itself." It is unlikely that such ideas would ever have be-come as popular as they are had they not been borne on the wings of powerful intellectual currents that have shaped the thinking of the modern church; and it is even less likely that theologians and religious activists of various stripes would have succumbed to them so readily had they not themselves been psychologically prepared for them by a long line of secu-lar thinkers who, knowing that they could not do away with Christianity altogether, sought to achieve their goal by redefining it in political terms or reducing it to a political phenomenon. Liberation theology is only one manifestation of the new syndrome, and perhaps not the most important one at that, for it too is but a bastardized product of the less visible ideo-logical forces that have been at work in our society for quite some time.

Rome's obvious uneasiness, not to say displeasure, with this type of theology and John Paul II's repeated warnings against the danger of mix-ing religion and politics suffice to remind us that not everybody in the Church is happy with the direction in which some of its lesser representa-tives have been moving. Yet the Pope's own stance and posture in regard to these matters often comes across as ambiguous. His public speeches in different parts of the globe are meant to have an impact on the lives of the nations at which they are aimed, and much of what he has said over the years in defense of social justice, world and local poverty, the role of labor, free trade unions, and the sharing of wealth has very definite polit-ical overtones. Because of the liberal thrust of many of these statements, some widely read analysts have gone so far as to brand him a socialist. Others find him inconsistent insofar as he strenuously objects to any cleri-cal meddling in politics while he himself never passes up an opportunity to lecture party chiefs and ruling juntas about violations of human rights, religious freedom, and human dignity. Still others have pointed to what they call his "geopolitical ambivalence," accusing him of using one set of standards when speaking to his fellow Poles and a different set when ad-dressing, say, Latin American audiences.

Much the same could be said of the American bishops, who have been outspoken in their criticism of the regime but refuse to admit that their pronouncements have anything specifically political about them. Many of them in fact were astonished and not a little dismayed when a few years ago, the *New York Times Magazine* ran a feature article boldly entitled

"America's Activist Bishops: Examining Capitalism."[1] To make matters worse, the front cover showed a group of three purple-clad bishops huddled in a private conference on which, judging from the grim look on their faces, the fate of the world seemed to hinge.

Are the ecclesiastical authorities confusing religion and politics when, in their official capacity, they lobby for or against defense systems, economic programs, and a host of domestic or foreign policies over which the country is divided at the present moment; or is there some higher ground on which their intervention in temporal affairs conceivably could be justified? What finally distinguishes the political from the nonpolitical, and where does one draw the line between them? This question, which is the one I would like to address, has yet to receive a clear answer. My tentative suggestion is that the Church's present stand on this matter makes complete sense only within the context of our typically modern understanding of society and the peculiar brand of apolitical politics that goes hand in hand with it. I shall begin with a brief account of what I take to be the fundamentally nonpolitical orientation of the New Testament, turn from there to a discussion of the manner in which its teaching was reinterpreted in the modern period, and finally take up the larger issue of the relationship between church and state as it presents itself in the light of recent developments.

THE NEW TESTAMENT AND THE POLITICAL LIFE

Anyone who reads the New Testament from the perspective of an outsider, without any previous knowledge of the centuries-old tradition that came out of it, is bound to be struck by its blatant disregard for the realities of the political life. The gospel all but opens and closes with a denial of the temporal or political nature of Jesus's messiahship. The temptation scene in the desert, which constitutes the immediate preparation for everything that follows in the text, shows Jesus refusing to change stones into bread and rejecting Satan's offer to place him at the head of all the kingdoms of the world if he would but worship him (Matt. 4:1-2). And when, toward the very end, Jesus is summoned by Pilate to respond to the charge that he had tried to make himself king, his answer takes the form of an emphatic proclamation to the effect that his kingship is not of this world (John 18:36).

Much of what transpires between these two scenes is a concerted effort to show that Jesus had come not to restore the old kingdom of Israel but to establish a new kind of kingdom, one that was spiritual in nature and totally divorced from the political concerns that animated his disciples. By way of example, it might help to take a quick look at Mark's account of the Transfiguration on Mount Tabor, where, for the only time in the gospel, Jesus reveals himself in all of the radiant splendor that belongs to him as the son of the Father (Mark 9:2-10). Characteristically, the event is set within the context of the celebration of the Feast of Tents, which lasted six days and during which, in accordance with Jewish custom, the Messiah to come was symbolically enthroned. Next to him appear Moses and Elijah, who represent the law and the prophets respectively, and thus sum up the entire pre-Christian dispensation. Contrary to all expectations, however, Moses follows Elijah and takes second place to him. It is not Moses, the founder of the nation and the mediator of its covenant, than whom none was thought to be greater (Deut. 34:10; cf. Num. 12:1-8), but the persecuted and suffering prophet who prefigures most directly the mission of the new Messiah.

This was a radical novelty, and that is why explaining it to the disciples proved to be such an arduous task. There is no evidence that any among them understood it during Christ's lifetime, and even when he appears to them after his resurrection, we still find them asking, "Lord, is it now that you are going to restore the kingdom to Israel?" (Acts 1:6). Unlike the Judaism in which it finds its preparation, Christianity is in essence not a political religion but an altogether different type of religion —one in which the individual's relationship to the transcendent deity is mediated not by a divine law or one's participation in a covenant that binds a group of people together as members of an earthly community, but by faith in the person of the living Christ. Nobody to my knowledge has stated the crucial difference between the two religions more succinctly and more perspicaciously than Gershom Scholem:

> Judaism, in all its forms and manifestations, has always maintained a concept of redemption as an event which takes place publicly, on the stage of history and within the community. It is an occurrence which takes place in the visible world and which cannot be conceived apart from such visible appearance. In contrast, Christianity conceives of redemption as an event in the spiritual and unseen realm, an event which is reflected in the soul, in the private world of each individual, and which effects an inner transformation which need not correspond to anything outside. Even the *civitas Dei* of Augustine, which within the confines of Christian dogmatics and in the interest of the Church has made the most far-reaching attempt both to retain and to reinterpret the Jewish categories of redemption, is a community of the mysteriously redeemed within an unredeemed world. What for the one stood unconditionally at the end of history as its most distant aim was for the other the true center of the historical process, even if that process was henceforth decked out as *Heilsgeschichte*. The Church was convinced that by perceiving redemption in this way it had overcome an external conception that was bound to the material world, and it had counterposed a new con-

ception that possessed higher dignity. But it was just this conviction that always seemed to Judaism to be anything but progress. The reinterpretation of the prophetic promises of the Bible to refer to a realm of inwardness, which seemed as remote as possible from any contents of these prophecies, always seemed to the religious thinkers of Judaism to be an illegitimate anticipation of something which could at best be seen as the interior side of an event basically taking place in the external world but could never be cut off from the event itself. What appeared to the Christians as a deeper apprehension of the external realm appeared to the Jews as its liquidation and as a flight which sought to escape verification of the Messianic claim within its most empirical categories by means of a non-existent pure inwardness.[2]

Despite its apolitical nature, the New Testament presupposes the existence of civil society and recognizes its necessity. Since Christians were not to form a nation of their own and were not given a law by which such a nation might be governed, it was assumed that they would continue to live under the laws of the particular societies to which they happened to belong and in accordance with them. In this sense, the New Testament view is perhaps best described as transpolitical rather than apolitical. Christians were not asked to turn their backs on their fellow human beings and could not reasonably be accused, as they often were by their pagan critics, of misanthropy or hatred of the world. On the contrary, it was their God-given duty to dedicate themselves to the welfare of others, with whom they were expected to share a common life even though they themselves were called to a loftier personal ideal. The knowledge acquired through revelation was inseparable from the love of neighbor. Therein lay the problem, for it was inevitable that at some point or other the demands made on them by their faith would come into conflict with those of a civil society that is always less than perfect and whose laws are never completely just.

On that score, the New Testament has no firm or fully developed teaching. In some instances, it takes a positive view of civil society, enjoining Christians to obey their rulers, "honor the emperor," pay taxes, be "subject for the Lord's sake to every human institution," and "maintain good conduct among the Gentiles." Since temporal rulers have been divinely ordained for the punishment of wrongdoers, to resist them was to resist the will of God himself, at the risk of incurring his wrath (Rom. 13:1-7; cf. I Pet. 2:11-17). Other texts strike a different, more negative note. They remind these same Christians that they are not to be "conformed to this world" (Rom. 12:2) and that they have to "obey God rather than men" (Acts 5:29). Some of them even take an extremely dim view of the Roman Empire, identifying it with the harlot "seated upon the seven hills" and "drunk with the blood of the saints and the martyrs of Jesus"

(Rev. 17:9 and 18:24).

How these conflicting strains could be harmonized is a question with which the New Testament never expressly deals. This does not appear to have been a major concern of the sacred writers, partly, one surmises, because all or most of them were convinced that the end-time was near. If the world as we know it was about to pass away, and if Christians had to "deal with it as though they had no dealings with it" (I Cor. 7:31), there was no need to be preoccupied with one's ambiguous relationship to political institutions, however good or bad they may have been.

This explains why it has always been so difficult to derive a coherent plan of action from the pages of the New Testament alone. The pagan Celsus, the first great philosophic critic of the new faith, concluded that, if Christians were consistent, they would go into the wilderness and die without having had any children,[3] which is exactly what Antony, the father of Christian monasticism, did after walking into a church one day and hearing the words: "Go, sell what you possess and give to the poor, and you will have treasure in heaven; and come, follow me" (Matt. 19: 21). Others, like the Spaniards of the Golden Age, chose instead to conquer the world, making "disciples of all nations [and] baptizing them in the name of the Father and of the Son and of the Holy Spirit" (Matt. 28:19), whether the baptized liked it or not.

The one text that seems to offer at least the beginnings of a solution to the problem at hand is the famous maxim "Render to Caesar what is Caesar's and to God what is God's" (Matt. 22:21; Mark 12:17; Luke 20: 25), which became a locus classicus of sorts in the political literature of the Middle Ages, but its scope is limited by the narrow context in which it occurs. Christ obviously had no intention of enunciating a general principle that could be applied to a variety of similar situations. His sole purpose was to elude the trap laid for him by his enemies, in this instance the Pharisees and the Herodians, who, although bitter rivals, had found an opportunity to make common cause against him. Their specific question was whether Jews ought to pay the tribute demanded by the emperor. If Christ answered yes, he was a collaborationist and hence a traitor to the Jewish nation; if no, he was a zealot who deserved to be punished for his refusal to comply with Roman law. The rejoinder was a clever one. It reminded the Pharisees that by using imperial currency they were availing themselves of the benefits of Roman rule and were thus bound to give the emperor his due; and it reminded the Herodians that their collusion with the Romans did not dispense them from the service owed to God. From an argument such as this one, few general conclusions can be drawn. Far from resolving the issue of church-state relations, it brings it into sharper focus by highlighting the inevitable tension between divine and political authority. Oscar Cullmann's observation that Christ was merely telling the state that it should not be "more than the state"[4] misses the point altogether, since the state was conceived as a unitary whole commanding the undivided allegiance of its citizens. To say that one should render to Caesar what was Caesar's meant nothing to someone who was convinced that everything was Caesar's. Such a claim may not have been fully enforceable, for history knows of no higher regime that has ever realized its ambition of becoming a total way of life; but it was a claim that the Roman rulers were not ready to give up.

For all its far-reaching political implications, the New Testament has no genuine political teaching of its own and, while it is not unaware of the harsh necessities of the social life, it evinces hardly any interest at all in them. As Thomas Aquinas would later say, the justice of which it speaks is not the justice of this world, sufficiently known through the first principles of natural reason, but the "justice of the faith," *iustitia fidei*,[5] which has been given to us not to solve the problems of civil society but to lead us to the blessedness of eternal life.

It is therefore no mere accident that the internal history of Christianity, in contrast to that of Judaism or Islam, traditionally has been dominated by doctrinal rather than legal or juridical preoccupations. Justification was achieved not through the performance of such righteous deeds as might be prescribed by the law but through faith. In the absence of a divinely mandated social organization, Christian unity was secured by a commonality of belief. Orthodoxy was thought to be more important than orthopraxy, and what one held as a believer took precedence over one's actual way of life, which could vary greatly from one place to another. Accordingly, theology and not jurisprudence became the highest science within the community and the locus of its liveliest debates, all of which tended to focus on points of doctrine rather than on points of law. As has rightly been observed, no other religious group has placed a greater premium on the purity of doctrine or has been so much on its guard against heresy. For this reason, the Church's authority was always understood to be first and foremost a spiritual authority. Nowhere in the Christian tradition does one encounter the kind of concern for the perfect law or the perfect social order that is so prominently displayed in both Judaism and Islam.

THE POLITICIZATION OF CHRISTIANITY

In light of everything that has been said thus far, it comes as something of a surprise that the leaders of the Christian churches suddenly should be so eager to invade the political domain not simply to demand the freedom required for the exercise of their proper functions, which they have always enjoyed in this country, but to influence public legislation and governmental policy on the premise that work for social justice is a "constitutive dimension" of their ministry. And it comes as an even greater surprise that they should insist on grounding their own proposals in the New Testament, which is mostly silent about such matters. A good case in point is the Catholic bishops' much publicized 1983 pastoral letter on war and peace and their more recent (1986) letter on the American economy.[6] I leave it to others to take issue with the specific positions taken in these two documents, as well they might, and concentrate only on the dangers inherent in the attempt to develop from the sacred text a doctrine that has no solid basis in it.

The simple fact of the matter is that the New Testament has no clear or consistent teaching on war, makes no effort to deal with any of the complex moral issues that it raises, and does not even show any distinct awareness of the existence of these issues. It is tempting, of course, to argue that war is ruled out by the commandment to love, but that inference is not necessarily warranted and is never drawn by the sacred writers themselves. Indeed, from the same commandment one could just as easily come to the opposite conclusion; it is surely not a sign of love to allow unjust aggressors to oppress or tyrannize one's friends, fellow countrymen, and others for whose safety one is responsible. Moreover, there appears to be at least an implicit recognition of the legitimacy of war under certain circumstances in the Letter to the Romans, which makes it plain that rulers have been ordained by God for the express purpose of inflicting his wrath on evildoers (Rom. 13:4)—something they could hardly do without resorting to force and, if need be, military force.

To be sure, Christians are urged to "turn the other cheek" (Matt. 5:39) and to requite evil with good (cf. Rom 12:17-21), which is not necessarily a bad idea; but these principles too, take on a different meaning when viewed within the context of one's relations with all of one's fellow human beings. A slap on the cheek with the back of one's hand was a common insult that ordinarily invited some form of retaliation. In such instances, the offended person might well choose to forgive rather than stand on his honor, especially if he has only himself to think about; but it is easy to imagine other instances in which the same forbearance would do more harm than good. Significantly, the situations envisaged in the New Testament are one-on-one situations rather than situations in which the welfare of third parties, to say nothing of entire communities, is at stake. They have nothing to do with the mode of behavior that is in order when conflicts of a general nature threaten to erupt. It is surely a sign of the fundamental ambiguity of the New Testament concerning these matters that, with seemingly equal right, radicals at both ends of the political spectrum have been able to appeal to it to defend pacifism or to justify their involvement in revolutionary activities.

Similar remarks could be made about the New Testament attitude toward wealth and poverty. It would be strange, to say the least, if Christ, who had no money of his own and who seems to have depended so much on the hospitality of his rich friends, had been opposed to riches as such. The gospel frequently alludes to the poor, but its overwhelming emphasis is on spiritual rather than material poverty, in keeping with a tendency that was already present in the Old Testament and that stretches back to at least the seventh century B.C. The key text on this subject is Zephaniah 3:12, where, perhaps for the first time, the word *poor,* which originally designated a social class, is invested with a religious and ethical meaning. The poor in this new sense are the pious and humble people who place their trust in God rather than in political schemes or worldly goods. Neither in this tradition nor in the New Testament tradition that grows out of it is there any suggestion that economic poverty could be abolished or that the lower classes ought to be elevated at the expense of the wealthier ones. In any event, most of the New Testament passages relating to poverty, the Canticle of Mary among them (Luke 1:46-55), are barely more than paraphrases of Old Testament texts whose meaning has been further spiritualized. The parable of Lazarus and the Rich Man (Luke 16: 19-31) is just another example of the same mentality. If the gospel writer had been intent on condemning wealth, he could hardly have missed this unique occasion to do so. Yet one looks in vain for any such condemnation in the text, which instead leaves us with the impression that there is a good and a bad use of wealth, just as there is a good and a bad use of poverty. The poor person who lives in envy of his rich neighbor does not thereby prove his moral superiority to him. Besides, if wealth itself were the issue, one would have a hard time explaining why, after his death, Lazarus was rewarded by being placed in the bosom of Abraham, who is said to have been "very rich in cattle, in silver, and in gold" (Gen. 13:2).

Finally, even when the New Testament does speak about the economically poor, it is less from their point of view than from the point of view of those who are called upon to help them. The New Testament attempts to persuade people both to cultivate an attitude of gratitude to God for his gifts and to benefit the poor by just and merciful deeds. It doesn't give detailed descriptions of unjust social conditions.[7]

Precisely because the New Testament has no political agenda of its own, Christian theologians concluded long ago that the knowledge required to deal with such problems would have to come from other sources, and that is why they turned to political philosophy. Decisions of a political nature became matters of prudential judgment, duly informed by general principles of justice and a proper regard for the demands of the common good. Few if any absolutes could be invoked to resolve the complex problems political leaders are likely to confront. In each instance, one had to look for the solution that was most conducive to the general welfare. The assumption was that what was best in itself was not necessarily best at this or that particular moment, or for this or that particular group, and, hence, no single solution or plan of action would ever accommodate all of the unforeseen and unforeseeable circumstances in the midst of which decisions affecting the life of a nation have to be taken. The highest court of appeal was the natural law, which gained new prominence in the twelfth century, once the task of sorting out ecclesiastical laws from civil laws had been accomplished. As originally conceived, however, the natural law was a subtle instrument, to be used in a manner that took into account the high degree of contingency that surrounds human actions. The inflexibility associated with natural law in the minds of our contemporaries is a more recent phenomenon, traceable to the changes it underwent in the sixteenth and seventeenth centuries under the impact of the new philosophic and scientific theories of that period. To appreciate the distance that separates its earlier exponents from the later ones, one has only to observe that, by the time we reach Grotius, the four short articles that Thomas Aquinas devotes to the problem of war in the *Summa theologiae* had expanded into a mammoth 800-page quarto.

If, then, the New Testament contains so few political imperatives, and so little that bears any direct relationship to the political life, what impels so many Church leaders to look to it for practical guidance in these matters? What accounts for the sudden surge of interest on their part in such mundane matters as military strategy, government planning, economic rights, and the like, often at the expense of other topics that were formerly considered more worthy of their attention and more appropriate to their calling? As I have already intimated, part of the answer may be found in a series of earlier developments, whose full impact has only recently been felt within the churches. Two authors in particular deserve to be singled out as having had a determining influence in this respect. One is Spinoza, who denied that there was any theoretical teaching whatsoever to be extracted from the Bible and reduced the whole of its content to a moral teaching. That moral teaching is summed up in two words: justice and charity. Charity itself is taken only in its crudest and most superficial sense as that virtue by which one is inclined to help the needy. It is not intrinsically linked to those moral virtues that perfect individuals in themselves and concerns only their relations with others. The argument was all the more clever as it served a twofold purpose: once accepted, it could help put an end to religious persecution, which Spinoza attributed to the traditional Christian emphasis on doctrine as distinguished from morality; and, more importantly in Spinoza's view, it could be used to reclaim for philosophy the independence it had lost as a consequence of its subordination to theology.

The second great fountainhead of our currently most fashionable ideas is Rousseau, the author most responsible for setting modernity on the course it was to follow with all kinds of detours down to our time. Only two related facets of Rousseau's thought need to be retained for present purposes: his reinstatement of religion as the true bond of society and the source of its vitality and his so-called politics of compassion, both of which are of a piece with his radical egalitarianism. As we know from the "Profession of Faith of the Savoyard Vicar" and other texts, the religion in question is a religion without dogmas, or with only the simplest of dogmas. It is essentially a civil religion, but one that differs from the civil theology about which the authors of antiquity had spoken in that it is meant to close off any access to a higher or nonpolitical realm. The politics of compassion serves the same goal by calling for a greater equalization of social conditions through a restructuring of the entire political order. People were no longer to be rewarded on the sole basis of what they contributed to society; their poverty was a sufficient title to whatever benefits they could expect to receive from it. The neediest elements of society were the ones most deserving of its attention, and every disadvantaged citizen had an immediate claim on the compassion and benevolence of his fellow citizens. In the name of egalitarian democracy, the poor had become, to use a more recent expression, the object of a "preferential option."[8]

The dramatic impact of Rousseau's views is observable throughout the vast literary, sociological, and philosophic production of the nineteenth century. In Saint-Simon, the founder of French socialism, the interest in the poor had grown to the point of excluding virtually every other concern. Taking his cue from Rousseau, Saint-Simon started from the belief in God but cleared theology of all dogma. His famous essay *The New Christianity* is predicated on the principle that "the whole of society ought to strive toward the amelioration of the moral and physical welfare of the poorest class" and that it should "organize itself in the way best adapted to the attainment of this end."[9] A similar note is struck in Victor Hugo's poetic masterpiece, *The Legend of the Centuries,* whose sketch of the progress of the human conscience from its biblical origins onward climaxes in the glorification of a family of lowly fishermen, "The Poor People," in mid-nineteenth-century Brittany.[10]

The rise of Neo-Thomism as the dominant force in Roman Catholic theology did much to stem the tide of modernity within the Church and gave a more conservative tone to the bulk of its theological speculation. However, in its initial stages the return to Thomas Aquinas was motivated primarily by practical rather than theoretical reasons. What rendered the Thomistic position attractive to Christians who were struggling to adjust to the conditions of post-revolutionary European society is that it recognized the naturalness of the political order and was thus able to supply the philosophical premises by which the Church could relate to it. From the use of Thomas Aquinas by nineteenth-century theologians came what is now referred to as the "social teachings" of the Catholic Church. Yet there is ample evidence to show that the solution these theologians finally came up with was more a compromise with modern thought than a genuine re-appropriation of the principles that had governed premodern Christian thought.

Be that as it may, the collapse of Neo-Thomism in the wake of Vatican II paved the way for the emergence of a variety of more recent schools, Christian Marxism and liberation theology among them, in which the characteristic nineteenth-century amalgamation of religious and social concerns attains its most acute form. Both of these movements share with earlier Christian thought the view that the gospel alone is incapable of furnishing the guidance needed for the successful management of political affairs. The difference is that the new theology looks to Marx rather than to classical philosophy for that guidance. This is certainly true of Christian Marxism, but it is equally true of liberation theology, however eager it may be to reject the abstruse philosophical speculations associated with European thought and to pass itself off as a native Latin American product. As the late Gaston Fessard has so well shown, there is a straight line that goes from the works of the defrocked Italian priest Giulio Girardi to those of Gustavo Gutierrez, the presumed originator of liberation theology and still its most respected spokesman.[11] Granted, Girardi is cited only twice in Gutierrez's *A Theology of Liberation*, but a careful scrutiny of this work, such as Fessard's, reveals a much closer connection between the two authors than Gutierrez himself has thus far been willing to acknowledge.[12] The point at issue is not whether the liberationists are right or wrong in criticizing the abuses of governmental power or the exploitation of the masses by powerful and selfish elites, but whether the systematic use of Marxist categories to diagnose the evils of contemporary society does not implicitly commit the user not only to the narrow conception of justice embodied in these categories but to the philosophical materialism from which they appear to be inseparable. This, I take it, is the ground of the reservations expressed, for example, in Cardinal Josef Ratzinger's *Instruction on Certain Aspects of the Theology of Liberation*.[13] Baptizing Plato and Aristotle, which is what the Church Fathers and their medieval disciples attempted to do, not without some difficulty, is one thing; baptizing Marx may turn out to be quite another.

THE CHURCH AND SOCIAL JUSTICE

This still leaves us without an answer to the more general question:

How can the Church censure these and similar movements as misguided attempts to politicize Christianity even as it continues to issue statements that are fraught with unmistakable and sometimes momentous political implications? Is the Church contradicting itself, or do its leaders really believe that their own incursions into this troubled area are situated on a plane that transcends that of politics?

One convenient way to approach the problem is through the notion of social justice, which in the last fifty years or so has become a standard feature of official church documents as well as of much of the theological literature of our time. Unfortunately, the history of that novel expression has yet to be written, and its meaning remains vague. Taparelli d'Azeglio, the Roman Catholic theologian who is credited with having coined it, is anything but explicit in his discussion of it in his *Saggio teoretico di dritto naturale appoggiato sul fatto* (*Theoretical Essay on Natural Right*).[14] Even a cursory glance at the book reveals that its author was more learned than profound, more eclectic than consistent, and more muddled than clear-sighted. In fairness, one has to admit that he inherited an unenviable situation when, at the age of fifty and without any previous experience, he was appointed to a chair of moral theology in Sicily. Seminary life had been severely disrupted by the French Revolution, scholasticism had ceased to be a living tradition, and the only manual at his disposal was that of the Swiss jurist Burlamaqui, of Rousseauean fame, which was ill suited to his purpose. The only alternative was to write his own manual, which turned out to be a blend of Thomistic, Lockean, and Traditionalist ideas in the midst of which there appears, seemingly out of nowhere, something labeled "social justice."

Taparelli's nineteenth-century followers, who by then had acquired a better knowledge of the premodern tradition, were understandably bewildered. To which, if any, of the three hitherto known forms of justice—legal, distributive, and commutative—was social justice related? The shrewdest among them leaned toward the first, but with some reluctance since the resonances did not appear to be the same. Besides, to call justice "social" seemed redundant, inasmuch as justice had always been regarded as the social virtue par excellence—unless, as proved to be the case, it could be shown to designate a reality that is not covered by any of the older views of justice. Still, it is doubtful whether the notion would have stuck had it not fallen on soil that was already well prepared to receive it.

Justice in this new sense, one is tempted to say, is not a virtue at all, in that it has more to do with social structures than with the internal dispositions of the moral agent. Its proper object is not the right order of the soul but the right order of society as a whole. It shares with early modern liberalism the view that society exists for the protection of certain basic and prepolitical rights, and it radicalizes that view by combining it with an emphasis, stemming ultimately from Rousseau, on the need for a greater equalization of social conditions as a means of guaranteeing the exercise of those rights. It thus takes for granted that social reform is at least as important as personal reform and that the just social order depends as much on institutions as on moral character. It keeps the Lockean notion of labor as the origin of property but looks upon the accumulation of wealth as a form of exploitation of the poor by the rich few and, hence, as the root cause of the inequities that pervade modern Western society. Accordingly, it calls for either a radical redistribution of material resources or, short of that, the establishment of a system that reduces as much as possible the distance separating the social classes. Its immediate goal, in short, is to produce happy rather than good human beings. Taparelli himself, who in this instance at least is fairly consistent, went so far as to claim that all human beings had a right to happiness and not just to the pursuit of virtue. In the final analysis, there is one and only one just social order, whose broad outlines are prescribed in advance and therefore are not a proper object of deliberation on the part of wise and prudent legislators.

Are the churches or their representatives engaging in politics in the guise of religion when, in the name of the gospel, they call for the eradication of the systemic injustices of society through the reform and, if need be, the overthrow of the structures in which these evils are supposedly embedded? For better or for worse, this is what some theologians who would reduce the Christian faith to a program of political and social liberation are in fact doing. Whether it is also what the American bishops are up to is doubtful. Yet they too, it seems, have unwittingly or uncritically acquiesced in the new trend and think they can exclude from the realm of politics proper certain decisions that were once regarded as preeminently political. It used to be that there were two noble alternatives to the political life: revealed religion and philosophy. We are now given to understand that there is a third, specifically moral and thisworldly in character and all the more attractive because it addresses itself to all decent human beings, whether or not they are religiously inclined.

At the risk of ending with what may appear to be a hopelessly vague generalization, I would simply ask whether the modern rights theory that lies at the root of so much of what our Church leaders have been saying lately is compatible with the stress on duties or virtue that is typical of the older approach to these matters. The controversies over this question in recent years suggest that the answer to it is not clear or that there is a certain lack of clarity in the minds of those who have been trying to answer it. There is, nevertheless, some comfort to be taken in the thought that in all but the rarest of instances, clarity about theoretical issues has never been the hallmark of our political and religious leaders.

NOTES

1. Eugene Kennedy, "America's Activist Bishops: Examining Capitalism," *The New York Times Magazine*, 12 August 1984.

2. Gershom Scholem, *The Messianic Idea in Judaism* (New York: Schocken Books, 1971), 1-2.

3. Origen, *Contra celsum*, translated with an introduction and notes by Henry Chadwick (Cambridge: Cambridge University Press, 1953), VIII.55.493.

4. Oscar Cullman, *The State in the New Testament* (New York: Scribners, 1956), 90.

5. Thomas Aquinas, *De veritate*, qu. 12, a. 3, ad 11.

6. National Conference of Catholic Bishops, "The Challenge of Peace: God's Promise and Our Response," *Origins* 13, no. 1 (1983): 1-32; "Economic Justice for All: Catholic Social Teaching and the U.S. Economy," *Origins* 16, no. 24 (1986): 409-55.

7. Cf. Caesarius of Arles, Sermon 25 in *Sermons*, vol. 31 (Washington, DC: The Catholic University of America Press, 1956), 127-28; and p. 299 of this volume.

8. See, for example, the Catholic bishops pastoral on the U.S. economy, notes 16, 52, and 85-91, pp. 411, 418, and 421-22.

9. Claude-Henri de Saint-Simon, *Le nouveau christianisme, Oeuvres complètes*, vol. 3 (Geneva: Slatkine Reprints, 1977), 117, 177.

10. Victor Hugo, "Les pauvres gens" in *La légende des siècles* (Paris: Garnier, 1964), 700-7.

11. Gaston Fessard, *Chrétiens marxistes et théologie de la libération* (Paris: Lethielleux, 1978), esp. 413-21.

12. Gustavo Gutierrez, *A Theology of Liberation* (Maryknoll, NY: Orbis Books), 973, 271, 275.

13. The text of the *Instruction,* issued by the Congregation for the Doctrine of the Faith, may be found in *Origins* 14, no. 13 (1984): 193-204.

14. Luigi Taparelli d'Azeglio, *Saggio teoretico di dritto naturale appoggiato sul fatto* (Palermo, 1840, and Rome: La Civiltà Cattolica, 1855), vol. I, bk. ii, ch. 3: "Nozioni del dritto e della giustizia sociale," Rome edition, 220-32.

THEOLOGICAL REFLECTIONS ON
THE CHALLENGE OF PEACE: GOD'S PROMISE
AND OUR RESPONSE,
A Pastoral Letter on War and Peace
(May 3, 1983)

We owe a deep debt of gratitude to our bishops for their splendid pastoral letter and to Fr. Bryan Hehir[1] for his brilliant analysis of the genesis and content of that letter. The final draft strikes me as being vastly superior to the first two and serves as a fine example of what can be accomplished through patient discussion, charitable debate, a genuine openness to opposing viewpoints, and a gentle nudge from on high, by which I mean of course, the Holy Spirit! The letter has the great merit of reminding us that politics and especially nuclear politics should not and cannot be divorced from considerations of morality, as it so often is in the minds of our contemporaries. It also has the merit of restoring a measure of unity among the bishops themselves, who in the recent past have given the impression of being at odds with one another, at the risk of confusing some of the faithful who expect them to speak with roughly the same voice at least on fundamental issues. Here at last is a statement to which virtually all of them could subscribe, which is more in line with the teaching of the universal church, and which is bound to have a considerable impact on the nation as a whole. It is by far the most ambitious statement to come out of the National Conference of Catholic Bishops in the last fifteen years or so and it fully deserves the acclaim with which for the most part it has been greeted by Catholics and non-Catholics alike. Since I mostly agree with the positions taken in the letter, I shall limit myself to a few brief remarks about some points that call for further discussion or clarification.

The first of these has to do with the so-called "limited" or "just war" theory. The letter, as I read it, endorses that theory and rightly credits St. Augustine with having supplied it with its central insight, which is that the New Testament command of love must be understood in such a way as to include the defense of the innocent. Simply put, to allow one's family, one's friends, or others for whom one is responsible to be tyrannized or massacred by an unjust aggressor is not an act of love. Unless I am mistaken, however, the letter's account of the just war doctrine owes more to the Spanish and Dutch theorists of the sixteenth and seventeenth centuries than it does to any of their predecessors. The result is a conception of the just war that is more juridical or legalistic in tone than the one advocated, for example, by Augustine or Thomas Aquinas, both of whom seriously doubted whether matters as complex as these could be legislated with any degree of precision, if only because decisions pertaining to them are necessarily affected by the immense web of ever-changing and largely unforeseeable circumstances in the midst of which they have to be taken. The Spanish theorists obviously thought they could be so legislated, but then they were faced with the problem of justifying Spanish imperialism on the one hand and of curbing its excesses on the other. Well intentioned as it may have been, their theory runs the twofold risk of granting to war a stronger legal sanction than it might otherwise have and of depriving statesmen of the flexibility required for the discharge of what Cardinal Casaroli recently called their "awesome responsibilities."

Let me illustrate this by means of a simple example. The letter expresses "profound skepticism" about the morality of any nuclear attack, whether it be by way of a first strike or by way of retaliation. The reason alleged is that the new weapons probably could not be used without endangering the lives of millions of noncombatants and hence without violating one of the basic tenets of the just war theory. At the close of the May meeting at which the final draft was emended and approved, the current president of the National Conference of Bishops was quoted as saying that he could not conceive of any situation in which the use of nuclear weapons would be morally justified. Neither can I, right now. It is nevertheless hard to believe that Catholic teaching can be adequately formulated on the basis of what anyone, even a bishop, is or is not able to imagine at any given moment. Say that a country is being attacked by a powerful enemy and that its existence is threatened by a fleet of nuclear submarines which can be destroyed in mid-ocean, or by an orbiting nuclear missile which can be destroyed in outer space, but only by another nuclear device. Would the same restrictions still apply? No one can answer that question in advance for the simple reason that the answer would depend on one's assessment of the total picture at the time of the emergency. This is why Thomas thought that warfare was first and foremost a matter, not so much of moral and legal principle, but of political prudence, duly informed by a proper regard for the requirements of the common good.[2] If time permitted I would argue that the original just war theory does greater justice to the nonviolent position than does the present one, even though the Christian tradition has never accepted nonviolence as a universally valid option.

In fairness, it should be added that the bishops' remarks are hedged in with all kinds of "whereases," so many of them in fact that assessing their exact meaning is like trying to pin a tail on a vanishing donkey. The letter states flatly that our "no" to nuclear war must be "definitive" and "decisive." Yet, when asked point-blank whether the bishops had condemned the use of nuclear weapons altogether, Archbishop Roach admitted that he was not sure. All he could say was they had certainly moved "closer to" that position. I admire his candor but I also understand his predicament. There are moments when the letter almost sounds like the Athanasian Creed, which is supposed to be the only nonheretical statement ever to be written about the Trinity because every time it affirms something, it immediately proceeds to deny it. Maybe the bishops could ask Fr. Hehir to tell them what they voted for!

My second point concerns the distinction between "principle" and "policy," which is more sharply drawn in the final draft, in response to some legitimate criticisms elicited from above and below by the earlier drafts. Accordingly, the bishops make it clear that the general principles enunciated in the first part of the letter carry more weight than the policy judgments contained in the second part. Unfortunately, the lines often crisscross, and ordinary lay people, who lack the theological sophistication of the bishops, cannot always tell which hat they are wearing when they make this or that statement. Such at any rate has been my experience in talking with them.

This may be only a minor problem, however. The crucial question is whether bishops really need to be as specific as ours have chosen to be in formulating proposals about nuclear policy. One can appreciate their desire for concreteness, since by their very nature moral principles are meant to be applied; but this still leaves open the question of the level of application, about which opinions tend to vary. The letter strongly favors nuclear disarmament and takes a rather dim view of nuclear deterrence. What if its recommendations, once implemented, were to lead to a nuclear holocaust? There are, after all, some highly reputable analysts who hold different views on this matter and who seem to think that, given the present world situation, the prospects for a lasting peace would be diminished rather than enhanced by the adoption of the bishops' proposals—that, until a better solution can be worked out, we should look instead for a more reliable balance of nuclear power, even if this means conducting further tests such as those to which the bishops want to call a halt. The sad fact of the matter is that no one today, not even the greatest experts, as they themselves will testify, knows what the best solution to the problem might be.

If, in spite of all their disclaimers, the bishops insist on speaking as strategists, they must expect to be judged by criteria other than the ones that normally apply to episcopal pronouncements. It is ironic, but not at all surprising, that one of the latest articles on the subject should be entitled precisely, "Bishops, Statesmen, and Other Strategists on the Bombing of Innocents."[3] I mention this only to indicate that it may not be in the bishops' interest to allow themselves to be cast in that role and have their views discussed on a par with those of, say, McGeorge Bundy, Stanley Hoffman (whose name they cannot seem to spell right), and Albert Wohlstetter, to cite only three authors who have dealt with this issue at the same time as the Catholic bishops.

No one denies that the Church has a vital stake in the political life of the nation, and it is regrettable that, through no fault of its own, it has been removed from it for so long. But to say, as the letter does, that it "should become involved in politics" is not the most prudent way to voice that concern, especially in a country such as ours. It does not suffice to reply that, as private citizens, the bishops have every right to make their opinions known to others. A statement issued in the name of a national hierarchy will always be perceived as having some sort of official status and claim more respect than the opinions of a private individual. Bishops are at their best and are most convincing when they speak as theologians, not as military strategists. Unless they can show that their views on nuclear deterrence are self-evidently or demonstrably superior to other possible choices, they should exercise the greatest restraint in articulating them in a public document such as this one. The trouble with the letter, if I may say so, is that it does not take the threat of nuclear annihilation seriously enough, just as in my opinion it is not sensitive enough to the impulse at work in the nonviolent option. Its lofty talk about respect for human life would have been even more compelling if greater attention had been paid to the possible outcome of some of the strategies that it outlines. As Cardinal Lustiger, the Archbishop of Paris, suggested not long ago in a different but related context, the position of the American bishops is that of "people who know or believe that nothing will happen to them anyway."

The usual objection to this line of questioning is that a mere rehash of the old Catholic teaching would have been trivial at this point, that more was needed in order to catch the dilemma of the age, that a specific program of action had to be laid out if the bishops were to make a positive contribution to the national debate and influence government policy while it is still in the process of being formed. All well and good. Still, the nuclear crisis has been with us for quite a while now, decades in fact, and there are those among us who have been striving to alert students to its gravity for a quarter of a century or more, with little support from the ecclesiastical establishment at a time when that support would have been most useful. Better a belated statement than no statement at all, I suppose, but one cannot help wondering why it was produced at this time and not earlier. The answer is fairly obvious. Without the prodding of a powerful and still growing antinuclear movement, paradoxically summoned into existence by the deployment of Soviet missiles in Europe, the bishops would never have dared to speak out as loudly as they did.

The same answer accounts for the political preferences evinced in the letter. Nobody expected or even wanted a pure and simple reiteration of the timeless principles of Catholic moral theology. The difficulty is that, by following those principles to their logical conclusion, one does not necessarily come out on the bishops' side of the political agenda. It is no mere coincidence that the policy which the bishops would like to shape has already been shaped to a large extent by "prophetic" forces outside the Church and without the benefit of its assistance. One emerges from a reading of the letter with the odd feeling that the bishops as a bloc are not really haunted by the specter of a nuclear war, all protestations to the contrary notwithstanding. To outsiders, their recent conversion will undoubtedly appear too sudden to carry much conviction. Besides, there are just too many inwardly more pressing issues which they have yet to address and on which they seem somewhat reluctant to pronounce themselves. Having read the "signs of the times" (as they call them), they sense very well that an equally "courageous" statement on those issues would not muster half as much support. Clearly, there are times when silence is golden. It must not be easy these days to preside over a Church which stands for so much that runs counter to popular sentiment.

In all of this I have not said anything that is not already said or hinted at in the letter. I have merely tried to refocus the issue by stating it in slightly different terms and from a slightly different perspective. My real regret is that there is so little about peace in a document published under the heading, *The Challenge of Peace.* Since our bishops have already done so much however, it would be cruel to reproach them with not having done more, and I for one am not about to do so.

NOTES

1. The Rev. J. Bryan Hehir was an advisor and employee of the U.S. Bishops at the time the pastoral letter was issued. He played a significant role in shaping the various drafts of the letter.

2. Thomas Aquinas, *Summa theologiae* II-II, qu. 50, a. 4.

3. Albert Wohlstetter, "Bishops, Statesmen and other Strategists on the Bombing of Innocents," *Commentary* 79, no. 6 (June 1983): 15-35.

23

CHRISTIANITY AND THE JUST WAR THEORY

In a world that seems to be held together by little more than a common fear of a common destruction, it was practically inevitable that the limited or just war theory should regain some of the credit that it lost when, following the breakdown of medieval Christendom and the emergence of the modern sovereign state, it was largely abandoned in favor of what now goes under the name of international law. The presumed advantage of the new approach to the problem of foreign relations was that it did not depend for its efficacy on the moral character of the nation or the virtuous dispositions of its rulers. Since few people can be trusted to behave reasonably when their private interests are at stake, one might, indeed, be tempted to look for a solution whose successful implementation was not contingent for the most part on the acquisition of virtue. Reliable institutions, designed without regard for the needs of others and for the sole purpose of promoting the welfare and prosperity of one's own nation, would automatically bring about a world order from which the specter of war would be banished altogether. Human beings could remain as they are as long as the laws under which they lived were what they ought to be. To help others they had only to think of themselves. In the interest of morality, but without the benefit of its assistance, nature had seen to it that a lasting peace would one day be achieved regardless of whether or not people were ready to curb their aggressiveness or renounce their worldly ambitions. Progress in the direction of that morally desirable goal was assured, not by the deliberate pursuit of moral purposes, but by the free interplay of essentially immoral and self-regarding passions. As a result of their incessant preying on one another, human beings had created a precarious situation from which they could extricate themselves only by desisting from any further attempt at aggrandizement at the expense of their fellow human beings. Wars would diminish in frequency and intensity, not because their irrationality offended mankind's moral sense, but because they were proving ever more costly, destructive, and detrimental to the material well-being of the warring nations. A hidden teleology at work in the historical process was sufficient to guarantee that the antagonisms which characterize human intercourse would constrain even the worst offenders to adopt a civic constitution recognizing everyone's rights and granting to each individual and nation as much freedom as is consistent with the freedom of others. As Kant finally put it:

> The friction among men, the inevitable antagonism, which is a mark of even the largest societies and political bodies, is used by nature as a means to establish a condition of quiet and security. Through war, through the taxing and never-ending accumulation of armament, through the want which any state, even in peacetime, must suffer internally, nature forces them to make at first inadequate and tentative attempts; finally, after devastations, revolutions, and even complete exhaustion, she brings them to that which reason could have told them at the beginning and with far less sad experience, to wit, to step from the lawless condition of savages into a league of nations. In a league of nations, even the smallest state could expect security and justice, not from its own power and by its own decrees, but only from this great league of nations, from a united power acting under decisions reached under the laws of their united will . . . All wars are accordingly so many attempts (not in the intention of man but in the intention of nature) to establish new relations among states, and through the destruction or at least the dismemberment of all of them to create new political bodies, which, again, either internally or externally, cannot maintain themselves and which must thus suffer like revolutions; until, finally, through the best possible civic constitution and common agreement and legislation in external affairs, a state is created which, like a civic commonwealth, can maintain itself after the fashion of an automaton.[1]

Chastened by two world wars and the persistent threat of a nuclear conflagration, many of our more thoughtful contemporaries have begun to entertain serious doubts about the practicality of the modern position. In that connection, the much publicized pastoral letter on nuclear deterrence

(*The Challenge of Peace*: *God's Promise and Our Response*) issued May 3, 1983 by the Roman Catholic hierarchy comes as a timely reminder that matters pertaining to war and peace cannot safely be divorced from considerations of morality. If the lively debate sparked by the bishops' letter suggests anything, however, it is that its noble attempt to reaccredit the just war theory is not without difficulties of its own, which need to be clarified before a genuine consensus among all interested parties can be reached.

One of these difficulties stems from the fact that the argument that the letter develops, and through which it seeks to influence public policy while it is still in the process of being shaped, is of specifically Christian origin and is liable to appear to impartial observers as a matter of party allegiance. The classical philosophers occasionally spoke of wars as being just or unjust but never made any real effort to elaborate what could properly be described as a just war "theory." Their main focus was on domestic policy, where principles of justice are both easier to recognize and more readily applicable. Although fully aware of the constant impingement of foreign affairs on the internal life of the nation, they seem to have concluded that, most of the time, necessity rather than choice was the controlling factor in the polis's dealings with other nations. It was more important to them that the citizens of a particular city develop such habits as would help them to moderate their appetite for conquest. Plato even went so far as to suggest that a decent city differs from a gang of robbers, not in how it acquires its property or its wealth, but in what it does with them once they have been acquired. Accordingly, external matters were most often left to the discretion of wise and prudent rulers. The solution was chancy at best, inasmuch as one can never be sure that the city will have such rulers at its head when it most needs them. It was, nevertheless, regarded as not only the best but the only viable solution to the nagging problem of international relations. Neither in Plato nor in Aristotle do we find anything that quite compares with, say, the famous question "On War" in Thomas Aquinas's *Summa theologiae*.[2]

The one possible exception to this rule among the ancient philosophers is Cicero, who as a statesman as well as a philosopher, could hardly avoid coming to grips with what had long been the dominant fact of Roman political life, namely, its military conquests, and whose works do contain something like a just war theory. Not surprisingly, it is from him that the Church Fathers inherited the notion of the just war, although their use of it went well beyond what he originally intended by it. It is doubtful whether Cicero himself was convinced of the unimpeachable justice of Rome's conduct toward other nations. He certainly knew that there was little historical evidence to support the contention that her wars were waged only in self-defense or for the defense of her allies. If such had been the case, these allies would have been returned to independent status as soon as the danger was over and Rome would never have established herself as the mistress of the civilized world. Anyone who reads Cicero's *Republic* carefully comes away with the impression that the just war theory that it expounds was nothing more than a subtle critique of Roman expansionism, written by someone who appears to have been more concerned with what the conquest of the world would do to the conqueror than with what it would do to the conquered nations.[3]

Second, as explained in the previous chapter, the Catholic bishops' version of the just war theory represents a significant departure not only from the teachings of the classical philosophers but from the spirit and the letter of the original Christian doctrine elaborated by St. Augustine. The bishops' version of the just war theory, influenced more by the Spanish and Dutch theologians of the sixteenth and seventeenth centuries, is more legalistic and juridical in tone than the theory put forth by Augustine and Aquinas.

This much being said, one has to admit that the problem of warfare has always been fraught with greater urgency for the Christian theologian than it was for any of the philosophers of classical antiquity. Given the force of the biblical teaching concerning the sacredness of life, it is understandable that from the beginning some Christians should have been reluctant to engage in bloodshed or condone the use of arms. When the ghost of Hamlet's father urges his son to "revenge his foul and most un-natural murder,"[4] he is careful to add: "murder most foul, as in the best it is," thereby calling attention to the peculiarly loathsome character of any deed that has as its object the taking of another person's life. Killing is, of course, not the same thing as murder and one does not normally look upon a soldier who kills in the legitimate exercise of his duties as a murderer. Yet, in view of the fact that war inevitably results in the death of persons who do not deserve to die and are not meant to die, the line between the two cannot always be drawn with absolute clarity. True, the Hebrew Scriptures contain more than their share of violence and bloody massacres; but these were holy wars, undertaken at God's command and for the sake of overcoming his enemies. The issue was whether the Christian, living as he did under a new dispensation, was free to indulge in a practice that was no longer ordered by God or ostensibly carried out in fulfillment of his mysterious designs. Is an act of war in any way compatible with Christian love and, thus, at least permissible within Christian ethics?

As far as the early Church is concerned, the record is spotty and probably less clear than the bishops seem to indicate. It is perhaps unfortunate that the drafters of the letter should have had no choice but to rely for their information on a variety of works published during the 1960s and early 1970s, often by scholars whose evaluation of the materials at hand bears traces of their pacifist leanings. More recent studies have done much to redress the balance, and there is reason to think that work currently in progress will corroborate their findings.

Be that as it may, one looks in vain for any thematic treatment of the question in the literature of the pre-Constantinian period, and all signals point to the fact that for most people the issue was somewhat theoretical. For one thing, there was no regular involuntary conscription in the Roman Empire; for another, many of the early Christians belonged to the class of slaves or freedmen, none of whom was eligible for military service. The situation changed slightly when pagans began to convert to the new faith in larger numbers, but even then no uniform solution prevailed. Some Christian writers, Tertullian and Origen in particular, voiced strong opposition to warfare and advocated or leaned toward pacifism as the only legitimate Christian response to aggression, albeit for different reasons. Violence is one of them, but only one. Equally prominent were idolatry or emperor-worship (which was generally restricted to superior officers) and the danger of immorality, a phenomenon not uncommonly associated with the military life. Others adopted a more moderate stance, forbidding Christians to enlist in the army but allowing professional soldiers to remain in it even after their conversion to the faith, since it would not have been feasible for most of them to do otherwise. Desertion was punishable by death and the soldier who abandoned his service forfeited all the benefits of property or money to which he was entitled upon retirement. Besides, military service did not necessarily involve one in acts of violence. For long stretches of time the Empire was at rest or compelled to wage war only sporadically and on a limited scale. More often than not, soldiers worked at peaceful jobs. They doubled as police officers and served as messengers, ushers, or accountants for their local commanders. One of them even brags about having done nothing but write during his entire twenty-five years of service. Nor for that matter was there any great incentive to join the army since the pay was not very good. Interestingly enough, the enrollment jumped the moment salaries were increased.

I will make one final remark on this subject, which has to do with the discrepancy that one detects between what Christians said and what they actually did. Tertullian, whose stand on the issue is based on the biblical injunction against killing, nevertheless looked upon participation in war as a potentially virtuous act and is himself a major witness for the presence of Christians in the army. Christians are known to have been included in the famous Thundering Legion under Marcus Aurelius, for otherwise the claim that they were responsible for its successes could not have been made with any degree of credibility. Agbar, the king of Edessa in Syria, converted to Christianity and made it the official religion of his realm, something that would hardly have been possible if it had deprived him of all recourse to military power. Diocletian singled out the Christians in his army as the prime target of his persecution, another indication of the fact that Christians served as soldiers, small as their number must have been. In short, the opposition to military service, to the extent it existed, was not adhered to in practice. Throughout that period, it was considered normal to profess one's loyalty to the Empire, but the implications of that loyalty were never made fully clear. One gathers from these observations that the Christian position on war, assuming that one can even speak of a Christian position, was not so much contradictory as immature or undeveloped.[5] What was needed was a more thorough examination of the question, of the kind that later writers would provide.

The first such attempt dates from the end of the third and the beginning of the fourth centuries and takes the form of what is sometimes referred to as the Christian idea of progress.[6] It developed the idea that the advent of Christianity had signaled a decisive turning point in the political as well as the spiritual history of the human race. With the spread of the new faith and its eventual recognition by the imperial rulers, harmony would slowly be restored throughout the world and human beings would never again be faced with the excruciating dilemma of having to take up arms against one another. As Isaiah had predicted, swords would be turned into plowshares and justice and peace would at last forge a lasting alliance (Isaiah 2:4). The accession of Constantine to the imperial throne, his toleration of Christianity, and his own subsequent conversion to it were all signs pointing in the same happy direction. Under the aegis of the new emperors, the Kingdom of God was about to be inaugurated not just in heaven, as some less worldly-minded apologists for the Christian faith had announced, but here on earth.[7]

The argument was as appealing as it was shallow. Its naiveté was exposed less than a century later when Rome was suddenly faced with a new and more powerful threat stemming, not from the centrifugal forces at work within it, but from mounting pressures on its borders. The death blow came in 476 when the empire was finally overrun by barbarians, but its extreme vulnerability was brought home to everyone in 410 when the city of Rome itself was occupied by Alaric and his Goths for a brief period of two weeks. The reaction of the pagan elite was swift and to the point. The blame for the disaster was promptly laid at the door of Christianity, which was held to be inimical to the welfare of the state on the ground that it disparaged military valor and deprived the city of its only reliable means of defense against its enemies. Rome, whose security had always rested on its military prowess, could no longer count on the strength of its once invincible armies. Its present plight was no mere quirk of chance. It could plausibly be ascribed to the rapid growth of a peace-loving religion and its recent proclamation as the official cult of a nation that had formerly been "dedicated to Mars."[8]

The challenge was a serious one. It struck at the very heart of the Christian faith and brought into full daylight a problem that had been submerged rather than solved by its earlier defenders. There was no denying the transworldly character of the Christian Scriptures, their profound silence on problems of a properly political nature, their manifest indifference to the distinction between regimes, their all but total disregard for the necessities of social life, and their emphasis on an elusive Kingdom of God that was supposedly taking shape within history but whose promises could only be fulfilled outside of it. For anyone who took their message at face value, even the most glorious human achievements paled into insignificance. Others might be preoccupied with the fate of empires; the Christian, who had a better world to look forward to, could afford to "let the world slip."[9] In the words of a prominent fourth-century Greek Father:

> Neither renown of ancestry, nor strength of body, nor beauty, nor stature, nor honors bestowed by all mankind, nor kingship itself, nor any other human attribute that one might mention, do we judge great; nay, we do not even look with admiration upon those who possess them, but our hopes lead us forward to a more distant time, and everything we do is by way of preparation for the next life.[10]

Little wonder that the new religion, which did so much to discourage public-spiritedness or at any rate so little to encourage it, should have been targeted as the prime cause of the deterioration of the empire's political fortunes and of its present inability to hold its enemies at bay.

Augustine's just war theory is part of a larger endeavor, the aim of which was to counter the pagan objection by demonstrating that, for all its seemingly radical apoliticism, the New Testament ideal was by no means irreconcilable with the requirements of dutiful citizenship. The premise on which the argument rests is that Christianity, unlike Judaism, presents itself first and foremost not as law but as a faith or a set of teachings that do not of themselves call for or encourage the formation of any kind of political community. It simply presupposes the existence of civil society and takes for granted that Christians will continue to structure their temporal lives in accordance with its needs. Precisely because it had no political program of its own, Christianity could adapt itself to any regime. Admittedly, they were all imperfect, but they could be improved; and by living in them, one could contribute to that improvement. This explains why Christians were to be found everywhere, mingling freely with the rest of the population, sharing their customs, their dress, and, within prescribed limits, their general way of life.[11] The only practices to which they were opposed were the ones that reason itself denounces as vicious or immoral and that are the real cause of the decline of states. By making civil obedience a religious as well as a moral duty (Rom. 13:1), their faith fostered law-abidingness, patriotism, and every form of public virtue.

Nor was it fair to say that Christianity condemns war and breeds contempt for military valor. As Augustine rightly saw, there was no consistent teaching on war to be found anywhere in the New Testament, for the simple reason that the question is never directly addressed by any of the sacred writers. The letter to the Romans did make it clear, however, not only that Christians had to obey their rulers but that these rulers had been ordained by God for the express purpose of inflicting his wrath on evildoers (Rom. 13:4). If, as one must suppose, their duties often entail the use of force, the Christian who carries out his orders is bound to become involved in acts of violence.

The few texts that could be invoked in support of nonviolence were either beside the point or subject to qualification in the light of other texts that run counter to them. Christ, who occasionally speaks with soldiers, does not rebuke them for their profession or urge them to lay down their arms, but actually commends them for their righteousness. As was mentioned earlier, the commandment "Thou shalt not kill" is really a modified version of the more precise "Thou shalt not murder" and blurs the distinction between these two notions, which the original text is careful to maintain. The soldier, unlike the murderer, is not someone who takes the law into his own hands. He does not exercise the right of private judgment or try to determine for himself who ought or ought not to be allowed to live. There is a difference. Christ does say elsewhere that "all who take the sword will perish by the sword" (Matt. 26:52), but he could not have intended this as a universally valid maxim inasmuch as, without ever having held a sword, he himself died a violent death.

As for the famous injunction to requite evil with good, it must be interpreted within the general context of one's relations with all of one's fellow human beings. The situations envisaged in the New Testament are typically one-on-one situations rather than situations in which the welfare of a third party is at stake. One ought to be ready to forgive personal offenses (Augustine, who is stricter than Thomas Aquinas on this point, goes so far as to forbid killing in self-defense), but one is surely not required to love the criminal more than his innocent victims, especially when one bears a special responsibility for their safety.

One could infer from all of this that the New Testament is less concerned with external actions than with the internal dispositions with which they are to be performed. It seeks to ensure that war, if it must be waged, will be carried out with a benevolent design and without undue harshness. Human beings are compelled at all times to do what is most likely to benefit others. In some instances, peace and the correction of wrongdoers are more readily and more perfectly achieved by forgiveness than by castigation, whereas in other instances one would only confirm the wicked in their evil ways by giving free rein to injustice and allowing crimes to go unpunished. What Christianity reproves is not war itself but the evils of war, such as love of violence, revengeful cruelty, fierce and implacable hatreds, wild resistance, and the lust for power. By yielding to these evils, human beings lose a good that is far more precious than any of the earthly possessions an enemy could take from them. Instead of increasing the number of the good, they merely add themselves to the number of the wicked. Just wars are therefore permissible, even though they must be undertaken only out of necessity and for the sake of peace.

On its own grounds, the foregoing theory is open to at least two serious objections to which there are unfortunately no neat answers. The first is directly related to the question of self-defense. The clearest instance of a just war is that of the war embarked upon for the sole purpose of protecting oneself or one's borders against an unmerited attack. This obviously presupposes that the borders to be defended are just in the first place. Yet, in light of what Augustine says about the origin of nations and empires, it is hard to believe that such borders are ever completely just or natural. There was surely nothing sacred about the borders of the Roman Empire, which had been secured by force of arms and at the price of untold miseries visited upon weaker or less belligerent nations. Rome itself was but a mammoth larceny, long unpunished only because of its might. From this larger perspective, its position was not morally superior to that of its enemies. If, in addition to that, one bears in mind that the state of war inevitably gives rise to situations in which the ordinary rules of justice are suspended, one might well conclude, as Augustine does, that war is part of an order of things that, no matter how one looks at it, leaves much to be desired and is suited only to the condition of a wounded or inherently imperfect nature.

The second problem concerns the tacit but questionable assumption that a war can be just on one side only. This much would seem to be implied in Augustine's statement to the effect that "when we wage a just war, our adversaries must be sinning."[12] If Rome's cause was just, then that of the invader was unjust. Still, there was plenty of evidence to show that the barbarians were often acting out of necessity and under pressure from powerful hordes to the east before whom they had little choice but to flee. Besides, in the course of the negotiations, Rome apparently had few qualms about breaking faith with them and even indulging in the most unspeakable treacheries. Thousands of barbarians who had been serving in the Roman army were mercilessly butchered once the reaction against them set in, and others, who had been promised asylum, were reduced to the most abject slavery upon their arrival.[13] Rome, too, bore its share of the responsibility for the evils that had recently befallen it. Augustine all but says that it was only getting what it deserved. The problem is not unlike the one that came up centuries later in the wake of the Spanish conquest of the New World. Even assuming that wars of civilization are permissible and that the Spaniards had every right to wage war on the Native Americans, one hesitates to fault the original inhabitants of the conquered lands for defending themselves.

All in all, Augustine's theory left much unsaid, although probably not unthought. If anything can be said to have motivated it, it is the conviction, not that wars can ever be completely just, but that under certain more favorable circumstances they might become a trifle less unjust. There are limits to how far one can go in establishing a nation's right to the territory over which it rules or in laying down rules for the defense of that territory. As Thomas Aquinas would later say, the art of warfare (*militaris*) belongs preeminently to the sphere of political prudence, duly informed by a proper regard for the requirements of the common good. The contingency of its matter is such that its exercise can never be governed by universal moral and legal principles, save of the most general kind.[14]

For Augustine, the choice was between civilization and barbarism, and it was in the light of this distinction that the decision to support one side or the other had to be made. No one, not even Cicero, had fewer illusions about the justice of the Roman Empire. If his heart was still with it, it is because he thought that the prospects for justice, slim as they always are, were greater within it than outside of it. No doubt, individual human beings are capable of a higher degree of moral perfection than society at large, but their chances of attaining that perfection will be enhanced in proportion as the life of their society is itself guided by principles of justice and moderation. Without justice in the soul, there will never be justice in the city or in the city's relations with other cities and nations. The paradox with which we are finally confronted was all too familiar to the political thinkers of the premodern period: good government makes for good citizens, but it is itself made by good citizens. That is the ultimate reason for which so many human societies had neither.

It was left to the great theorists of the modern age to come up with what was supposed to be a more workable solution to the problem by removing ethics from the realm of politics and by attempting to derive the common good from the enlightened self-interest of the individual members of society. Kant, in whose works the new view achieved perhaps its finest expression, went further than anyone else in proclaiming that the perfectly just society was possible. The paradox this time was that the perfect society could theoretically be made up entirely of "devils," albeit "intelligent" ones.[15]

It took nothing less than the horrifying prospect of a nuclear holocaust to cast doubts on the wisdom of the modern approach and to reawaken our interest in the discarded alternative to it. One can be grateful to the U.S. bishops for their unique effort to remoralize the issue of warfare and rethink it in terms of the situation created by the proliferation of nuclear weapons. The novelty of their letter, however, is not so much that it revives and updates the old just war theory as that it seeks to combine it with its opposite. This much is evident from the enormous stress that the letter places on matters of legality and strategy. As noted in the previous chapter, it comes as something of a surprise that the U.S. bishops would say so little about peace in a pastoral letter entitled *The Challenge of Peace*. To be sure, the letter contains a long section on peace, but that section is itself devoted mainly to problems of arms negotiation, arms control, arms reduction, the establishment of a national peace academy, "reverence" for the United Nations, the improvement of relations between the superpowers, the positioning of missiles, and a host of other similarly mundane, procedural, or technical matters. One searches in vain for a genuine discussion of peace, based on a profound analysis of the human soul, such as the one that accompanies Augustine's treatment of the just war. In defense of their position, the bishops have argued that the old just war theory, with its one-sided emphasis on morality, is inadequate to the needs of the age and that it must be supplemented by considerations of a more political nature. If so, one might be inclined to think that, in its own way, their letter does as much to hallow the divorce between ethics and politics as it does to overcome it. The bishops clearly want the best of both worlds. It remains to be seen whether these two originally antithetical worlds are suited to each other or how well they can get along together.

NOTES

1. Immanuel Kant, "Idea for a Universal History with a Cosmopolitan Intent," translated by L. W. Beck, *On History* (Indianapolis: Bobbs-Merrill, 1963), 18-19.

2. Thomas Aquinas, *Summa theologiae* II-II, qu. 40.

3. For a brief but careful and penetrating analysis of the classical position, see T. L. Pangle, "The Moral Basis of National Security: Four Historical Perspectives," in *Historical Dimensions of National Security Problems*, edited by K. Knorr (Wichita: University Press of Kansas, 1976), esp. 308-17.

4. William Shakespeare, *Hamlet*, act 1, scene 5, lines 25, 27-28.

5. See, on these and related matters, L. Swift, "War and the Christian Conscience I: The Early Years," in *Aufstieg und Niedergang der römischen Welt*, vol. II, 23, 1, edited by H. Temporini and W. Haase (Berlin and New York: Walter De Gruyter, 1979), 835-68.

6. See T. E. Mommsen, "St. Augustine and the Christian Idea of Progress," reprinted in *Medieval and Renaissance Studies* (Ithaca: Cornell University Press, 1959): 265-98.

7. For a summary of the recent discussions of this issue, see E. L. Fortin, "Augustine's *City of God* and the Modern Historical Consciousness" in *Classical Christianity and the Political Order: Reflections on the Theologico-Political Problem*, edited by J. Brian Benestad (Lanham, MD: Rowman and Littlefield, 1996), 117-36.

8. See Augustine, *De civitate Dei* (*City of God*) IV.29.

9. William Shakespeare, *The Taming of the Shrew*, Induction, scene 2.line 146.

10. Basil the Great, "Address to Young Men on Reading Greek Literature," in *The Wisdom of Catholicism*, edited by A. Pegis (New York: Random House, 1949), 10.

11. See Augustine, *City of God* XIX.17.

12. Ibid., XIX.15.

13. R. Bainton, *Christian Attitudes toward War and Peace* (New York and Nashville: Abingdon Press, 1960), 99-100.

14. See Aquinas, *Summa theologiae* II-II, qu. 50, a. 4.

15. Immanuel Kant, "Perpetual Peace," First Supplement, in *On History*, translated by L. W. Beck (Indianapolis: Bobbs-Merrill, 1963), 112.

24

CATHOLIC SOCIAL TEACHING
AND THE ECONOMY:
CRITERIA FOR A PASTORAL LETTER

This chapter is directed to the problem of the relationship between Christianity and the social order, particularly as it bears on matters of economic policy. The importance of this problem, to which no thoughtful Christian can be indifferent, has recently been brought to our attention by the bishops' decision to issue a pastoral letter dealing with the American economy from a Roman Catholic perspective.[1] That decision and the round of discussions initiated for the purpose of implementing it have again stirred the hopes and fears of a substantial segment of the Catholic community, not to mention others outside of it. The hopes correspond to a desire, shared by all decent human beings, to see justice promoted in our society. The fears have to do with the eventual content of the letter, which some think could be unduly politicizing. Both the hopes and the fears are probably exaggerated. Although it is hard to measure the impact of such documents, recent experience has shown that their persuasiveness tends to diminish in proportion as the bishops are perceived to be talking about subjects that do not fall within their jurisdiction or their acknowledged sphere of competence. Even the *National Catholic Reporter* admitted in bold type that the pastoral on nuclear deterrence, published on May 3, 1983, had already "sunk into *divine* oblivion."[2] The editors meant "benign" oblivion, or so one gathers from the rest of the article. I leave it to others to judge whether the slip was more divine than benign.

Few people would challenge the bishops' right to intervene in the current debate insofar as it involves general moral principles which, in their capacity as the official teachers of the Church, they have the duty to expound and bring to bear on contemporary issues. The crucial question concerns the level at which their intervention could most appropriately and most fruitfully take place. How they themselves propose to deal with that thorny question is still anybody's guess. Strangely enough, the chairman of their ad hoc committee has gone out of his way to indicate that the first draft of the projected statement would not be released until after the November election, lest it should appear to have a direct bearing on it. While one can be grateful for the promise and trust that it will be kept, one is almost forced to interpret it as a tacit admission that the bishops are not averse to taking sides on matters of personal political judgment or preference. As for the claim that in so doing they are merely exercising a right guaranteed to everyone under the Constitution, it is not likely to enhance their stature in the eyes of their readers, for it conveys the vague but lingering impression that they intend to speak as private citizens rather than as bishops. There are better uses to which the prestige of their office can be put.

This said, one can sympathize with their plight, which, when it comes to problems of this sort, is not and has never been an enviable one. The difficulty lies in part with the enormous and ever growing complexity of modern society, but its real roots are to be found in the nature of Christianity itself, which is not first and foremost a political religion. Anyone who reads the New Testament Scriptures carefully from this point of view cannot help being struck by their all but total indifference to questions of a properly political nature. Unlike the Hebrew Scriptures, they do not call for or encourage the formation of a particular political community or lay down a set of laws by which such a community might be governed. Nowhere in them do we find any proposals concerning the structures of civil society, public legislation, the administration of justice, or the production, management and distribution of material goods. Their ruling principle is not justice as ordinarily understood, that is to say, general or legal justice, but love: "This is my commandment, that you love one another as I have loved you" (John 15:22). Granted, love is a powerful human motive, and there is no reason to think that it cannot inform all of our actions, political or otherwise; but it does not specify the content of those actions save in the most general way and, hence, fits into the category of what used to be called "common" as distinguished from "proper" principles.[3] It is significant that the situations envisaged in the gospel are typically one-on-one situations from which there are few definite conclusions to be drawn regarding the behavior that is in order when the security and welfare of the larger community are at stake. "Love your enemy," "Be merciful," "Turn the other cheek," and the like may be valid maxims for the person who prefers forgiveness to punishment, prizes mercy more than justice, and has only himself to think about. They are, however, less readily applicable to multilateral situations involving third parties for whom one is responsible and whom one also has the duty to love.

The other side of the story, and it is no less important than the first, is that, in marked contrast to the Gnostic sects of late antiquity, the New Testament does not preach withdrawal from society or demand that its fol- lowers turn their backs on it. It simply assumes that Christians will continue to organize their temporal existence in accordance with the re- quirements of the society to which they happen to belong. Yet it does not dwell on the practical implications of their involvement in the social life or make any effort to explain how, concretely, the lofty moral ideal of the Sermon on the Mount can be reconciled with the duties of citizenship in a society that is always less than perfectly just. Its teaching in that respect, if it can be said to have any, is at best ambivalent. Christians are told to obey their rulers (Rom. 13:1), and at the same time they are reminded that they are to obey God rather than men (Acts 5:29). That the New Testa- ment writers should not have been overly preoccupied with this dilemma comes as no surprise since all or most of them were convinced that the end of the world was near. In the interim, Christians, who were still only a handful anyway, had more urgent things to worry about than the reform of the Roman Empire or the rooting out of its "systemic injustices." To quote St. Paul once again, "The appointed time has grown very short. From now on, let those who have wives live as though they had none..... and those who buy as though they had no goods, and those who deal with the world as though they had no dealings with it" (I Cor. 7:29-31). So much in a nutshell for the political and social program of the New Testa- ment: it simply doesn't provide specific prescriptions for public policy in the contemporary period or in any other period.

Later generations of Christians, who had to face the fact that the world was not about to end and who belonged to a religious community that had since grown to sizable proportions, were therefore compelled to look else- where for the practical guidance that the New Testament neglected to pro- vide. They tried the Old Testament for a while and, when that failed, they turned to classical philosophy. The solution that they came up with was later systematized by the great theologians of the Middle Ages and forms the basis of what is now referred to as the "social teaching" of the Church, a teaching which in its present shape dates back to the last part of the nineteenth century.

Adapting that teaching to the contingencies of our time has proved to be an uncommonly difficult task, however, for the simple reason that modern liberal society is founded on principles that are neither specifically religious nor particularly moral. To be sure, the leading theorists and advocates of modern liberalism were not necessarily hostile to religion and morality; but, having concluded that the unity of society had little chance of being restored on the basis of either one, they looked for a scheme that could function successfully even without them. The solution, as we saw in previous chapters, was to restructure the whole of society in such a way as to render its well-being less dependent on the moral character of its members. Once the right social structures were in place, people would best be able to serve their fellow human beings by pursuing their own selfish interests. Any detriment resulting from the depreciation of virtue would be amply compensated for by the untold blessings conferred upon us by what Hamilton called the spirit of unbridled commercial enterprise.[4] The greatest benefactors of humanity, the true heroes of the coming age, were not the dedicated parish priest or the saintly Christian toiling selflessly for the good of others, but the new captains of commerce and industry who, by enriching themselves, would enrich everyone else as well. Subsequent efforts to temper the heartlessness of the new society by introducing an element of compassion into it—more, as mentioned, under the influence of Hume and Rousseau, the true originators of the "politics of compassion," than under that of the gospel—have only partially succeeded in removing the inequities to which in time it gave rise. Not only for the sake of justice but in the interest of a more noble or dignified democracy, something more needs to be done.

These, I take it, are the concerns that motivate the bishops and they are concerns that we can all appreciate. There are nevertheless limits to how far one can go in determining *the* proper Christian response to any given social or economic problem. I, for one, do not know for example whether in the long run the nation would be better off if the combined employer-employee Social Security tax bite were raised beyond the 14.1 percent level scheduled to be reached on January 1, 1996 in order to meet the needs of the elderly. Nor do I know whether the top 40 percent of our taxpayers should be made to contribute more than the present seven out of every eight dollars collected by the IRS. And I know even less how much of the nation's disposable capital ought to be reinvested rather than spent for the purpose of generating the funds needed to maintain and improve our welfare programs. These are the types of issues that divide political parties and they are issues over which well informed and well intentioned elected officials and citizens are bound to disagree.

This is not to say that some solutions are not better than others, but only that, when the bulk of the Christian tradition is not clearly on one side, one should think twice before affixing the seal of divine approval on any of them. As I intimated earlier, the New Testament has no definite social or economic agenda of its own. Even though it warns against injustice and evinces a special concern for the poor, its overwhelming emphasis is on spiritual rather than material poverty. Never to my knowledge does it venture to propose that the lower classes be elevated at the expense of the wealthier ones. Moreover, even when it does speak about the sociologically poor, it is less from their point of view than from the point of view of those who are called upon to help them. The New Testament, it seems, is more interested in the internal dispositions of the doer (or non-doer) of the just or merciful deed than in the social condition of its recipient. In the words of an ancient Christian writer, who was merely echoing what others before him had said, "What kind of people are we if, having received everything from God, we refuse to give anything to others?"[5] Succinctly stated, one does not bear witness to the love of God by closing one's heart to the victims of misfortune or the worthy poor. The remark merits pondering, inasmuch as it calls attention to the proper theological ground of the Christian position on the sharing of earthly goods. This ground, needless to say, is very different from that of the pagan philosophers, for whom the most noble deeds were prompted by nothing more than a desire to please one's better self.

By way of a footnote, I might add that, perhaps because of the inherent apoliticism inherited from its origins, the Christian tradition has frequently been inclined to take a dimmer view of the accumulation of wealth than other religious traditions. The long fight against the sole use of money to produce more money—"usury," as it was termed—bears ample testimony to that fact, although the rationale for the opposition seems to have come more from Aristotle than from the Bible. The famous and seemingly endless late medieval debate on the subject of wealth and poverty is another case in point. It is interesting to note, however, that in this instance the attacks on wealth were mostly directed against a tithe-exacting Church that cost the villagers a lot of money but did not take care of their souls in return.

Once again, one cannot fault the bishops for wishing to reformulate the social doctrine of the Church in terms that reflect the economic realities of the hour, as long as they do not confuse that doctrine with the private opinions of the carefully selected *periti* on whose advice they apparently plan to rely. In this connection, the distinction between principle and policy, convenient as it may be as a rule of thumb, has a tendency to break down beyond a certain point and ceases to be operative in areas where one shades into the other. What one has finally to decide is whether the principle and the policy in question are so closely intertwined as to be inseparable from each other, or whether their relationship is such as to leave room for a variety of opinions among which one remains free to choose. If the Church were never allowed to pronounce itself on matters of immediate practical concern, it would be severely restricted in its ability to denounce some of the most flagrant violations of justice and right. Some refinement of that otherwise useful distinction would thus appear to be in order. The problem is not exactly new. It is already implicit in Thomas Aquinas, who shrewdly refrained from drawing too sharp a line of demarcation between the natural law and its derivative, the human law—a wise judgment for which he has often been unfairly criticized.

This still leaves us with the question of the criterion on the basis of which one distinguishes between a general principle and a more or less contingent policy. Here I would simply caution against an excessive reliance on the highly ambiguous notion of social justice. As one well known journalist recently mentioned in conversation, "We cannot talk about social justice any more, it just does not ring a bell." My hunch is that the time has come to lay the ghost of this ill-fated avatar permanently to rest. Recent popes, beginning with Pius XI in *Quadragesimo anno*, have of course used it, but only sparingly and without giving it anything like the prominence that it has lately acquired in the minds of our zealous and increasingly vocal social reformers.[6]

The fact remains that general moral principles, if they are to become effective, must be translated into practical programs for the benefit of those who are required to act on them. Such programs are formerly thought to be the particular province of experienced, well educated, and dedicated lay persons, who, by reason of their status, are in a position to take a reasonable stand on controversial issues without directly engaging the authority of the Church. This, unless I am mistaken, is precisely the task that the Lay Commission on Catholic Social Teaching and the U.S. Economy undertook to perform, to the obvious if somewhat muted displeasure of some of our bishops.[7] As I see it, there are two ways in which it can proceed. It may wish to counter the anticipated "liberalism" of the bishops by making a strong case for a more moderate position—a case based among other things on a comparison between the undeniable merits of liberal democracy and the now well-publicized failures of modern socialism; or else it can do what the bishops perhaps ought to be doing but as yet have given no evidence of wanting to do, namely, stress, the need for virtue and the importance of moral character as the true foundations not only of liberal democracy but of any legitimate regime. Liberal democracy may be strictly utilitarian in its deepest roots, but it does not follow that the people who live under it fully endorse its materialistic orientation. The more plausible supposition is that, like most human beings, they prefer to harmonize heaven and earth and are usually happier when, to borrow a phrase from Tocqueville, they can combine material well-being with moral delights. By building on the element of idealism that lies just beneath the surface of American life, one might eventually be able to do more for the poor as well as the rich of our country than by issuing a series of impassioned pleas and counter-pleas for either the maintenance or the transformation of the status quo.

NOTES

1. The bishops published the final version of their pastoral letter in November, 1986 under the title, *Economic Justice for All: Catholic Social Teaching and the U.S. Economy* (Washington, D.C.: United States Catholic Conference, 1986).

2. *National Catholic Reporter*, 20 July 1984, p. 16.

3. Cf. Thomas Aquinas, *Summa theologiae* I-II, qu. 94, a. 4.

4. Cf. *Federalist Papers*, no. 7.

5. Caesarius of Arles, Sermon 25 in *Sermons*, vol. 31 (Washington, D.C.: The Catholic University of America Press, 1956), 127-28.

6. J. Brian Benestad, "The Catholic Concept of Social Justice: A Historical Perspective," *Communio* 11, no. 4 (1984): 364-81.

7. Cf. Lay Commission on Catholic Social Teaching and the U. S. Economy, *Toward the Future: Catholic Social Thought and the U. S. Economy* (New York: Lay Commission on Catholic Social Teaching and the U. S. Economy, 1984).

25

THE TROUBLE with CATHOLIC
SOCIAL THOUGHT

Toward the close of Vatican II, a distinguished Protestant theologian, himself an observer at the Council, remarked that, in its endeavor to come to terms with the modern world, the Roman Catholic Church was trying to complete in less than five years what Protestantism had failed to accomplish in well over a century. The statement was intended as a friendly warning that the domain into which the council fathers had ventured was strewn with pitfalls that the euphoria generated by their deliberations could easily cause them to ignore.

The truth of that statement has been amply borne out by more recent developments, not the least notable of which is the Church's unprecedented involvement in problems of a directly political nature.

No one denies that many of the bishops' statements,[1] and above all the widely publicized pastorals on nuclear warfare and the American economy, represent a substantial contribution to the national debate, which the bishops are eager to elevate by injecting a strong moral note. What is less obvious or less often said is that the same documents bring into sharper focus a number of tensions inherent in the Church's current position on social matters. My aim is not to defend or criticize that position but to indicate wherein, as I see it, the tensions lie and in what sense the chord sounded in the pastorals and related texts differs from the one to which Catholics were formerly accustomed. With all due respect to the authority of the bishops, I shall argue that their latest teaching is implicitly grounded in a bifurcated anthropology; that it simultaneously pursues two sets of goals that are not easily reconcilable with each other; that it combines traditional Christian doctrine with ideas that once were and may still be fundamentally antithetical to it; in short, that it suffers from a latent bifocalism that puts it at odds with itself and thereby weakens it to a considerable extent.

Anyone who reads the new documents against the background of the older ones cannot help being struck by the overwhelming emphasis that they place on the notion of human rights, the hallmark of modern ethical theory. What once presented itself as first and foremost a doctrine of duties and hence of virtue or dedication to the common good of one's society now takes its bearings, not from what human beings owe to their fellow human beings, but from what they can claim for themselves. This observation is only apparently belied by the fact that chapter 2, section B, of the letter on the economy opens with an "outline of the *duties* all people have to each other and to the whole community"[2]; the rest of the text makes it clear that these duties are rooted in pre-existing rights which everyone is obliged in conscience to honor and which must, therefore, be regarded as the primary moral phenomenon.

At first glance, the difference between the two views might be looked upon as one of approach rather than of genuine substance, and the more so as rights and duties are to some extent correlative. If I have a duty to do something, I must also have the right to do it, although the converse need not be true. Moreover, there is no reason to think that the use of different approaches to the same problem leads of necessity to substantially different results. Climb to the top of Mt. Everest from the north or from the south, and you arrive at the same destination, even if one side turns out to be easier or otherwise more suitable than the other. Differences of this kind are not uncommon in the Christian tradition and are in part responsible for the vitality that it has demonstrated across the centuries. The early Church Fathers were attracted to Plato and found him better adapted to their purpose, which was to implant Christianity in a world that was as yet unfamiliar with it. For reasons of their own, the medievals preferred Aristotle, whose philosophy is eminently suited to a situation in which Christianity was already fully accredited. No one would describe the two groups as being at loggerheads with each other.

Such is not always the case, however. If one's method of procedure has been deliberately chosen with a view to a different end, rather than as a different means to the same end, its employment is bound to have an impact on what is accomplished. It is significant that the notion of universal human rights, understood as rights that inhere in each human being by reason of the fact that he or she is a human being, does not occur anywhere in premodern thought and, until very recent times, only sparingly in Roman Catholic thought. The Bible itself, which shares to some degree the perspective of classical philosophy on this point, does not, as we saw in chapter 2, promulgate a Bill of Rights, of which it knows nothing; instead, it issues a set of commandments. For centuries, the cornerstone of Catholic moral theology was not the natural or human *rights* doctrine but something quite different, called the natural *law*. Rights, to the extent that they were mentioned at least by implication, were contingent on the fulfillment of prior duties. Far from being absolute or inalienable, they could be forfeited and were so forfeited by the individual who failed to abide by the law that guaranteed them. Simply stated, what the church taught and tried to inculcate was an ethic of virtue as distinct from an ethic of rights.

To the best of my knowledge, the true originator of the rights doctrine is Hobbes, from whom it was taken over by virtually all of the great early modern thinkers, Spinoza, Locke, and Rousseau foremost among them. That doctrine emerged by way of a reaction against premodern thought and signals a radical departure from it (cf. chapter 12, "Thoughts on Modernity"). Its underlying premise is that, contrary to what had been previously assumed, human beings are not intrinsically ordered to a natural end, in the attainment of which they find their happiness or perfection. In Hobbes's own words, "there is no such *finis ultimus,* utmost aim, nor *summum bonum,* greatest good, as is spoken of in the books of the old moral philosophers."[3] Human beings are universally actuated, not by a desire for the good of reason, but by an amoral passion, and not the most noble one at that, to wit, the fear of violent death, which constitutes the sole foundation on which a viable theory of justice can be erected.[4] For, as Hobbes observes elsewhere, "as often as reason is against a man, so often will a man be against reason."[5] To be sure, reason is still involved in the process, but only in an instrumental capacity, that is to say, as a faculty whose function is to "reckon consequences"[6] and, by so doing, insure one's self-preservation.

This is not to suggest for one moment that the bishops are pure and simple Hobbesians; far from it. Like most contemporary ethicists, they are vastly more influenced by Kant and his latter-day disciples, who managed to give the original rights theory a more exalted status by grounding it, not in a selfish passion, but in practical reason or the dignity of the individual person as an autonomous moral agent. Yet in the final analysis, Kant's theory is still only a modification of the Hobbesian theory, whose nonteleological orientation it preserves and with which it has more in common than it does with classical moral philosophy. Rights remain paramount and the actualization of the just social order is made to depend more on institutions than on moral character. Virtue, although desirable, is not essential to the scheme, and one need not acquire it in order to reap its benefits. Human beings may stay as they are as long as the laws under which they live are as they ought to be. The only condition is that these laws be devised in such a way as to be equally favorable to everyone, to scoundrels no less than to saints. Wars will disappear, not because their atrocities offend one's moral sense, but because with the progress of modern science they are rapidly becoming too costly, if not downright suicidal.[7] Commerce will curb the spread of religious fanaticism and bind people together more closely than ever before.[8] A strong league of nations will insure the proper relations among the peoples of the world,[9] and a day is coming when no one will have to worry about sacrificing himself for the good of others. In Kant's famous phrase, the perfectly just society does not require that the bulk of its citizens be angels or even minimally decent human beings; it can be made up entirely of "devils," on condition that they be intelligent.[10] The key to its proper functioning is not genuine moral virtue, on which one can rarely depend, but enlightened self-interest.

I am well aware that the preceding account of the relationship between premodern and modern ethical thought runs counter to what is still far and away the most fashionable view among present-day scholars, the majority of whom look upon the rights doctrine as nothing more than the perfected version of the old natural law doctrine. According to that interpretation, there is no real breach of continuity between premodern and modern thought. The latter is a child of the former, whose intention it does not oppose but rather seeks to fulfill more effectively. Such was the position taken a short generation ago by Jacques Maritain and John Courtney Murray, and such also is the one taken in our day by John Finnis and Felicien Rousseau, to name only a few Catholic thinkers who speak for a host of others in this regard.[11] It is emphatically not the position taken by the early modern thinkers themselves, all of whom were firmly convinced that, like Columbus, they had discovered a new continent and stood on fundamentally different ground.[12] Modernity, they thought, had its own independent principle and was not to be viewed as a mere prolongation of what had gone on before. At the deepest level, it owed nothing of importance to either the classical or Christian tradition, from which it was separated by an abyss. The issue was stated with the utmost clarity, for the last time or just about, by Jonathan Swift, who pictured the two camps as solidly entrenched, each on its own summit, which neither was willing to surrender to the other.[13] It is only with Hegel and, some years later, with Nietzsche, that a powerful attempt was made to bridge the chasm between them by showing that the modern world was nothing other than the culmination of a process that had been unfolding since the dawn of history or, in Nietzsche's case, since the dawn of philosophic consciousness. And it is from them that the unitary conception of the development of Western thought was passed on to assorted philosophers and historians down to our time.[14]

The point to note, however, is that recent years have witnessed a spectacular return on the part of some prominent thinkers to the pre-Hegelian or pre-Nietzschean view. How this shift came about is a question that need not concern us here, except to say that the incentive was provided by Nietzsche's and especially Heidegger's all-out critique of the philosophers of classical antiquity, which swept away the rubble of a centuries-old tradition of Scholastic interpretation and paved the way for a fresh insight into the nature of their thoughts. Accordingly, scholars in growing numbers are once again inclined to view the history of Western thought as one that is marked by a sharp break that was consummated in the course of the sixteenth and seventeenth centuries and in the light of which virtually all of our basic problems must be reassessed. The phenomenon, I might add, is not limited to the critics of modernity. It is equally visible, as we have seen, in Hans Blumenberg's much discussed book, *The Legitimacy of the Modern Age*, which comes out resolutely on the side of the Moderns.[15]

We shall gain a clearer understanding of what is at stake in all of this if, by way of illustration, we glance ever so briefly at a few of the specific issues with which the bishops have been confronted or to which they have been forced to address themselves. One of these is abortion, the case against which, not surprisingly, they now base on the rights of the unborn. Earlier theologians would have begun by asking, not what abortion does to the foetus, but what it does to the person who performs it or undergoes it. Duties took precedence over rights and determined the conduct that was appropriate in a given set of circumstances. The more complex the case, the more it called for deliberation in the light of certain general principles which remain the same at all times but whose degree of applicability varies greatly from one instance to another. Considerable room was thus left for prudence, through the exercise of which one could hope to arrive at a sensible decision. The same cannot always be said of the argument from rights, which deprives us of the ability to determine in a principled way which of the two supposedly unconditional rights, that of the mother or that of the unborn child, has the green light when they come into conflict with each other.

Nor is this all. On the ground that all human life is sacred, some bishops have been invoking the same principle to ban not only abortion but nuclear warfare and capital punishment as well. Actions as disparate as the taking of an innocent life, the execution of a common criminal, and the defense of one's country against a foreign invader have been lumped together and condemned in the name of a would-be "consistent ethic of life" or an abstraction derived from modern science, which in large part is what I take the rights doctrine to be. Perhaps there are no ready-made solutions to any of these problems, but even a morally sensitive person may find it strange to see them treated as if they were identical.

A similar problem arises in connection with the Church's efforts to deal with theologians who dissent from its official teaching. Having endorsed the principle of freedom of conscience and freedom of expression, the bishops have been hard pressed to defend the sanctions imposed by Rome on some of the dissenters. One cannot publicly proclaim the rights of the individual conscience and then take action against the person who exercises them and thus appears to have justice on his side. It might be replied that the Church, too, has its rights, which need to be defended when they are infringed. But, never having been spelled out with the same degree of precision, these rights remain somewhat vague or less compelling than the ones claimed by and guaranteed to the individual person. The problem becomes even more acute when one considers that the dissenter is himself a member of the organization whose rights are being urged against him. All of this is to say that the Church is having to pay a price for espousing the principles of the Enlightenment along with their hidden premise, the ideology of progress. Only a stronger notion of the common good, coupled with a more subtle formulation of the notion of individual rights, could justify its stance and posture in regard to this thorny problem.

My last series of comments concerns the bishops' teaching on the economy and their so-called "preferential option for the poor," one of the pivots of that teaching,[16] where the argument is again made predominantly in terms of rights. One can hardly fault the bishops for wanting to do something about the scandalous persistence of pauperism in the midst of a society that is daily becoming more prosperous. On this score, they are merely heeding the injunctions of the Bible, which, as they correctly point out, evinces a good deal of concern for the poor, the afflicted, and the underprivileged. Unlike the bishops, however, the Bible never goes so far as to propose that something might be owed to them simply by reason of their poverty. Here as elsewhere, it looks at the problem from the standpoint of the doer of the righteous or merciful deed rather than that of its beneficiary. The thrust of the New Testament message is that the Christian, who has received everything from God, must imitate him by sharing his superfluous goods with those who lack the necessities of life. The poverty about which it speaks is more often than not spiritual poverty, which sometimes goes hand in hand with material poverty but which cannot be equated with it. God wants all people to be saved, the rich as well as the poor.[17]

It is true, as the bishops likewise remind us, that Jesus himself lived in a ten-thousandfold poverty and did not have a stone on which to rest his head; but it is also true that he had some fairly rich friends, whom he does not condemn and of whose hospitality he was quite willing to avail himself. And he apparently had no qualms about allowing his equally improvident disciples to steal a few ears of corn from neighboring fields from time to time. Imagine what human society would be like if everybody were to live like that! On balance, one is almost tempted to say that the Bible is more concerned with the rich than with the poor. From its perspective, they are the ones who need help. Besides, if the poor are really closer to God, I suppose one should think twice before robbing them of their poverty.

Even more striking is the fact that the New Testament has nothing at all to say about the reform of social structures or the establishment of institutions geared to the relief of human misery. (This part has been a theme recurring in several chapters.) What the bishops and others today call "social justice," as distinguished from, say, legal justice, as we have seen, is a nineteenth-century invention usually attributed to the Roman Catholic theologian Taparelli d'Azeglio.[18] The pedigree of that ill fated notion can easily be traced to Rousseau, who did call for the reform of society along egalitarian lines. The change was a momentous one. Prior to that time, everybody took it for granted that one was rewarded by society in the measure of one's services to it. Widows, orphans, and the victims of undeserved misfortune were obviously to be cared for, but they presented a special case demanding special treatment. Rousseau went further, stipulating that rewards were to be meted out on the basis of need rather than merit. The less one was in a position to contribute to society, the more one could expect from it. Compassion rather than reason became the stamp of human dignity and would henceforth dictate one's comportment toward the needy.[19] The battle cry was taken up by countless authors throughout the century that followed. In the *Legend of the Centuries*, Victor Hugo could say of a horse that steps aside to avoid trampling a toad: "Wiser than Socrates and greater than Plato"—*Plus sage que Socrate et plus grand que Platon.* But compassion, for all its attractiveness, is still, as its name indicates, a passion. It is only distantly related to what the Christian tradition calls "mercy" and "charity," both of which, as virtues, require the control of reason. John Stuart Mill was not far off the mark when he noted a century and a half ago that the new mood among Christians owed more to Rousseau's sentimental deism than it did to the spirit of the gospel, for which neither he nor Rousseau had much use.[20]

But if social, rather than personal, reform is the key to the problems of civil society, training in virtue loses much of its importance. There is no real need to dwell on it, and, in fact, the bishops show relatively little interest in it. Like their modern predecessors, they appear to have more faith in institutions, or so one gathers, for example, from the list of concrete proposals to which more than half of their pastoral on the economy is dedicated, and which includes such specific items as tax reform, the reform of the international economic system, the right to employment, trade unions, farm policy, food distribution, cooperation with firms and industries, national and international cooperation, welfare programs, job training and job creation programs, and foreign investments.

Part of the danger is that the vast amount of time, energy, and resources devoted to practical issues of this sort risks being interpreted as a tacit acquiescence in what Robert Nisbet has described as the demise of transcendental Christianity.[21] One begins to wonder whether the bishops have not abandoned their most important task for the sake of a possibly more urgent one. That it should sometimes be necessary to do this goes without saying. If my house is burning, I had better put the fire out, even if that means postponing any attempt to figure out what the bishops are up to. Unfortunately, there are always urgent matters to be taken care of; but if I am constantly preoccupied with them, I shall never be able to attend to the others. What is more, paying the proper attention to important matters at the right time is often the best way to forestall emergencies.

It is curious, to say the least, that bishops, whose first duty is to preach the gospel, should suddenly have become so obsessed with problems that Christianity has always identified as the proper province of the civil authorities. As a professor of Catholic theology, I find it regrettable that I cannot recommend their texts to my students without inviting a minor disaster. In the vast majority of cases, these students come away from them with the feeling that they can do better by reading the works of experts on the subject. Whatever gems of Christian wisdom may be imbedded in them are lost in a maze of academic detail and references that only a specialist or a born masochist would care to look up. By contrast, the same students respond with enthusiasm to books such as Allan Bloom's *The Closing of the American Mind,* which give them an inkling of what the world would be like if we were all better human beings or of what they could see in it if they themselves were to become better human beings. It surprises them that the bishops, to whom they still look up, should spend so much time talking about self-preservation (as they do in the letter on nuclear warfare) and about comfortable self-preservation (as they do in the letter on the economy), and so little time talking about the love of the good and the beautiful. Few object strenuously to any of the specific points made in those letters, which they are quite willing to accept as one opinion among others, but they instinctively resist what they perceive as an attempt to reduce Christianity to a teaching on rights or a form of deontological moralism.

I began by suggesting that the bishops' teaching was predicated on a bifurcated anthropology, one part of which reflects a teleological and the other a nonteleological view of nature. It might be more accurate to say that it suffers from the lack of any definite anthropology. A pastoral letter is admittedly not the best place to engage in remote philosophic disquisitions about the nature of the human being, but one would nonetheless expect its authors to exhibit a greater awareness of the theoretical implications of the doctrines expounded in it. It is no accident that the bishops have failed to win the wholehearted support of at least two groups of sympathetic critics within their own fold: the Neo-Conservatives, who agree with them about the end or ends pursued in their letters, but disagree about the means to those ends,[22] and a smaller or less noisy band of scholars who perversely insist that there can be no fruitful discussion of the means unless the ends themselves are first clarified.

My point, and it is the only one I have tried to make, is that the bishops may have confused some of their readers by using language that looks in two different directions at once: that of rights or freedom on the one hand, and of virtue, character formation, and the common good on the other. They would certainly be ill advised to give up their vigorous defense of rights, especially since the pseudomorphic collapse of Neo-Thomism in the wake of Vatican II has left them without any alternative on which to fall back; but they have yet to tell us, or tell us more clearly, how the two ends are supposed to meet. Reading their letters reminded me of a little ditty my gang used to sing when I was a kid:

> *I love Carolina,*
> *I love Angelina too.*
> *I can't marry both,*
> *So what I gonna do?*

NOTES

1. For an analysis of these documents up to the year 1980, see J. Brian Benestad, *The Pursuit of a Just Social Order: Policy Statements of the U. S. Catholic Bishops, 1966-1980* (Washington, D.C.: Ethics and Public Policy Center, 1982).

2. *Economic Justice for All: Pastoral Letter on Catholic Social Teaching and the U. S. Economy*, no. 62. The text of the pastoral may be found in *Origins* 16, no. 24 (1986).

3. Hobbes, *Leviathan*, ch. 11, init.

4. Cf. *De cive* I.1.7: "Therefore the first foundation of natural right is this, that every man as much as in him lies endeavor to protect his life and members." Also *Leviathan*, ch. 13-14.

5. *De hominis natura*, epist. dedic.

6. Cf. *Leviathan*, ch. 4.

7. Kant, *Idea for a Universal History*, 7th thesis.

8. Ibid., 8th thesis.

9. *Perpetual Peace*, Section II, Second Definitive Article.

10. Ibid., First Supplement.

11. See esp. J. Finnis, *Natural Law and Natural Rights* (Oxford: Clarendon Press, 1980), and F. Rousseau, *La croissance solidaire des droits de l'homme* (Paris and Montreal: Desclée et Cie, 1982). Rousseau's book is of special interest in that it offers an unusually penetrating account of the Thomistic natural law theory and promptly goes on to equate that theory with the modern rights theory. The argument is all the more intriguing as Rousseau is one of the few authors who still looks to Aquinas for the guidance that we need in these matters. I notice only that, as might have been expected, none of the quotes from Aquinas, with which the lengthy chapters devoted to him abound makes any mention of "rights."

12. See, for example, Machiavelli, *Discourses*, first book, Introduction, and F. Bacon, *New Organon*, book 1, Aph. 92.

13. Jonathan Swift, *The Battle of the Books, ca. init.* As is well known, the opposition between Ancients and Moderns is the theme of *Gulliver's Travels*, where it is explored at much greater length.

14. A. J. and R. W. Carlyle's monumental *History of Mediaeval Political Theory in the West*, 2d edition (Edinburgh and London, 1927), did much to popularize the Hegelian thesis among historians of twentieth-century political thought. As A. J. Carlyle writes (vol. I, p. 2): "Just as it is now recognized that modern civilization has grown out of the ancient, even so we think it will be found that modern political theory has arisen by a slow process of development out of the political theory of the ancient world—that, at least from the lawyers of the second century to the theorists of the French Revolution, the history of political thought is continuous, changing in form, modified in content, but still the same in its fundamental conceptions."

15. H. Blumenberg, *The Legitimacy of the Modern Age*, translated by R. M. Wallace (Cambridge, MA, 1983). Cf. chapter 11 of this volume for an extended discussion of this book.

16. *Economic Justice for All*, nn. 16, 87, et passim. For a balanced assessment of the bishops' teaching on this and related issues, see Avery Dulles, S.J., "The Gospel, the Church and Politics," *Origins* 16, no. 36 (1987): 637-646.

17. The two New Testament texts most frequently invoked by theologians in this connection are Matt. 25:31-46, where, speaking of those who will have fed the hungry and clothed the naked, Jesus declares: "Truly, I say to you, as you did it to one of the least of my brothers, you did it to me," and Luke 4:16-19, where, quoting from Isaiah

58:6 and 61:1-2, Jesus announces that he has come to "preach the good news to the poor," "proclaim release to captives," and "set at liberty those who are oppressed." It is by no means certain, however, that a universal teaching relative to poverty can be drawn from either text. The "brothers" spoken of in Matthew are not the poor in general but the members of the Christian community and, more specifically, the ministers of the gospel. Such at any rate is the interpretation suggested by Augustine, who remarks that in the New Testament " 'brothers' without any qualification always refers to Christians" (*Enarr. in Psalm.* [*Exposition on the Psalms*], 32.29). The same interpretation is supported by some modern biblical scholars. As for Luke 4, its message seems to be revealed by the deliberate omission of any reference to what follows in Isaiah's text, to the effect that divine vengeance is about to be wrought on those who persist in their injustice. In simple words, the God of the New Testament wants the conversion of the sinner and not his punishment. The biblical notion of spiritual poverty (or "humility"), as distinct from material poverty, goes back to Zephaniah 3:12-13, which is generally thought to be the oldest text in which the term designating the poor (*anawim*) is used in a specifically religious sense. For a concise statement of the Church's classic under-standing of this problem, see Leo the Great, *Sermon* 95.2, in *A Select Library of Nicene and Post-Nicene Fathers*, Second Series, edited by P. Schaff, vol. 12, 203.

18. L. Taparelli d'Azeglio, *Saggio teoretico di dritto naturale appogiato sul fatto* (Palermo, 1840, and Rome: La Civiltà Cattolica, 1855), vol. I, book ii, ch. 3: "Nozioni del diritto e della giustizia naturale," Rome edition, 220-32.

19. See on this subject C. Orwin, "Compassion," *The American Scholar* 48 (1980): 309-33.

20. Cf. *Three Essays on Religion*, in *The Philosophy of John Stuart Mill*, edited by Marshall Cohen (New York: Modern Library, 1961), 450.

21. Cf. *The New York Times Review of Books*, 28 April 1985, apropos of James Turner's book, *Without God, Without Creed: The Origins of Unbelief in America* (Baltimore: John Hopkins University Press, 1985).

22. For a clear statement of the Catholic 'Neo-Conservative' position, see *Toward the Future: Catholic Social Thought and the U. S. Economy* (New York: Lay Commission on Catholic Social Teaching and the U. S. Economy, 1984).

26

FRIEND and TEACHER: ALLAN BLOOM'S OBSESSION with the MYSTERY of the SOUL

With the untimely death of Allan Bloom on October 7, 1992, America lost one of its finest, most brilliant, and most successful teachers. At the same time and by more than mere coincidence, American higher education lost its most courageous and incisive critic. Others have already said this and much else about Allan, both here and on a previous occasion. I, for one, can only add a personal footnote to their magnificent tributes.

My own friendship with Allan goes back to the mid-fifties when we were in Paris together, he as an exchange student from the University of Chicago and I as a doctoral candidate at the Sorbonne. We met for the first time at a seminar on Plato's *Laws* that the distinguished French Dominican and classicist A.-J. Festugière happened to be teaching at the Ecole Pratique des Hautes Etudes. As the only Americans in the group, we inevitably gravitated toward each other and were soon having tea after class in nearby sidewalk cafés, going out for an occasional meal (never a very lavish one in those days), or attending some play or other in one of the city's many theaters.

A typical adventure was our going to see Pirandello's *Six Characters in Search of an Author* in a small playhouse not far from the celebrated

Montmartre district. The fifties were the heyday of French existentialism, and Pirandello, whose plays mirrored the cultural confusion of the age, was in vogue. Since the Pirandello play was relatively short, a short one-act play had been tacked on to round out the program. We did not know what to expect and, quite frankly, never quite figured out what we got. When the curtain rose for the second play, neither of us even suspected that the first one had ended. I have never seen Bloom with a more sheepish look on his face. It was as if someone had played a trick on him. He decided on the spot that, whatever it may have meant, and its meaning was still being hotly debated, existentialism was not for him.

This accounts among other things for his endless questions to me about contemporary Roman Catholics, some of whom he had befriended in Paris. One such person was Père Maydieu, the editor of a well-known review called *La Vie Intellectuelle*, who each week on Tuesday afternoon received anyone who cared to drop in on him. People from all over the world were likely to be met there on any given day. What interested Bloom was not the positions they took on this or that issue but what moved them to take such a position. What were these people really like beneath the appearances? Here were members of an organization that once had connections with the great centers of power and culture, an organization that, unlike most groups in the modern world, stood for some high principle, however weakened it had become, intellectually and otherwise. What was left of it may have been only fragments, but they were fragments of a once coherent and immensely impressive whole.

Funny things almost always occurred during our student days in Paris, about which we were still laughing years later. One evening, in a small family restaurant on the Boulevard Saint-Germain, not far from where Allan lived, we spotted frogs legs on the menu and promptly ordered them as an appetizer. Sitting next to us were two older ladies who kept eyeing us suspiciously, for what reason I could not fathom. Perhaps, I thought, because they resented the presence of a couple of loudmouths intruding on their privacy and only too eager to offend their sensibilities by speaking a language they did not understand. But no, it was not that at all. They were just curious to know what we were eating. Imagine two barbarians savoring a local delicacy in the hub of the civilized world and being asked by the natives themselves what we were munching on. Hilarious!

This was Bloom's first triumph in Paris, but it was not to be his last. The incident was still fresh in his memory when, following the publication of the French edition of *The Closing of the American Mind*, photos of him, often against the background of the Eiffel Tower or the dome of the Invalides, appeared in virtually all of France's major journals and newspapers. For an obscure midwesterner far removed by his origins from the glamour of one of the world's great capitals, the new triumph was an unbelievable achievement.

It was on informal occasions such as these that, instinctively and without my being aware of it, Bloom took over my education. Reluctant as I was to admit it, I found myself learning more about Plato from him than from anyone else around, including our common teacher, Festugière, of whom we were both fond because he liked Americans and did us a lot of favors. To say that Allan defended his ideas with vigor would be an egregious understatement. His method was to hit you over the head with them, a practice that he raised to the level of a fine art. Clarity about fundamental issues was paramount. One had to be made to see things. Philosophers hate nothing more than the penumbra of the proverbial cave or the night in which all cows are black. Resolving conflicts by the favorite American technique of splitting the difference between the conflicting positions rather than on the basis of principle was loathsome to him. Being a sensible man, he sometimes retreated from his most extreme stances, but only after having assured himself that this was not just a compromise born of cowardice.

One of the things that surprised him was that I knew next to nothing about politics and showed no interest in it. In this I was not alone. Few Catholics of my generation fared much better in that regard. Our teachers dutifully repeated what Aristotle says to the effect that politics is the queen of the practical sciences and the one in which all the others culminate; after which they proceeded to ignore it completely. The situation, though slightly improved, still leaves much to be desired.

For the twenty-three year old Bloom, learning political theory meant reading Leo Strauss, some of whose books and unpublished articles he had brought with him from America. I still have some of them, which he sold to me when, as not infrequently happened, he was broke. Here all of a sudden were possible answers to many of the questions that kept coming up in my work. Strauss knew something that neither I nor my Sorbonne professors, world-famous scholars all of them, knew. A new world had opened up with which it would take me a long time to become familiar. The rewiring had barely begun.

Bloom's efforts paid off in other ways as well. I began to glimpse what education meant for him—a consuming, lifelong, and all-encompassing enterprise. Aristotle thought that on it depended the fate not only of individuals but of cities and nations. Allan later became a master at it. As a teacher, he had the knack of identifying the most promising students and of provoking in them the root-and-branch change that one must undergo at that age or forgo the possibility of ever undergoing it. Americans tend to go through school like a letter in the mail, the same at the end as they were before and with nothing but a diploma—a cancelled stamp, if you wish—to show for their efforts. Not so with Allan's students, who experienced at first hand what Plato's *Republic* describes as a *periagôgê*, a conversion or turning-around of the soul. The underlying image is that of Athena twisting Odysseus's head and forcing him to look in another direction. As a result, they were able to see on their own that true education had nothing to do with what was being peddled under that name by our then most popular gurus.

Effecting this kind of conversion in another person is not an easy task. Allan had been taught by Plato and soon discovered for himself that in order to learn one has to begin by unlearning, that is to say, by shedding one after another all of the partially false opinions with which one grows up. The problem he encountered is that today's students, brought up as most of them are in an ultra-liberal atmosphere, rarely carry with them any firm convictions from the tyranny of which they can be freed. A new pedagogical strategy was required, one that consists in inculcating a set of prejudices in their largely empty heads as a means of recreating the conditions under which the liberating experience is apt to take place. It dawned on me that this is what lay behind Allan's endearing habit of driving his points home with a sledge hammer.

The process was greatly facilitated by his love of literature, which he used as a tool both to combat the abstractions of modern thought and enable students to gain self-knowledge, something they were prodded to do, not by narcissistically gazing at themselves on somebody's couch, but by looking outside of themselves and allowing themselves to be mirrored in the literary characters with whom they were coming into contact. I have no idea of how many of Allan's students found themselves by reading Shakespeare, Swift, Flaubert, or Stendhal under his guidance, but my guess is that the number is quite large. Literature was indeed a privileged instrument with which principles could be taught and taste developed. The two went together. Principles without taste, he said, are crude; taste without principles is trivial. The trouble with the modern academy was that it had neither.

For a man who surrounded himself with books, Allan was anything but a voracious reader, a fact for which he blamed his dyslexia. This did not prevent him from posing as an authority on all sorts of books when summoned to do so. One day in my presence a student asked him whether he had read a particular book. He replied good-naturedly and with his usual aplomb, "No, but I'll tell you whatever you want to know about it."

When he did read books, it was with extreme care and an uncommon sensitivity to the nuances of the text. Three authors define the coordinates of his intellectual orientation: Plato, the exponent par excellence of classical education; Rousseau, the greatest authority on modern education; and Nietzsche, whose own educational program is an effort to rescue the thought of his illustrious predecessors from the clutches of modern science, but not without reinterpreting it in the light of his own extraordinary project. All three authors have something in common: they are obsessed by the mystery of the human soul and, through an intensely personal approach to its study, manage to enlist the reader as an accomplice in their endeavor to penetrate its recesses. One either likes them or dislikes them with a passion, but one cannot remain indifferent to them. Few authors in our tradition belong to that category. Bloom was attracted to them because of an inborn affinity with them. He himself confessed to an "insane" fascination with the soul and its endlessly varied epiphanies through the lives and actions of human beings.

It is no accident that the books for which he is best known are his translations and interpretations of Plato's *Republic* and Rousseau's *Emile*, along with his surprise bestseller, *The Closing of the American Mind*, written so to speak in the shadow of Nietzsche. Because he understood Nietzsche so well, Allan has been mistaken for a nihilist by people who have forgotten or never knew that his first and last love was Plato. His book *Love and Friendship*, completed just before his death, sets the record straight on that score.

For all its astonishing popularity, *The Closing* was not well understood by most readers. Reviewers responded to the culture criticism with which it begins and ends but skipped over its difficult and highly original central section, which contains the most acute diagnosis by any contemporary writer of the intellectual crisis of our time. I refer to the crisis precipitated by Nietzsche's proclamation of the death of God and characterized by a keen sense of the terribleness of life in the absence of any compelling metaphysical or religious horizon. Rousseau's *Emile* was a heroic attempt to reconstruct the Platonic soul with all its longings and aspirations, and, by so doing, to restore some meaning to human life on the basis of our modern scientific understanding of the universe. The attempt, later pursued and supposedly brought to completion by German Idealism, had failed. As Nietzsche pointed out with implacable logic, it was predicated on the false assumption that one can jettison classical speculative thought and preserve the morality that is rooted in it, or get rid of the architect and keep the building. The jig was up. All of our most cherished ideals could be seen to be at risk. Sad to say, hardly anyone was willing to face up to the situation. Instead, an unconscious, debonair, or, as Bloom put it, "laid-back" nihilism had settled over the land.

Not surprisingly, the cultural Left was quick to declare the author of *The Closing* guilty of racism, sexism, and elitism, the three great sins of the modern world, all of them sins against equality. Undaunted by this blistering assault, Bloom remained steadfast in his conviction that greatness is our true vocation and that no amount of propaganda or social conditioning will ever succeed in obliterating from our consciousness the natural differences that distinguish us from one another. In this he spoke as a true liberal. The fanatics, he said half jokingly, "are all on the other side." They are the ones who, in the name of freedom, inflict on students a politically sanitized curriculum against which they rebel inwardly, and ostracize them if they fail to conform to their wishes. The educator in Bloom was appalled to see that, just as the weakest among us had once been openly discriminated against by society, so now the best had become the victims of a vastly more subtle brand of discrimination.

His job, as he saw it, was not to arouse the moral indignation of his charges—at Cornell in the late sixties he had seen where that could lead, and there is nothing he detested more—but to enlighten their minds and lead them out of the cave of an ever more ideologically driven public opinion. It was to imbue them with a spirit of moderation and show them, as Swift had once done, how one can live in one's time without sharing the principles of that time.

His own deepest satisfactions in life were not satisfactions that could be taken away from him. They were bound up with the life of the mind and hence of the kind that can be enjoyed even in the midst of great disappointments. And enjoy them he did. His exuberance, his love of life, his irrepressible good humor, above all his extraordinary ability to inspire students and friends alike have placed us forever in his debt. Much as he liked Rousseau, he thought there was something "crabbed" about him. There was nothing crabbed or ungenerous about Allan. He always gave more than he got in return.

Thomas Aquinas, admittedly more my mentor than his, explains somewhere that a teacher can never receive adequate payment for his services because what he offers to others, namely, truth, is incommensurable with any material good. Today, we can do no more than lament Allan's passing and, without any possibility of repaying him, express our gratitude for all that he has given us. I know of no more appropriate way to honor his memory than by continuing to do with our meager talents what he himself was able to do so much better while he was still with us.

INDEX

abortion 226, 253, 307
Abou Masar 121
Abraham 140, 202, 237, 269
Abrahams, I. 129n3
Achilles 233
Adam 112-13
Adams, James Luther 172n1,
   172n18
Adler, Mortimer 50, 56n1, 180
Adorno 180
Aeneas 26
agapism 52
Agbar, king of Edessa 287
Alamain, Jacques 219n46
Alaric 288
Alexander 234
Alexander, F. S. 219n43
Althusser, Louis 105
Ambrose, St. 20, 232
American economy 9, 268, 295,
   303
American founders 30
American founding 17, 193
Antonazzi 214n15, 214n17,
   215n22

Antony of the desert, St. 52,
   266
apocalypticism 137, 247
apologists 4, 14, 120, 255, 288
Aquinas, St. Thomas 20, 59,
   64-65, 68, 71n8, 133n49,
   155, 161, 163, 164n4, 167,
   169-70, 196-97, 202, 207,
   210-11, 219n47, 220nn55-
   56, 221n63, 224, 227, 237,
   240n9, 253, 256, 267, 270-
   72, 275n5, 278-79, 282n2,
   285-86, 290, 292, 293n2,
   294n14, 300, 301n3, 312n11,
   320
Arendt, Hanna 193
Arianism 141
Aristotelianism 68, 119-20,
   123, 164n1
Aristotle 20, 26, 59, 100n34,
   100n36, 115-19, 121, 123,
   126-27, 129nn4-7, 129nn10-
   12, 133nn49-50, 133n49,
   161, 164n5, 177-78, 184,
   187-88, 192, 196, 200, 207,

ABOUT THE AUTHOR

A native of Rhode Island, Ernest L. Fortin, A.A., received his B.A. degree from Assumption College (Worcester, MA) in 1946, his Licentiate in Theology from the Angelicum (Rome) in 1950, and his Doctorate in Letters from the Sorbonne in 1955. He has also done post-doctoral work at the Ecole Pratique des Hautes Etudes (Paris) and the University of Chicago. He taught at Assumption College from 1955 to 1970, and as a part-time visiting professor of philosophy at Laval University from 1965 to 1972. Since 1971 he has been teaching theology and political theory at Boston College, where he also co-directs (with C. Bruell) the Institute for the Study of Politics and Religion. He has lectured widely to scholarly audiences both in America and in Europe. His publications include *Christianisme et culture philosophique au cinquième siècle: la querelle de l' âme humaine en Occident* (Paris, 1959); *Medieval Political Philosophy: a Sourcebook*, edited with M. Mahdi and R. Lerner (New York, 1963); *Dissidence et philosophie au moyen âge: Dante et ses antécédents* (Paris and Montreal, 1981); *Dantes Göttliche Komödie als Utopie* (Munich, 1991), and *Augustine: Political Writings*, edited with D. Kries (Indianapolis, 1994). His articles, review articles, and book reviews have appeared in a wide variety of professional journals and symposia. An English translation, with an introduction and notes, of Thomas Aquinas's *Commentary on the Politics of Aristotle* is scheduled to appear in 1997.

ABOUT THE EDITOR

J. Brian Benestad is professor of theology at the University of Scranton, a Jesuit University in Northeastern Pennsylvania. He has been teaching at Scranton since the fall of 1976. A native of New York City, he received his B.A. from Assumption College in 1963, a Licentiate in Theology from the Gregorian University (Rome) in 1968, and a Ph.D. in political science from Boston College in 1979. In 1981 he co-edited a collection of the U.S. bishops' policy statements issued between 1966 and 1980, and authored *The Pursuit of a Just Social Order* (1982). Most recently he completed an article entitled "Ordinary Virtue as Heroism," published in *Seedbeds of Virtue*, edited by Mary Ann Glendon and David Blankenhorn.